DFG

Occupational Toxicants

DFG Deutsche Forschungsgemeinschaft

Occupational Toxicants

Critical Data Evaluation for MAK Values and Classification of Carcinogens

Volume 19

Edited by Helmut Greim

Commission for the Investigation of Health Hazards of Chemical Compounds in the Work Area

WILEY-VCH

WILEY-VCH GmbH & Co. KGaA

Prof. Dr. Helmut Greim
Senatskommission
zur Prüfung gesundheitsschädlicher Arbeitsstoffe
der Deutschen Forschungsgemeinschaft
Technische Universität München
Hohenbacherstraße 15-17
85354 Freising

Translation and layout: Dr. A. E. Wild and J. Handwerker-Sharman

Library of Congress Card No. applied for.

A catalogue record for this book is available from the British Library.

Deutsche Bibliothek Cataloguing-in-Publication Data:
Occupational toxicants : critical data evaluation for MAK values and classification of carcinogens / DFG, Deutsche Forschungsgemeinschaft ; Commission for the Investigation of Health Hazards of Chemical Compounds in the Work Area.
- Weinheim ; New York ; Chichester ; Brisbane ; Singapore ; Toronto : Wiley-VCH
Einheitssacht : Gesundheitsschädliche Arbeitsstoffe <engl.>
Erscheint unregelmäßig. - Aufnahme nach Vol. 1 (1991)
Vol. 1 (1991) -

Printed in the Federal Republic of Germany.
Printed on acid-free paper.
Printing: betz druck GmbH, Darmstadt
Bookbinding: Litges & Dopf Buchbinderei GmbH, Heppenheim

Note

In the present volume reference is frequently made to the various sections of the *List of MAK and BAT Values** (MAK List). The MAK List was introduced and its sections described in detail in the first volume of this series.

---

* Available from the publisher, WILEY-VCH Verlag GmbH & Co. KGaA, Weinheim, both in the original German edition and as an English translation.

Contents of Volume 19

Acetonitrile

MAK value (2001)	**20 ml/m^3 (ppm) ≙ 34 mg/m^3**
Peak limitation (2001)	**Category II, excursion factor 2**
Absorption through the skin (2001)	**H**
Sensitization	**–**
Carcinogenicity	**–**
Prenatal toxicity (2001)	**Pregnancy risk group C**
Germ cell mutagenicity	**–**
BAT value	**–**
Synonyms	cyanomethane ethanenitrile methyl cyanide
Chemical name (CAS)	acetonitrile
CAS number	75-05-8
Structural formula	CH_3CN
Molecular formula	C_2H_3N
Molecular weight	41.05
Melting point	–43°C
Boiling point	81.6°C
Density at 20°C	0.786 g/cm^3
Vapour pressure at 20°C	96.6 hPa
log P_{OW}*	–0.34
1 ml/m^3 (ppm) ≙ 1.70 mg/m^3	**1 mg/m^3 ≙ 0.59 ml/m^3 (ppm)**

Acetonitrile is a by-product of acrylonitrile production. It is a volatile, polar solvent and has many uses, for example, for the extraction of fatty acids from plant fats and oils and for extractive distillation of organic petrochemicals (fractionation of olefin-diolefin mixtures). Acetonitrile is used as a solvent in high performance liquid chromatography, in DNA synthesis and in the sequencing of peptides. The main impurity in commercial acetonitrile is propionitrile.

---

* *n*-octanol/water distribution coefficient

The present document is based in part on the review published by the WHO (1993) and the risk assessment of the EU (Spanish Ministry of Health 2000).

1 Toxic Effects and Mode of Action

Acetonitrile is readily absorbed through the gastrointestinal tract, the skin and the lungs and is distributed uniformly in the whole body in both man and experimental animals. *In vivo*, the radioactivity from 2-^{14}C-acetonitrile is associated with the fractions containing proteins, nucleic acids and lipids; differentiation between covalent binding to and metabolic incorporation of the ^{14}C atom into the macromolecules has not been carried out. The extent to which acetonitrile is metabolized is not known. The biotransformation involves oxidation to formaldehyde cyanohydrin which is then broken down enzymatically to yield cyanide and formaldehyde (the latter has not yet been identified as a metabolite). In the organism, cyanide is mostly converted to thiocyanate. The elimination half-time for acetonitrile in blood is about 30 hours in man and about 8 hours in the mouse. Acetonitrile, thiocyanate and a very small fraction of the cyanide are excreted in the urine. After intravenous injection of a sublethal dose of acetonitrile into monkeys, about 12 % of the dose was recovered in the urine in the form of thiocyanate. In a drinking water study, rats excreted about 70 % of the absorbed acetonitrile as thiocyanate. A little of the cyanide is eliminated in the exhaled air. Complete elimination of a sublethal dose of acetonitrile can take several days in man. Because the half-life of the substance is so long, accumulation during the working week must be assumed.

In man, acute acetonitrile intoxication produces symptoms of cyanide poisoning which appear after a delay because the metabolism is slow: a feeling of tightness in the chest and chest pain, restlessness, tachycardia, dizziness, headaches, vomiting, reduced blood pressure, shallow breathing, metabolic acidosis, pulmonary oedema, loss of consciousness, extension spasms and respiratory paralysis.

The effects on animals of short-term and long-term inhalation of acetonitrile are described below. Effects are seen on body weights and organ weights, histological changes in the liver, lungs, stomach and kidneys and coordination disorders. In addition, the substance has irritant effects on the eyes and the respiratory tract including emphysema, bronchitis and pneumonia. Oedema and haemorrhage are also seen in the brain. Mortality was seen in mice in a 13-week study at concentrations of 400 ml/m^3 or more. Long-term toxicity in mice is characterized by epithelial hyperplasia in the forestomach. This is seen in female mice exposed to concentrations of 50 ml/m^3 or more and in male mice from 200 ml/m^3, but the significance of the effect for man is not clear. In a 2-year study, the incidence of basophilic liver foci was increased, but not significantly, in male rats exposed to 100 ml/m^3 and a slight increase (significant trend) in the incidence of benign and malignant liver tumours was seen in male rats exposed to 400 ml/m^3.

Mutagenicity tests with various *Salmonella typhimurium* strains in the presence and absence of rat liver S9 fraction yielded negative results as did the hypoxanthine guanine phosphoribosyl transferase (HPRT) test and the mouse lymphoma test. In *Saccharomyces cerevisiae*, acetonitrile induced gene conversion and, at very high concentrations, also mitotic chromosome loss and duplication. Aneugenic effects were demonstrated in *Drosophila melanogaster*. Administered in very high doses to NMRI mice, acetonitrile induced micronuclei in the bone marrow. In a more reliable study, the incidence of micronuclei in the bone marrow cells and in peripheral blood cells was not increased.

In two valid developmental toxicity studies in which rats inhaled acetonitrile, in the concentration range between 1200 and 1500 ml/m^3 adverse effects were seen in the dams but not in the progeny. This exposure level corresponds to a dose of about 500 mg/kg body weight and day. An inhalation study with a second species is not available, but there is a study with oral administration to rabbits in which a maternally toxic dose of 15 mg/kg body weight was also shown to have no prenatal toxic effects. After exposure of mice and rats to acetonitrile concentrations up to 400 ml/m^3 for 13 weeks, no effects on testis weights or sperm motility were detected.

2 Mechanism of Action

Acetonitrile is hydroxlated by the cytochrome P450 monooxygenase system to yield formaldehyde cyanohydrin, the α-hydroxynitrile, which is broken down by catalase to yield hydrogen cyanide and formaldehyde (Figure 1). Free cyanide induces the symptoms of intoxication listed in Section 1 above.

For years it has been discussed in the literature whether or not other metabolites such as formaldehyde also contribute to the toxicity of acetonitrile. Formaldehyde and cyanide would be expected to be produced in equimolar amounts, but formaldehyde has not yet been detected as a metabolite of acetonitrile. Other metabolites such as formaldehyde cyanohydrin, thiocyanate, and cyanate could play a role in acetonitrile toxicity, as could the availability of sulfur-containing amino acids.

The reduced body weights seen in experimental animals exposed to high levels of acetonitrile have been put down to depletion of sulfur-containing amino acids, because detoxification of cyanide requires sulfur provided by cysteine or the essential amino acid methionine which are then no longer available for protein synthesis (Swenne *et al.* 1996). Rats given a diet free of methionine and cysteine also lost weight (Tor-Agbidye *et al.* 1998). This could be the mechanism responsible for the delayed body weight gains seen in animals given cyanide doses of about 10 mg/kg body weight or more (see the MAK documentation for Hydrogen cyanide (2001), this volume).

Reduced longitudinal growth has been observed in the Democratic Republic of Congo in children who lived on tubers of the tropical plant cassava (*Manihot esculenta* Crantz) (Banea-Mayambu *et al.* 2000). It has been estimated that 500 million people depend on cassava as their staple food or as part of their diet (Padmaja 1995). Cassava tubers contain the cyanoglycosides linamarin and lotaustralin which release hydrogen cyanide on drying in the sun. Consumption of insufficiently dried tubers (which still contain cyano-compounds) can result in chronic intoxication. As stated above, the organism which has taken up cyanide uses sulfur-containing amino acids as sulfur donors in detoxification.

The use of sulfur-containing amino acids to detoxify cyanide could therefore be a factor contributing to the delayed growth of the children (Banea-Mayambu *et al.* 2000). However, impairment of thyroid function, documented as increased goitre incidence associated with the consumption of cassava (Banea-Mayambu *et al.* 2000), could also delay physical development.

3 Toxicokinetics and Metabolism

3.1 Absorption, distribution, elimination

3.1.1 Man

Studies of smokers have revealed that about 90 % of the acetonitrile inhaled with cigarette smoke is taken up by the organism. When the smoke was retained in the mouth for 2 seconds, 74 % of the acetonitrile was retained (Dalhamn *et al.* 1968a, 1968b). The level of acetonitrile in blood can serve as a measure of the number of cigarettes smoked daily. Depending on cigarette consumption, the acetonitrile concentrations are in the range between 43.3 and 148.3 µg/l blood (1 and 3.7 µmol/l blood) (Houeto *et al.* 1997). In another study, urinary acetonitrile levels were determined for 40 persons who smoked at least 3 cigarettes per day and for 20 non-smokers. The average acetonitrile concentration in the urine of smokers was 117.6 µg/l (2.9 µmol/l), in the urine of non-smokers 2.9 µg/l (0.07 µmol/l) (McKee *et al.* 1962).

The levels of cyanide and thiocyanate in the blood of persons with acetonitrile poisoning are shown in Tables 1 and 2. In a case of lethal acetonitrile poisoning, 5 days after the exposure the highest concentrations of acetonitrile were in the liver (0.29 mmol/kg), spleen (0.23 mmol/kg) and kidneys (0.33 mmol/kg); the concentrations in the heart and lungs were lower, pancreas and bladder contained only traces of the substance. The concentrations of free cyanide were highest in the lungs (13 µmol/kg) and spleen (17 µmol/kg). It is, however, doubtful whether lower doses which do not saturate metabolism and elimination are distributed in a similar manner. Cyanide and acetonitrile were detected in the urine (Dequidt *et al.* 1974).

From the available data it is not possible to estimate the extent of biotransformation of acetonitrile in man. Although it has been observed that persons poisoned fatally with acetonitrile excreted very large amounts of the unchanged substance, it is not known how much of the low doses taken up during chronic exposure is converted to thiocyanate. A case of attempted suicide (ingestion of about 5 ml acetonitrile, that is about 64 mg/kg body weight, 1.56 mmol/kg body weight) demonstrates that acetonitrile is still detectable in the blood 100 hours after ingestion. From these data, half-lives of 32 hours for acetonitrile and 15 hours for cyanide were estimated (Michaelis *et al.* 1991). As cyanide is formed continuously from acetonitrile in metabolic equilibrium, it is difficult to conceive how it can have a shorter half-life than the starting substance. Perhaps the elimination of cyanide was affected by the thiosulfate therapy carried out in this patient.

3.1.2 Laboratory animals

Cyanide is released from acetonitrile more slowly than from other nitriles. Thus the levels of cyanide in the blood or brain of the rat 1 hour after administration of the oral LD_{50} of acetonitrile (2460 mg/kg body weight) are much lower than after administration of the LD_{50} of other nitriles (Ahmed and Farooqui 1982). The peak blood cyanide concentration is first attained 7.5 to 15 hours after intraperitoneal injection of acetonitrile (1470 mg/kg body weight). The serum acetonitrile concentration after 3 and 7.5 hours was about 1000 mg/l; the maximum cyanide concentration, in contrast, about 7.5 mg/l (Freeman and Hayes 1985).

Rats given acetonitrile in oral or intraperitoneal doses of 0.75 mmol/kg body weight (30.75 mg/kg) excreted within 24 hours 11.8 % and 4.4 % of the dose, respectively, as urinary thiocyanate. With other aliphatic nitriles, 27 % to 75 % of the dose was excreted within 24 hours (Silver *et al.* 1982), this result confirms that acetonitrile is metabolized relatively slowly.

A study in which three dogs were exposed for 4 hours to acetonitrile in a concentration of 16000 ml/m^3 yielded evidence of rapid pulmonary absorption. The blood cyanide concentrations reached values of 330 to 530 µg/l (12.7–20.4 µmol/l) after one hour. After three hours, the maximum concentrations of 3050 to 4330 µg/l (117–167 µmol/l) were attained. All three animals died about 14 hours after the start of exposure (Pozzani *et al.* 1959).

Mice were given single i.v. doses of 2-^{14}C-acetonitrile and the distribution of the radioactivity in lipids and in the macromolecular fraction (proteins and nucleic acids together) of various organs was determined. In the first hour after administration of the substance in practically all tissues studied, 5 % to 10 % of the total ^{14}C was detected bound to the protein and nucleic acid fraction, 6 % to 16 % bound to the lipids. However, the method used did not permit differentiation between direct binding and metabolic incorporation of the ^{14}C atom. During the 48 hours after administration, the absolute amount of radioactivity in the tissues decreased, the relative amount of bound radioactivity, however, increased. The findings after 48 hours are summarized below. Of the total radioactivity found in the individual tissues, 25 % to 45 % was associated with the protein and nucleic acid fraction, 30 % to 50 % with the lipids. The highest levels of bound radioactivity were found in the liver and pancreas. The accumulation of radioactivity in various tissues after i.v. administration of 2-^{14}C-acetonitrile was followed by means of whole body autoradiography. After 5 minutes the radioactivity was particularly high in the liver, the kidneys, the contents of the upper gastrointestinal tract, the contents of the gall bladder and urinary bladder, the oral mucosa, the skin, the nasal secretion and the cartilage. After 24 hours, accumulation was also detected in the preputial gland and in the bone marrow. After 48 hours the tissues listed above contained less radioactivity; the relatively highest amounts remaining were in the liver, gastrointestinal mucosa, submucosa and muscularis, the testes, the preputial gland and the prostate. The elimination half-times after a single intravenous 2-^{14}C-acetonitrile dose of 2.46 mg/kg body weight differed for the different organs and were in the range of 5 to 8 hours. The values were 8.5 hours for blood, 6.8 hours for brain, 7.3 hours for lung, 5.5 hours for liver, 6.4 hours for kidney and 7.5 hours for testis. The terminal half-times were, for example, 50 hours for the liver, 90 hours for the epididymis and 540 hours for the small intestine (Ahmed *et al.* 1992).

Sublethal doses of acetonitrile (0.1 ml/kg body weight, about 80 mg/kg body weight, 1.9 mmol/kg body weight) were administered by i.v. injection to three rhesus monkeys and the amount of thiocyanate excreted daily in the urine determined for a period of 4 to 7 days. In this experiment, a total of 12 % of the acetonitrile was recovered as thiocyanate. Two months later the animals were used again, this time to determine how much of an intravenous dose of sodium thiocyanate (about 500 mg thiocyanate) appeared in urine as thiocyanate. Again, the urine was analyzed daily until the thiocyanate level had returned to control values. Only 55 % of the injected sodium thiocyanate was recovered. Therefore, in the experiment described above, more than 12 % of the acetonitrile could have been metabolized (Pozzani *et al.* 1959).

Rats given an intraperitoneal injection of 200 mg acetonitrile excreted 3 % of the dose as acetonitrile and traces of hydrogen cyanide, and 2.3 % as "bound cyanide" (sulfur-containing cyano compounds, cyanohydrins, cyanocobalamin) in the urine during the 11 days *post applicationem* (Haguenoer *et al.* 1975). Because of the low recovery, the only possible conclusion from this study is that acetonitrile and hydrogen cyanide appeared in the urine.

Groups of 15 male and 15 female Wistar rats were exposed to acetonitrile vapour in concentrations of 166, 330 or 655 ml/m^3, 7 hours daily, 5 days per week for 90 days. From the animals of the groups exposed to 166 and 330 ml/m^3, 16-hour urine samples were collected during days 59 to 66 and the thiocyanate concentrations were determined. These concentrations were in the range between 170 and 790 mg/l (2.9–13.6 mmol/l) for both groups, independent of the exposure level. The urinary thiocyanate levels returned to zero during the 2-day exposure break at weekends, but not between the daily exposures (Pozzani *et al.* 1959).

Dogs and rhesus monkeys were also exposed to acetonitrile in a concentration of 350 ml/m^3, 7 hours/day on 5 days/week for 91 days. At the beginning of week 7 after the 2-day weekend exposure break, no cyanide could be detected in the dogs' blood. On the last of the five exposure days, in contrast, the cyanide concentrations were 76 to 92 µg/l (2.9–3.5 µmol/l). In week 7 of the study on five consecutive days, 16-hour urine samples were also collected. The average thiocyanate concentrations in the urine were between 69 and 252 mg/l (1.2 and 4.9 mmol/l). At the end of the subsequent 2-day exposure-free interval, the thiocyanate level in the urine of the only dog for which it was determined was 156 mg/l (2.7 mmol/l). In spite of the exposure-free interval, the level had not returned to normal. Blood samples taken from three rhesus monkeys after the 2-day exposure-free weekend before the 55th exposure contained cyanide concentrations of 18 to 29 µg/l (0.7–1.1 µmol/l), those taken on exposure day 59 contained 47 to 54 µg/l (1.8–2.1 µmol/l). The thiocyanate excretion was also determined on 5 consecutive days in 16-hour urine samples. The values were between 60 and 114 mg/l (1.0 and 2.0 mmol/l). After the 2 exposure-free days, the thiocyanate values were not lower, but had increased to an average level of 192 mg/l (3.3 mmol/l) (Pozzani *et al.* 1959). The data demonstrate that during a 5-day exposure of dogs and monkeys to acetonitrile, slight accumulation of thiocyanate in urine occurs. Allometric considerations suggest that rats might eliminate cyanide more rapidly than do dogs and monkeys and that especially monkeys are expected to excrete acetonitrile in the form of thiocyanate very slowly.

The effect of acetonitrile on the body weight development of SD rats (12–15 weeks old, weight 325–410 g) was studied. Six groups of six animals were formed; three groups were given standard diet with 18 % protein, the other three protein-reduced diet (5 % protein). Two of the groups on the normal diet and two on the protein-reduced diet were given drinking water (*ad libitum*) to which acetonitrile (40 mM or 80 mM) had been added. The resulting acetonitrile doses were about 1.5 to 1.6 and 2.7 to 3.2 mmol/kg body weight and day (about 65 and 120 mg/kg body weight and day) for the 40 and 80 mM groups, respectively. Water consumption by the animals given acetonitrile was about 50 % of the corresponding control values. The amount of food consumed by the animals was not stated. The animals were given acetonitrile for 3 to 4 weeks and some of them were then observed for another 2 weeks. Animals on a normal diet which drank water containing acetonitrile gained weight like the control animals during the first three weeks of the study but then lost weight until, after another three weeks, the body weights were like those at the beginning of the study. The levels of total sulfur compounds in the 24-hour urine were similar in the three groups of animals (about 1200 µmol/kg body weight) but the excretion of thiocyanate in the urine of animals treated with acetonitrile increased from 1.5 µmol/kg body weight and day (control) to 850 or 1120 µmol/kg body weight and day, while the inorganic sulfate decreased from 750 µmol/kg body weight and day (66 %; control) to 42 µmol/kg body weight (3 %) in the group given 40 mM acetonitrile and 109 µmol/kg body weight and day at 80 mM. In the urine of animals treated with acetonitrile, the thiocyanate-bound sulfur accounted for 67 % and 81 % of the total excreted as sulfur compounds (control 0.13 %). Thus the increase in the level of thiocyanate excretion was less than proportional to the dose of acetonitrile. Animals given a protein-reduced diet and normal drinking water lost weight markedly from the beginning of the study; the relative weight loss became gradually less over a period of 9 weeks until the severely reduced body weights remained practically constant at 73 % of the initial weights. The excretion of sulfur compounds in the urine was also reduced. Total urinary sulfur compounds amounted to 370 µmol/kg body weight and day; the thiocyanate level was very low at 0.6 µmol/kg body weight, inorganic sulfur accounted for 280 µmol/kg body weight and day. When the animals were not only given a protein-deficient diet but also acetonitrile in the drinking water, the body weight losses were so marked that they led to emaciation. The amount of sulfur compounds excreted in the 24-hour urine increased to more than twice that recorded for the acetonitrile-free control group (from 370 to 890 or 1136 µmol/kg body weight and day), the thiocyanate to more than 1000 times the control value (from 0.6 to 620 or 850 µmol/kg body weight and day) while inorganic sulfate was reduced from 280 to 70 or 45 µmol/kg body weight and day. Under these conditions, that is, in spite of the protein-deficient diet, 70 % to 77 % of the total sulfur excreted was in the form of thiocyanate (Swenne *et al.* 1996). The authors concluded from their results that cyanide detoxification uses up enormous amounts of protein and therefore can have adverse effects on body weight development. The formation of one molecule of thiocyanate or one molecule of 2-aminothiazoline-4-carboxylic acid consumes one molecule of cysteine or methionine, because methionine can be converted to cysteine. Methionine is an essential amino acid; cysteine, however, is only not essential while sufficient methionine is available. Persistent deprivation of these amino acids has the result that proteins containing cysteine or methionine are no longer formed. Thus the other available amino acids (more than 10 times the weight of cysteine and methionine) cannot be used for protein synthesis.

Cyanide administered for 14 days in concentrations of 1000 and 3000 µg/l drinking water also causes loss of weight. This effect, however, was put down to the reduced water consumption of the animals (NTP 1993) which was also observed by Swenne *et al.* (1996).

3.2 Metabolism

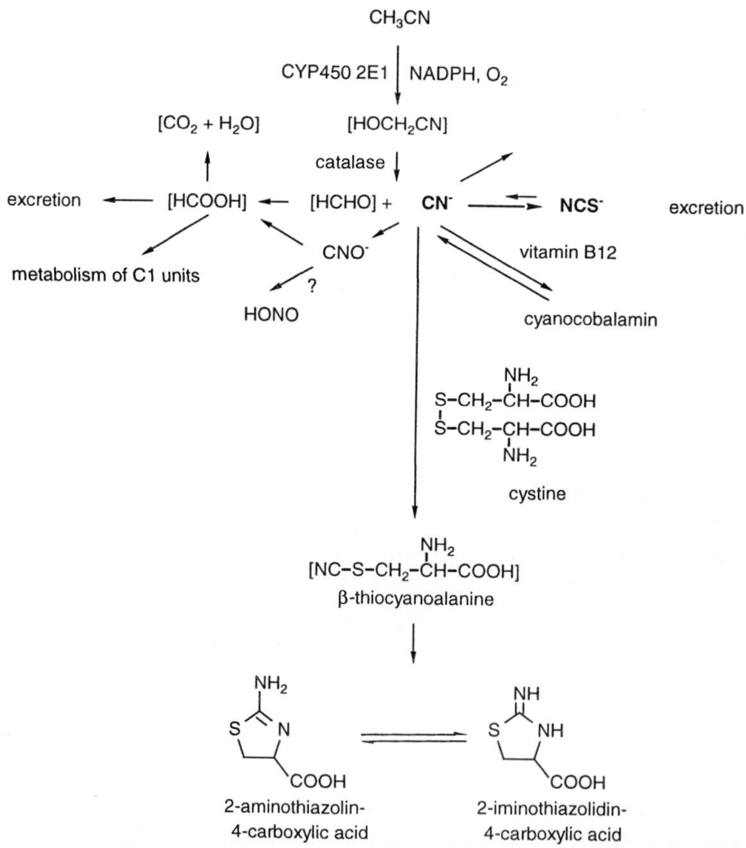

Figure 1. Biotransformation of acetonitrile in mammals (from NL Health Council 1997)

The biotransformation of acetonitrile has been studied in detail in various species of experimental animal or in their tissues and cells (WHO 1993). The conversion of acetonitrile to cyanide and formaldehyde is an O_2-dependent and NADPH-dependent reaction; *in vitro* it obeys classical Michaelis-Menten kinetics. The first step, hydroxylation of the α-C atom of the acetonitrile to yield formaldehyde cyanohydrin, takes place in the liver and is probably catalysed by CYP2E1 (Feierman and Cederbaum 1989, Freeman and Hayes 1988). In the second step the cyanohydrin is broken down to cyanide and formaldehyde, a catalase-dependent reaction (Feierman and Cederbaum 1989). However, formaldehyde cyanohydrin and formaldehyde have not yet been detected as metabolites (Figure 1).

Microsomes from rat liver and rat nasal mucosa produce cyanide from acetonitrile. The rates of metabolism have been shown to be 0.098 nmol cyanide per mg protein per minute in the liver and 0.9 nmol cyanide per mg protein per minute in the nasal mucosa (Dahl and Waruszewski 1989). The K_m and V_{max} values for these reactions in hepatocytes from female rats were 3.4 mM and 1.1 nmol cyanide per 10^6 cells and 10 minutes (Freeman and Hayes 1987). For mouse hepatocytes the values were 4.19 mM and 14.3 ng cyanide per mg protein and 15 minutes (Tanii and Hashimoto 1984).

Most of the cyanide is converted to thiocyanate by the enzyme thiosulfate:cyanide sulfurtransferase (EC: 2.8.1.1; trivial names: thiosulfate-sulfurtransferase, rhodanese) (Vennesland *et al.* 1982). *In vitro* the enzyme cystathionase-γ-lyase (EC: 4.4.1.1) was also found to be involved in the transsulfuration of cyanide. The main sulfur donors are thiosulfate and thiocystine (Porter *et al.* 1996). The enzyme 3-mercaptopyruvate:cyanide sulfurtransferase (EC: 2.8.1.2) also appears to make a significant contribution to the detoxification of cyanide (Nagahara *et al.* 1999). The enzymes and substrates involved in cyanide detoxification are shown in Figure 2.

The known physiological compounds containing sulfane sulfur include not only thiosulfate but also polythionates, polysulfides and albumin-bound elemental sulfur. Thiosulfate:cyanide sulfurtransferase can catalyze interconversion of these various sulfane sulfur compounds and also transfer their sulfur to cyanide (Vennesland *et al.* 1982). It has been estimated that in man about 80 % of the cyanide is converted to thiocyanate. The rate of detoxification of cyanide is limited by the availability of sulfane sulfur compounds (NL Health Council 1997).

The reverse reaction of thiocyanate to yield cyanide and thiosulfate is possible. It is catalysed by various peroxidases (NL Health Council 1997).

Other biotransformation pathways for cyanide include: a) the non-enzymatic reaction with cystine to form thiazoline or thiazolidine carboxylic acids (Swenne *et al.* 1996), b) the reaction with α-keto acids (e.g. pyruvic acid, α-ketoglutaric acid, oxaloacetic acid) or dihydroxyacetone to yield cyanohydrins (Huang *et al.* 1998, Porter and Baskin 1996), c) fixation of the cyano group in a cobalamin (Vennesland *et al.* 1982), d) the strong (but reversible) binding of the cyanide ion to methaemoglobin (this is the basis of therapy of cyanide or acetonitrile poisoning with a methaemoglobin-producer as antidote. However, the cyanide group is then released during reduction of methaemoglobin to haemoglobin), e) the breakdown of a small amount of cyanide (about 2 % in the mouse) to carbon dioxide; intermediates are cyanate and formate (McMahon and Birnbaum 1990).

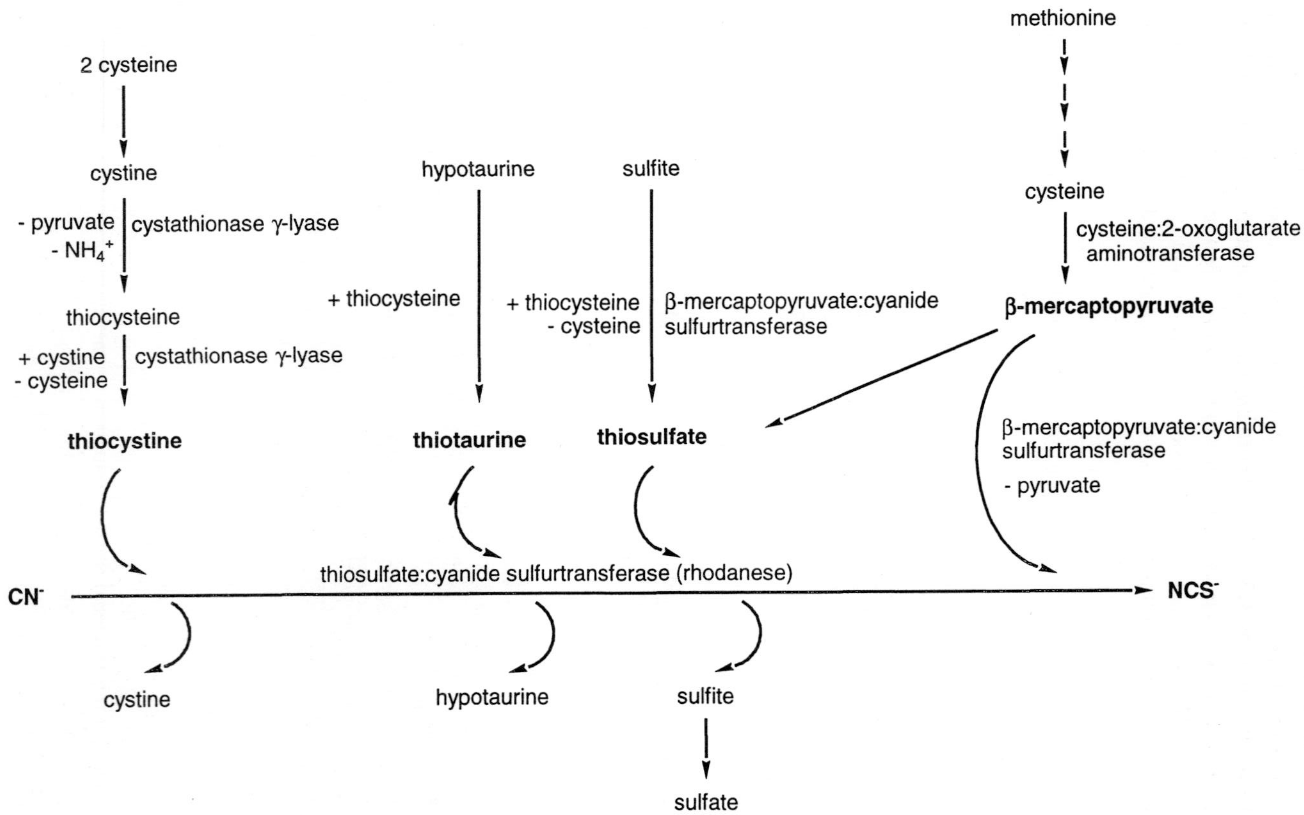

Figure 2. Biotransformation of cyanide (from Ahmed *et al.* 1992, Feierman and Cederbaum 1989, Huang *et al.* 1998, Luo and Horowitz 1994, Ohkawa *et al.* 1972, Porter *et al.* 1996, Vennesland *et al.* 1982)

Studies aimed at clarifying the fate of the formaldehyde formed in oxidation of acetonitrile have not been published. It is, however, assumed that in various tissues the formaldehyde is used in the synthesis of nucleic acids and proteins (Ahmed *et al.* 1992).

The metabolism of acetonitrile and thus also its toxicity has been reduced by inhibition of the microsomal oxygenases of the liver with carbon tetrachloride (Tanii and Hashimoto 1984, Willhite and Smith 1981). Similarly, simultaneous administration of acetone reduced the toxicity of acetonitrile (Freeman and Hayes 1985). Competitive inhibition of CYP2E1-dependent metabolism of acetonitrile is responsible for this effect (Freeman and Hayes 1988). Enzyme induction by pretreatment with acetone increased the toxicity of acetonitrile *in vivo* (Freeman and Hayes 1985).

4 Effects in Man

There are no data available for the effects of repeated exposure of man to acetonitrile nor for local effects on skin or mucous membranes, allergenicity or genotoxicity.

4.1 Single exposures

4.1.1 Laboratory exposures

In a toxicity study, three male volunteers (ages between 31 and 47 years, evidently three of the authors of the study) inhaled acetonitrile for 4 hours in a vapour concentration of 40 ml/m^3 in an 8 m^3 exposure chamber. The blood cyanide levels were determined 19 days and one day before the beginning of the study and immediately after the exposure. The thiocyanate levels in 24-hour urine were also determined on days 19, 18 and 1 before the study and immediately after the 4-hour exposure. The two older volunteers, one smoker and one non-smoker, did not notice any adverse effects either during or after the inhalation. In these two persons the blood cyanide level remained below the detection limit. Urine thiocyanate was not increased. The youngest volunteer, however, a non-smoker, noticed a slight tightness in the chest on the evening after the exposure and the next morning a feeling of coldness when inhaling, like that felt on inhaling menthol. This feeling of coldness persisted for about 24 hours. Cyanide was not detected in blood but the urine thiocyanate level was increased slightly. All three persons identified the smell of acetonitrile during the first two to three hours but after that did not notice it as much. One week after the exposure to 40 ml/m^3, the two older of the three volunteers subjected themselves to a higher level exposure. They inhaled acetonitrile in a concentration of 80 ml/m^3 for 4 hours. Again the two persons noticed no ill effects. In blood samples taken immediately before, 0.5, 3 and 24 hours after the exposure, no increase in the cyanide concentrations was detected. The results of thiocyanate analysis of urine samples collected one and two days after the exposure were not consistent. In one person, the smoker, the thiocyanate level in the urine collected before the exposure was higher than in that collected afterwards. The values for the other person remained constant. Nine days after the exposure to 80 ml/m^3, the two persons subjected themselves to an even higher level exposure. They spent 4 hours in an atmosphere containing an acetonitrile concentration of 160 ml/m^3. One of the two persons registered transient facial flushing 2 hours after the exposure and 5 hours later a slight feeling of bronchial tightness. The feeling of tightness was gone by the next morning. This person had suffered from increased nasal secretion before the exposure. The blood cyanide concentrations before and after exposure were not significantly different; the same applied for the amounts of thiocyanate excreted in the urine (sampling times as for the 80 ml/m^3 exposure). Both persons stated that they would not hesitate to repeat the 4-hour exposure to 160 ml/m^3 (Pozzani *et al.* 1959). It is difficult to decide whether the symptoms described by the one volunteer after the 40 ml/m^3 exposure can be interpreted as adverse because, even after exposure to 160 ml/m^3, the feeling of tightness described by another person was slight and this person would have been willing to repeat the exposure. It is questionable whether this effect was caused by cyanide because the blood cyanide levels were not increased, unlike in a case of cyanide poisoning in which a feeling of tightness in the chest was also described (Amdur 1959).

4.1.2 Case reports

In the literature, about 25 cases of acetonitrile poisoning are described, mostly caused by inhalation of the substance in industry (see Tables 1 and 2). A latency period of 4 to 12 hours between the exposure and the appearance of the first symptoms is characteristic. This time interval is evidently determined by the metabolic oxidation step which slowly releases cyanide from acetonitrile. The elimination of the cyanide is correspondingly protracted. Early symptoms include pains in the chest and upper abdomen, dizziness, and vomiting, and are followed later by convulsions, unconsciousness, acidosis and circulatory collapse.

Because of the delay in appearance of the symptoms, it is recommended that all cases of acetonitrile poisoning be admitted to hospital for observation (Geller *et al.* 1991). The preferred treatment is a combined therapy with a methaemoglobin-producer such as *p*-dimethylaminophenol and with sodium thiosulfate to promote metabolism of the cyanide.

4.2 Reproductive and developmental toxicity

The results of a case-control study (206 cases, 329 controls) indicated that there was no association between the incidence of spontaneous abortions in female laboratory assistants and the reported exposure to acetonitrile (odds ratio 1.4, 95 % confidence interval 0.4–4.7) (Taskinen *et al.* 1994).

Table 1. Clinical findings and cyanide and thiocyanate levels in persons suffering from acute occupational acetonitrile poisoning

Age	Work process	Symptoms	Course of the intoxication	Maximum blood cyanide	Maximum serum thiocyanate	References
23	painting the inside surface of a 19 m^3 tank with a paint containing acetonitrile for 12 h	ash-grey skin colour, chest pains, dizziness, vomiting, blood-stained sputum, convulsions, shallow irregular breathing, coma	death 14 hours after stopping work	7.9 mg/l (306 µM)	n.d.	Amdur 1959
35	painting the inside surface of a 19 m^3 tank with a paint containing acetonitrile for 3 h	extreme weakness, pallor, dizziness, vomiting bile-coloured stomach contents, initial tachycardia, later slow pulse (55/min), slow shallow breathing, absence of deep reflexes, spasmodic abdominal pain	went back to work after 11 days	3.06 mg/l (118 µM)	160 mg/l (2.8 mM)	
28	painting mainly the outside surface of a 19 m^3 tank by hand with a paint containing acetonitrile for 12 h	semi-conscious, slate grey skin colour, blood pressure 100/50 mm Hg, pulse 45/min, shallow intermittent breathing, feebleness, absence of deep reflexes	went back to work after 20 days	10.88 mg/l (418 µM)	230 mg/l (4.0 mM)	
28	painting the inside surface of a 19 m^3 tank with a paint containing acetonitrile for 2.5 h	headaches, feeling of a pulse in the neck, weariness, difficulty in swallowing	worked for only short periods in the first days after the exposure, regularly from day 18	0.72 mg/l (28 µM)	145 mg/l (2.5 mM)	
20	painting jobs during 2 days	headaches, shortness of breath (but pre-existing hyperventilation)	did not go back to work	0.58 mg/l (22 µM)	180 mg/l (3.1 mM)	
18	painting with paint containing acetonitrile for 8 h, sand-blast cleaning and mixing the above paint for 7 h	headaches, weakness, tightness in the chest, shortness of breath	recovery within a few days, shortness of breath persisted for one month	0.33 mg/l (13 µM)	100 mg/l (1.7 mM)	
42	mixing of paints containing acetonitrile	tiredness, dizziness; symptoms disappeared after about 24 h	went back to work after 7 days	0	135 mg/l (2.3 mM)	

Table 1. continued

Age	Work process	Symptoms	Course of the intoxication	Maximum blood cyanide	Maximum serum thiocyanate	References
25	painting with paint containing acetonitrile for 1 h, working for one day in a room in which the above paint was being used	severe chest and abdominal pain after 3 days, hepatomegaly, signs of liver insufficiency ("disability"), no icterus but "icteric index of 13.5 units"; transient albuminuria	went back to work after 2 weeks	0	60 mg/l (1.0 mM)	Amdur 1959
19	work in a photographic laboratory; acetonitrile was heated	after 4 h: upper abdominal pain, dizziness, vomiting; after 12 h: unconsciousness, convulsions, low blood pressure, vomiting blood; for several days: deep coma	death on day 5	1.1 mg/l blood (42 µM), 0.46 mg/l urine (acetonitrile: 11.7 mg/l blood, 31.1 mg/l urine)	n.d.	Dequidt *et al.* 1974

n.d. not determined

Table 2. Clinical findings in persons suffering from acute acetonitrile poisoning

Number of cases	Estimated dose	Latency period	Symptoms	Course of the intoxication	Acetonitrile and cyanide concentrations	References
dermal/inhalation						
1 2-year-old child, weight 12 kg	about 30 ml spilled over the body and the bed	8 h	little responsiveness, vomiting, lethargy, sounds over the lungs	the poisoned child survived	cyanide in blood: 231 µmol/l (6 mg/l) (after 12 h), 62 µmol/l (after 24 h), 69 µmol/l (after 48 h), 15 µmol/l (after 60 h)	Caravati and Litovitz 1988
oral						
1 16-month-old child, weight 11.8 kg	15 to 30 ml	several hours; first vomiting almost 20 h after ingestion	laboured breathing, blood pH 6.95, sinus tachycardia, pulmonary oedema	death after 12 h	cyanide in blood (3.1 mg/l) and in brain (0.2 mg/kg)	Caravati and Litovitz 1988
1	40 g (about 570 mg/kg body weight)	3 h	vomiting, convulsions, unconsciousness, respiratory failure, severe metabolic acidosis (pH 6.4), shock, twice cardiac arrest	coma for 6 days, severe anaemia, liver insufficiency, radial paralysis, muscular atrophy, recovery after 3 months	thiocyanate still detectable 20 days after the intoxication	Jaeger *et al.* 1977
1	59 ml	12 h	shallow breathing, metabolic acidosis, convulsions	recovery after 5 days	cyanide in blood: 3.1 mg/l (after 8 h), maximum 12.8 mg/l (after 59 h)	Turchen *et al.* 1991

Table 2. continued

Number of cases	Estimated dose	Latency period	Symptoms	Course of the intoxication	Acetonitrile and cyanide concentrations	References
1 3-year-old child, weight 17.2 kg	15 to 30 ml (about 800–1700 mg/kg body weight)	12 h	confusion, excitation, anxiety, vomiting	recovery after 24 h	cyanide in blood: 1.3 mg/l (after 3.45 h), thiocyanate: 11 mg/l	Geller *et al.* 1991
1 2-year-old child, weight 15.8 kg	5 to 10 ml	10 to 14 h	restlessness, vomiting, tonic-clonic cramps	recovery after 2 days	cyanide in blood: 1.82 mg/l	Kurt *et al.* 1991
1	5 ml plus 1 ml ammonia solution (about 64 mol/kg body weight)	5 h	nausea	recovery after 30 h	acetonitrile in blood: maximum 99.2 mg/l cyanide in blood: 15 mg/l	Michaelis *et al.* 1991
2	not known	not known	vomiting	death	acetonitrile in blood: 800 mg/l, in urine 1000 mg/l cyanide in blood: 2.4 and 4.5 mg/l	Jones *et al.* 1992

4.3 Carcinogenicity

In a case-control study of the relationship between exposure of factory workers to chemicals and the incidence of non-Hodgkin's lymphoma (52 cases), multiple myeloma (20), non-lymphocytic leukaemia (39) and lymphocytic leukaemia (18), the odds ratio for exposure to acetonitrile was 5.2 for non-Hodgkin's lymphoma (2 cases), 2.5 for non-lymphocytic leukaemia (1), and not increased for the other disorders. Because of the small numbers of cases and the fact that the persons were exposed to mixtures of substances, a carcinogenic effect of acetonitrile in man cannot be concluded from these data (Ott *et al.* 1989).

5 Animal Experiments and *in vitro* Studies

5.1 Acute toxicity

Dyspnoea, gasping for breath, trembling, convulsions and clouding of the cornea were seen in mice 1 to 5 hours after exposure to acetonitrile by inhalation (Willhite 1981) and intraperitoneal injection (Willhite and Smith 1981). Deaths were recorded as long as 3 days after exposure (Willhite 1981). This syndrome caused by acute acetonitrile poisoning could not be distinguished from cyanide poisoning (Willhite and Smith 1981).

Different species are very different in their sensitivity to acetonitrile. Mice are more sensitive than rats. Symptoms of intoxication which are common to all laboratory animals include irregular and rapid breathing, coordination disorders, stiffness and convulsions. Rabbits produce these typical symptoms after intraperitoneal injection of an acetonitrile dose as low as 90 to 150 mg/kg body weight (2.2–3.7 mmol/kg body weight).

A detailed review of LD_{50} values for various species and administration routes has been published by the WHO (1993).

5.1.1 Inhalation

The 2-hour LC_{50} values determined for the mouse are in the range between 2300 and 5700 ml/m^3. In a study carried out according to OECD guidelines, the 4-hour LC_{50} found for the mouse was 3544 ml/m^3 (Spanish Ministry of Health 2000). The LD_{50} values for the rat were between 7550 and 12400 ml/m^3 for 8-hour exposures and 16000 ml/m^3 for 4-hour exposures. The 4-hour LC_{50} values indicate that rabbits (2800 ml/m^3), mice and guinea pigs (5580 ml/m^3) are more sensitive to the effects of inhaled acetonitrile than are rats and dogs (7900–15800 ml/m^3). The symptoms of acetonitrile poisoning were similar in different animal species (WHO 1993). Monkeys reacted to inhalation of acetonitrile in a concentration of 2510 ml/m^3, after a delay, with coordination disorders and laboured breathing (Pozzani *et al.* 1959).

5.1.2 Ingestion

In a study carried out according to OECD guidelines, an oral LD_{50} of 617 mg/kg body weight was found for the mouse (Spanish Ministry of Health 2000). In other studies too, the mouse, with LD_{50} values below 600 mg/kg (WHO 1993), was more sensitive than the rat for which the oral LD_{50} was between 1327 (adult males) and 6762 mg/kg body weight (adult females) (Pozzani *et al.* 1959). Young animals are much more sensitive with an LD_{50} of 158 mg/kg body weight (Kimura *et al.* 1971). Diluted acetonitrile seems to be taken up more readily from the gastrointestinal tract of the rat and is therefore more toxic than undiluted acetonitrile for animals of both sexes (Pozzani *et al.* 1959). A clear dependence of the LD_{50} values on the administration route—oral administration compared with intravenous or intraperitoneal injection (see Section 5.1.4)—is not recognizable. This indicates that most of the ingested acetonitrile is absorbed.

5.1.3 Dermal absorption

The dermal LD_{50} for the rabbit after occlusive epicutaneous application of undiluted acetonitrile (984 mg/kg body weight) is about the same as the oral LD_{50} for the rat. The dermal LD_{50} for aqueous acetonitrile in the rabbit is 395 mg/kg body weight (Pozzani *et al.* 1959). In a limit test carried out according to OECD guidelines, acetonitrile doses of 2000 mg/kg body weight were applied epicutaneously (occlusion not specified) to 5 male and 5 female rabbits for 24 hours. All animals survived and no clinical signs of effects of the substance were seen, apart from reduced defecation by the female animals which was considered to be substance-related. Autopsy did not reveal substance-related gross pathological changes (Spanish Ministry of Health 2000).

5.1.4 Intravenous and intraperitoneal injection

Determination of the intraperitoneal LD_{50} for acetonitrile in the rat yielded values between 702 and 6281 mg/kg body weight, depending on whether diluted or undiluted acetonitrile was tested. The intravenous LD_{50} for acetonitrile in the rat was found to be 1320 mg/kg body weight (Pozzani *et al.* 1959). Guinea pigs and mice reacted to intraperitoneal injection of acetonitrile more sensitively than did the rat (Pozzani *et al.* 1959, Willhite and Smith 1981). Thus the intraperitoneal LD_{50} for the mouse (181–521 mg/kg body weight) is about one tenth of that for the rat.

5.2 Subacute, subchronic and chronic toxicity

5.1.1 Inhalation

The toxicity data for acetonitrile after repeated inhalation exposure are shown in Table 3.

Table 3. Effects of repeated exposure of experimental animals to acetonitrile

Species (strain) number/group	Concentration (mg/m^3)	Concentration: effects	References
mouse			
($B6C3F_1$) 10 ♂	100, 200, 400 6 h/day, 5 days/week, 13 weeks	**400 ml/m³:** no effects on weights of testes or *cauda epididymidis*, no effects on sperm motility, increased body weights	Morrissey *et al.* 1988
($B6C3F_1$) 10 ♂, 10 ♀	100, 200, 400, 800, 1600 6 h/day, 5 days/week, 13 weeks	**from 100 ml/m³:** ♂: relative liver weights increased **from 200 ml/m³:** ♂: absolute liver weights increased, ♀: forestomach hyperplasia **from 400 ml/m³:** ♀: relative liver weights increased, 1 death; ♂: forestomach hyperplasia, ♀/♂: cytoplasmic vacuolation in hepatocytes **800 ml/m³:** ♀: absolute liver weights increased, 4 deaths, ♂: 1 death, ♂/♀: hypoactivity in week 1 **1600 ml/m³:** ♂/♀: hypoactivity in week 1, all animals died	NTP 1996
($B6C3F_1$) 10 ♂, 10 ♀	50, 100, 200 6 h/day, 5 days/week, 15 months	**100 ml/m³: NOAEL** **200 ml/m³:** forestomach hyperplasia blood not studied	
rat			
(not specified) ♂, ♀	618, 1847, 6239 6 h/day, 5 days/week, 1 month	**618 ml/m³: NOAEL** **from 1847 ml/m³:** irritation of eyes and nose, effects on the nervous system, slight anaemia **6239 ml/m³:** reduced body weight gains, deaths	WHO 1993
(Carworth Farms Wistar), 15 ♂, 15 ♀	166, 330, 655 7 h/day, 5 days/week, 90 days	**166 ml/m³:** 1 animal with aggregations of histiocytes in the alveoli, another with atelectasis **330 ml/m³:** 3 of 26 surviving animals with abnormal findings: bronchitis, pneumonia, atelectasis, aggregations of histiocytes in the alveoli **655 ml/m³:** 10 animals with congestion of the alveolar capillaries, focal oedema, bronchial inflammation, exfoliation of the bronchial epithelium, hypersecretion of mucous, 8 animals with cloudy swelling of the renal tubules, 7 animals with cloudy swelling of the liver **all exposure groups:** no effects on body weight gains, relative kidney or liver weights in comparison with the control group; some animals died with infections	Pozzani *et al.* 1959

Table 3. continued

Species (strain) number/group	Concentration (mg/m^3)	Concentration: effects	References
(F-344/N), 10 ♂, 10 ♀	100, 200, 400, 800, 1600 6 h/day, 5 days/week, 13 weeks	**400 ml/m³: NOAEL** **800 ml/m³:** ♂: 1 animal with coordination disorders died **from 800 ml/m³:** ♂/♀: absolute and relative thymus weights decreased, ♀: anaemia, ♂: hypoactivity in week 1 **1600 ml/m³:** ♂: anaemia, 6 deaths, ♀: 3 deaths, absolute and relative weights of kidneys, heart, liver increased, hypoactivity in week 1, triiodothyronine decreased, ♂/♀: body weights decreased	NTP 1996
(F-344/N), 8 ♂, 8 ♀	100, 200, 400 6 h/day, 5 days/week, 15 months	**400 ml/m³:** ♂: increased erythrocyte counts, decreased cell volumes and haemoglobin, ♀: decreased haematocrit and haemoglobin, increased leukocyte counts not biologically relevant, **NOAEL**	NTP 1996
dog			
(mongrel) 2 ♂, 1 ♀	350 7 h/day, 5 days/week, 91 days	**350 ml/m³:** body weight reduced on days 3 and 5 and on 8 other days before week 72, haemoglobin and haematocrit values reduced from week 5; in 2 animals values returned to normal by day 91, no effects on erythrocyte, neutrophilic leukocyte or lymphocyte counts compared with the values determined before the exposure, focal emphysema, proliferation of the alveolar septa	Pozzani *et al.* 1959
monkey			
(rhesus) 3 ♂[a]	350 7 h/day, 5 days/week, 91 days	**350 ml/m³:** brain haemorrhage, focal emphysema, proliferation of the alveolar septa, swelling of the convoluted renal tubules	Pozzani *et al.* 1959
(rhesus) 1 ♂[a]	330 7 h/day, 5 days/week, 99 days	**330 ml/m³:** chronic pneumonia, excitability	
(rhesus) 2 ♀[a]	660 7 h/day, 5 days/week, 23 and 51 days	**660 ml/m³:** animals died on days 23 and 51; brain haemorrhage (in and under the *dura mater*), emphysema, atelectasis, proliferation of the alveolar septa, swelling of the convoluted renal tubules	

[a] for these studies no control animals were used

13-week inhalation studies have been carried out with $B6C3F_1$ mice and F344/N rats. As these were range-finding studies, only a limited number of biochemical parameters and clinical symptoms were monitored. Groups of 10 male and 10 female $B6C3F_1$ mice and F344 rats were exposed to acetonitrile in concentrations of 100, 200, 400, 800 or 1600 ml/m^3, 6 hours/day, 5 days/week for 13 weeks. The control animals, the males of the 800 ml/m^3 group and all animals of the 1600 ml/m^3 groups were subjected to a complete histological examination. In the lower concentration groups, other organs were also examined when lesions developed. Mortality was seen in the 400 ml/m^3 group. Lymphoid depletion and lymphocytolysis were seen in the thymus, spleen and bone marrow, and pulmonary congestion in the animals which died; these symptoms were considered to be non-specific disorders of moribund animals. Body weights were reduced only in male animals but the reductions were not significant and not concentration-dependent and thus probably not exposure-related. In the male mice, a significant concentration-dependent increase in relative liver weights was observed at concentrations of 100 ml/m^3 or more and in absolute liver weights from 200 ml/m^3. In the female mice, the relative liver weights were increased from 400 ml/m^3 and the absolute at 800 ml/m^3. Forestomach lesions were clearly associated with the exposure. The gross pathological examination revealed single and multiple foci in the forestomach mucosa of the females exposed to 200 to 1600 ml/m^3 and in the males exposed to 400 to 1600 ml/m^3. Histological examination of the foci revealed epithelial hyperplasia. The mucosa in these foci was sometimes highly thickened, folded and covered with a pronounced keratin layer (hyperkeratosis). In the adjacent submucosa, inflammatory infiltrates were found. In 7 animals, ulcers had formed in the area of the epithelial hyperplasia. In the male and female mice of the 400 and 800 ml/m^3 groups liver changes were also detected. Very frequently vacuoles formed in the hepatocyte cytoplasm; the authors considered these to be a slight enlargement of already present vacuoles ("clear spaces") and to indicate increased glycogen deposition (NTP 1996). A NOAEL for the mouse cannot be derived from this study.

Unlike the $B6C3F_1$ mice, the rats exposed in this 13-week study developed anomalies only after exposure to 800 and 1600 ml/m^3 and were less sensitive than the mice, as has already been shown in the acute toxicity studies. Deaths were seen only in the 800 and 1600 ml/m^3 groups. Coordination disorders and convulsions were seen in rats which died during the first week of exposure to 800 or 1600 ml/m^3. Pulmonary congestion, oedema and alveolar haemorrhage were detected in the animals which died and also brain haemorrhage, cell depletion in the bone marrow, thymus atrophy, lymphoid depletion in the spleen and the *corpora lutea* of the ovary. In the 1600 ml/m^3 groups, body weights were reduced. In the male and female animals of the 800 and 1600 ml/m^3 groups, the absolute and relative thymus weights were significantly lower than those of the control animals. Normochromic normocytic anaemia was diagnosed in the 1600 ml/m^3 group rats and in the female animals of the 800 ml/m^3 group. In the females of the 1600 ml/m^3 group, the concentration of serum triiodothyronine (T_3) was decreased whereas the thyroxine (T_4) and thyreotropic hormone concentrations remained constant (NTP 1996). From this 13-week study, a NOAEL of 400 ml/m^3 for the rat can be deduced.

An earlier NTP 13-week study which has not been published but is cited in several reviews of the toxicology of acetonitrile (Spanish Ministry of Health 2000, WHO 1993) was considered by the NTP itself to be inadequate. The results cited are the interpretation of the laboratory which carried out the study and not of the NTP (NTP 2001). They were therefore not used in the current assessment.

Further studies of the subchronic toxicity of inhaled acetonitrile for the rat, dog and monkey are shown in Table 3. The target organs were shown to be the lungs, liver, kidneys and brain. Monkeys proved to be more sensitive than dogs (Pozzani *et al.* 1959). The studies with dogs and monkeys are, however, of only limited usefulness because the control dogs were not subjected to histological examination and there were no control monkeys.

In a 2-year NTP carcinogenicity study (1966; see Section 5.7) groups of 8 rats and 10 mice of each sex were examined after exposure for 15 months (rats 100, 200, 400 ml/m^3, mice 50, 100, 200 ml/m^3). No significant concentration-dependent effects on the weights of body, liver, kidneys or lungs were found. Haematological changes seen at 400 ml/m^3 were not considered by the NTP to be biologically relevant. In the mice, the incidence of epithelial hyperplasia in the forestomach was increased at 200 ml/m^3. The incidence of basophilic and clear cell foci in the livers of male rats was not increased above that in the control animals but for both controls and exposed animals was much higher (about 75 %) than in the 2-year study (NTP 1996).

In the 2-year study (NTP 1996) male and female mice and rats were exposed to acetonitrile in concentrations up to 200 ml/m^3 (mice) and 400 ml/m^3 (rats) (Tables 6, 7). Survival and body weight development were not different from the control values. In fact, the male mice of the 200 ml/m^3 group lived significantly longer. The incidence of hyperplasia of the forestomach epithelium was significantly increased at 100 ml/m^3. The mechanism of development of this hyperplasia is not understood. It is conceivable that acetonitrile is transferred by mucociliary clearance to the stomach where it causes unspecific irritation. Apart from the increased incidence of basophilic liver foci in the male rats (see Section 5.7) and the forestomach hyperplasia in mice, no treatment-related effects were found in the organs and tissues examined. The forestomach hyperplasia in the mice could be the expression of irritant effects in the forestomach; however, irritation like that normally caused by irritant substances in the noses of the animals was not seen. In addition, the fact that no forestomach lesions developed in rats even at 8 times higher concentrations suggests that this is rather a species-specific finding in the mouse; it has also been reported in other NTP studies (NTP 1996). Thus this end point is considered to be irrelevant for man. The study yields a LOAEL of 100 ml/m^3, the concentration at which the incidence of basophilic liver foci was higher, but not significantly higher, in exposed male rats than in the control animals. The NOAEL in this study is thought to be about 50 ml/m^3.

Many of the findings of the 15-month study and the 2-year study contradict those of the 13-week study described above. This could be accounted for by proposing that the effects described in the 13-week study are somehow subject to adaptation or are chance findings; thus the 13-week study is not suitable for the derivation of a threshold concentration for the workplace.

5.2.2 Ingestion

Daily injection or oral administration of acetonitrile (0.1–0.25 ml subcutaneous or 0.5 ml oral) to 4-month-old rabbits (weights not given) for a period of 21 days induced marked hyperplasia of the thyroid in the male animals. After 3 weeks of treatment, the thyroid glands were at least twice their normal size and were very hyperaemic. Allyl cyanide and benzyl cyanide, and to a much lesser extent also phenyl cyanide and cyanamide induced thyroid hyperplasia but produced less organ enlargement. Young rabbits reacted to the individual cyanides with more pronounced thyroid enlargement than did older rabbits. Sodium thiocyanate, administered intraperitoneally in doses of 0.025 to 0.2 ml (no further details), had no effects (Marine *et al.* 1932).

5.3 Local effects on skin and mucous membranes

5.3.1 Skin

A study of irritation of rabbit skin carried out according to OECD guidelines revealed that after epicutaneous application of 0.5 ml acetonitrile for 4 hours, irritation was not detected at any of the examination times (Spanish Ministry of Health 2000).

5.3.2 Eyes

A study of irritation of the rabbit eye carried out according to OECD guidelines revealed that 0.1 ml acetonitrile was clearly irritating (Spanish Ministry of Health 2000). Eye irritation was also seen in hamsters which had inhaled acetonitrile for one hour in a concentration of 8000 ml/m^3 (Willhite 1983). Not only inhalation but also intraperitoneal injection of acetonitrile in acute toxicity studies resulted in corneal clouding in the mouse (Willhite 1981, Willhite and Smith 1981); thus this is not just a local irritative effect.

5.4 Allergenic effects

A Buehler sensitization test carried out with undiluted acetonitrile and 20 guinea pigs yielded negative results (BP Chemicals 1999).

5.5 Reproductive and developmental toxicity

5.5.1 Fertility

In a screening test for reproductive toxicity, inhalation of acetonitrile in concentrations of 100, 200 or 400 ml/m^3 for 13 weeks did not affect sperm motility or the weights of testis or *cauda epididymidis* in mice or rats, or epididymis weights in rats (Morrissey *et al.* 1988). However, as these results are from the 13-week study considered by the NTP to be inadequate (see Section 5.2.1), their interpretion as clearly negative is questionable.

5.5.2 Developmental toxicity

The developmental toxicity studies are listed in Table 4.

Groups of 33 SD rats were exposed to acetonitrile in concentrations of 100, 400 or 1200 ml/m^3 for 6 hours daily on days 6 to 19 of gestation. Another 10 pregnant animals per concentration were used to determine the blood acetonitrile and cyanide levels on days 8 and 18 of gestation. In addition, 10 not pregnant animals were exposed to the same concentrations. In the 400 ml/m^3 group 1/33 animals died; in the 1200 ml/m^3 group 2/33 pregnant and 1/10 not pregnant animals died. At 1200 ml/m^3 the animals were hypoactive. Neither maternal body weight development nor the reproduction parameters were affected by the exposure. Weighing and examination of the foetuses for gross pathological, visceral, skeletal and craniofacial microscopic malformations revealed no abnormal findings. The only effect was an increase in the incidence of supernumerary ribs which was not concentration-dependent and was statistically significant only at 100 ml/m^3. The NTP gave the maternal NOAEL as 100 ml/m^3 and that for developmental toxicity as 1200 ml/m^3. Only in the 1200 ml/m^3 group was the blood cyanide level increased on day 8 (2 µg/ml); on day 18 it was only about 0.5 µg/ml. The concentration of acetonitrile in blood was constant (NTP 1994). The study demonstrates that even under conditions of slightly increased mortality (1200 ml/m^3), no adverse effects on the foetuses could be detected. The maternal NOAEL could also be about 400 ml/m^3 because it is not clear that the death of the one animal was exposure-related. (For comparison: 400 ml/m^3 was tolerated by rats for 13 weeks without deaths (NTP 1996)). In the other studies, lethality is associated at least with a reduction in body weight gains. This effect was not seen here and so a chance finding cannot be excluded.

Groups of 20 SD rats were exposed to acetonitrile in concentrations of 900, 1200, 1500 or 1800 ml/m^3 for 6 hours daily on days 6 to 20 of gestation (analyzed concentrations given in Table 4). Of the dams exposed to 1800 ml/m^3, 40 % died. In dams exposed to 1500 ml/m^3 or more, body weight gains were reduced to 60 % of the control value. In the 1800 ml/m^3 group, the incidence of post implantation deaths and early resorptions was increased. In the animals exposed to 1500 ml/m^3, the number of implantation sites, the sex ratio and the foetal weights did not differ from the control values. The increase in the incidence of not ossified sternebrae and supernumerary lumbar ribs was not concentration-dependent and not higher than in other control groups in this study. There was no significant increase in visceral or skeletal variations in the exposed groups. However, the incidence of supernumerary cervical ribs was higher than in any of the control groups in the study. The authors did not consider this finding to be relevant. The maternal NOAEL was 1200 ml/m^3, the NOAEL for developmental toxicity 1500 ml/m^3 (Saillenfait *et al.* 1993). This concentration corresponds to an acetonitrile dose of about 500 mg/kg and day, assuming 10 l/h inhaled air volume and 300 g body weight.

A segment II study in which acetonitrile was administered orally to rats demonstrates that only after administration of lethal doses (275 mg/kg body weight and day) are embryotoxicity and foetotoxicity seen in the form of slight disorders of sternebral ossification. The NOAEL for maternal toxicity and developmental toxicity was 190 mg/kg body weight. The incidence of not ossified sternebrae was increased even in the lower dose groups (control: 7/121, 125 mg/kg body weight: 17/146, 190 mg/kg body weight:

20/153), but in two other control groups of the same authors the incidences of this finding were 19/146 and 21/163; therefore these findings in the lower dose groups are not relevant (Johannsen *et al.* 1986).

Rabbits were given oral acetonitrile doses of 2, 15 or 30 mg/kg body weight and day during gestation. Doses which were lethal for the dams caused embryotoxicity but no malformations. The dose of 15 mg/kg body weight caused slight maternal toxicity but no effects in the foetuses. The NOAEL for maternal toxicity was 2 mg/kg body weight (Spanish Ministry of Health 2000). Thus rabbits were much more sensitive to the toxic effects of acetonitrile than were rats.

A study with rats demonstrated that even after oral administration of acetonitrile doses which were lethal for the dams (300 and 500 mg/kg body weight), there were no effects on postnatal survival or body weight development of the progeny during the 42 days *post partum*. Teratogenicity was not studied (Smith *et al.* 1987).

With golden Syrian hamsters it was demonstrated that inhalation for one hour of lethal concentrations of acetonitrile induced malformations. The effects could be lessened by administration of thiosulfate (Willhite 1983). Lethal doses administered orally to hamsters had teratogenic effects. In some cases malformations were seen in the progeny of animals without severe symptoms of intoxication. The study in which acetonitrile was administered by intraperitoneal injection is difficult to interpret because the effects were neither statistically significant nor dose-dependent, and even when they were present were not severe. Even the authors considered that the progeny were normal (Willhite 1983). Administration of a single oral dose of 2000 mg acetonitrile to rats on day 10 of gestation caused weight loss and, in 9/34 embryos examined on day 12, misalignment of the allantois, the trunk and the caudal extremities (in 3/34 only the allantois affected). These effects are morphological anomalies which can also be induced with a dose of 6 mg sodium cyanide and also *in vitro* with sodium cyanide (see below). Sodium cyanide also caused mortality of the dams (Saillenfait and Sabaté 2000). Because the substance was administered as a single dose, the studies are not useful for the evaluation of risks of workplace exposure during pregnancy.

Seen as a whole, the relevant studies demonstrate that prenatal toxic effects of acetonitrile are only manifested after maternally toxic doses and that the dose-response relationship is mostly steep.

In vitro

It was shown with 10-day-old rat embryos cultivated for 46 hours *in vitro* that exposure to acetonitrile concentrations of 40 mM or more reduced growth rates, 120 mM caused malformations in the head area and 160 mM increased mortality. Addition of a metabolic system with NADPH did not change the results. The malformations in the head area and the reduced growth rates were also caused by 0.125 to 0.3 mM sodium cyanide. This treatment, however, also caused misalignment of the allantois, the trunk and the caudal extremities. Addition of a metabolic system with NADPH completely prevented the development of malformations. Cyanide was 400 to 1000 times more effective than acetonitrile (Saillenfait and Sabaté 2000).

Table 4. Developmental toxicity studies with acetonitrile

Animal species (administration route)	Dose/concentration (animals per group)	Effects on the dams	Effects on foetuses or new-born pups	References
Inhalation				
Sprague-Dawley rat (6 h daily on days 6–20 of gestation)	0, 100, 400, 1200 ml/m^3 (33 animals)	**400 ml/m^3: (NOAEL?),** 1/33 died **1200 ml/m^3:** 2/33 died, 14/33 hypoactivity, no effects on body weights	**0 ml/m^3:** supernumerary ribs 2.6 % **100 ml/m^3:** supernumerary ribs 9.0 % **400 ml/m^3:** supernumerary ribs 6.5 % **1200 ml/m^3: NOAEL**, supernumerary ribs 4.4 %, no effects on foetal weights, sex ratio, resorptions, litter size, number of implantations; no skeletal or visceral malformations	NTP 1994
Sprague-Dawley rat (6 h daily on days 6–20 of gestation) animals killed on day 21 after mating	0, 900, 1200, 1500, 1800 ml/m^3 (20–23 animals) analysed concentrations: 1000, 1287, 1592, 1827 ml/m^3	**1287 ml/m^3: NOAEL** **1592 ml/m^3:** body weight gains decreased [a] by 40 % **1827 ml/m^3:** 40 % mortality, body weight gains decreased [a] by 40 %	**0 ml/m^3:** supernumerary cervical ribs 1/101 (1 %) **1000 ml/m^3:** supernumerary cervical ribs 5/153 (3.3 %) **1287 ml/m^3:** supernumerary cervical ribs 11/125 (8.8 %) **1592 ml/m^3: NOAEL,** supernumerary cervical ribs 3/142 (2.1 %) **1827 ml/m^3:** supernumerary cervical ribs 0/29, significantly increased number of resorptions foetal weights unchanged, no significant increase in visceral or skeletal malformations or variations	Saillenfait *et al.* 1993
golden Syrian hamster (once for 1 h on day 8 after mating) animals killed on day 14 after mating	0 (10), 1800 (6), 3800 (6), 5000 (6), 8000 ml/m^3 (12)	**1800 ml/m^3: NOAEL** **3800 ml/m^3:** 1 animal with dyspnoea, tremor, ataxia, increased salivation, death after 3 h **5000 ml/m^3:** all animals irritated by the acetonitrile fumes, excessive salivation, 1 animal with dyspnoea, tremor, death after 5 h **8000 ml/m^3:** laboured breathing, snapping for breath, lethargy, ataxia, irritation of eyes and nose, 3 animals died after 90 min	**3800 ml/m^3: NOAEL** **5000 ml/m^3:** 6 of 53 foetuses with malformations, exencephaly (2 foetuses), encephalocele (2), fused ribs (2), missing renal primordium (1), increased number of resorptions **8000 ml/m^3:** 29 of 115 foetuses with: exencephaly (13 foetuses), encephalocele (14), fused ribs (9), extrathoracic ectopia of the spinal column (1), decreased foetal weights, no effects on resorptions, exencephaly and encephalocele could be prevented completely by administration of thiosulfate, incidence of fused ribs reduced	Willhite 1983

Table 4. continued

Animal species (administration route)	Dose/concentration (animals per group)	Effects on the dams	Effects on foetuses or new-born pups	References
Gavage				
Sprague-Dawley rat (daily on days 6–19 of gestation) animals killed on day 20 after mating	0, 125, 190, 275 mg/kg body weight (3.1, 4.6, 6.7 mmol/kg body weight) in distilled water (25 animals)	**190 mg/kg body weight: NOAEL** **275 mg/kg body weight:** 4 animals emaciated, 2 animals died, reduced body weight gains	**190 mg/kg body weight: NOAEL,** total number of implantations not reduced, no implantation losses, no reduction in the number of viable foetuses, foetal weights not reduced **275 mg/kg body weight:** increased implantation losses (early resorptions), average number of foetuses reduced, non-ossified sternebrae in 28/121	Johannsen *et al.* 1986
Long-Evans rat (daily on days 7–21 of gestation) animals killed one day after birth	0 (155), 50 (20), 150 (20), 300 (22), 500 mg/kg body weight (20), in tricaprylin, 1 ml/kg body weight	**50 mg/kg body weight:** mortality 10 % **150 mg/kg body weight:** **NOAEL**, mortality 0 % **300 mg/kg body weight:** mortality 55 % **500 mg/kg body weight:** mortality 80 %, 67 % of animals had viable pups; body weight gains decreased	**300 mg/kg body weight: NOAEL**; in ♀: liver weights significantly increased on days 41 and 42 *post partum*, in ♂: lung and spleen weights increased **500 mg/kg body weight:** body weight gain significantly increased on day 4 *post partum* external examination revealed no malformations, no effects on litter size, weights of new-born pups or on survival until day 4 *post partum*	Smith *et al.* 1987
Sprague-Dawley rat (single dose on day 10 of gestation), animals examined on day 12	0 (4), 2000 mg/kg body weight (3)	**2000 mg/kg body weight:** piloerection, prostration, tremor, maternal weight loss	**2000 mg/kg body weight:** 35 % of foetuses with misalignment of the allantois, the trunk and the caudal extremities, malformations identical with those caused by a sodium cyanide dose of 6 mg/kg body weight	Saillenfait and Sabaté 2000
New Zealand white rabbit (daily on days 6–18 of gestation)	0, 2, 15, 30 mg/kg body weight (49, 365, 731 µmol/kg body weight) (25 animals)	**2 mg/kg body weight: NOAEL** **15 mg/kg body weight:** decreased body weight gains **30 mg/kg body weight:** 5 animals died, stomach walls atrophically thin in the region of the cardia, little motor activity, ataxia, slower breathing, anorexia, decreased body weight gains	**15 mg/kg body weight: NOAEL** **30 mg/kg body weight:** decreased number of viable foetuses no effects on implantations, *corpora lutea*, malformations, foetal weights	Spanish Ministry of Health 2000

Table 4. continued

Animal species (administration route)	Dose/concentration (animals per group)	Effects on the dams	Effects on foetuses or new-born pups	References
golden Syrian hamster (one dose on day 8 after mating) animals killed on day 15 after mating	0 (6), 100 (6), 200 (6), 300 (6), 400 (12) mg/kg body weight (2.4, 4.9, 7.3, 9.7 mmol/kg body weight) in water	**< 100 mg/kg body weight: NOAEL** **from 100 mg/kg body weight:** decreased body weight gains **300 mg/kg body weight:** 1 animal died **400 mg/kg body weight:** 4 animals died, no effects on body weight gains	**0 mg/kg body weight:** no anomalies (0/44) **from 100 mg/kg body weight:** decreased foetal weights **100 mg/kg body weight:** encephalocele (2/58) **200 mg/kg body weight:** no malformations (0/38), increased number of resorptions **300 mg/kg body weight:** 12/65 with malformations: exencephaly (1), encephalocele (1), fused ribs (10) **400 mg/kg body weight:** 14/76 with malformations: encephalocele (1), fused ribs (13), increased number of resorptions **NOAEL < 100 mg/kg body weight**	Willhite 1983
Intraperitoneal injection				
golden Syrian hamster (one dose on day 8 after mating) animals killed on day 14 after mating	0 (6), 100 (6), 200 (6), 300 (5), 400 (12) mg/kg body weight (2.4, 4.9, 7.3, 9.7 mmol/kg body weight) in water	**400 mg/kg body weight: NOAEL**, no clinical signs of toxicity, no effects on body weight gains	**0 mg/kg body weight:** no anomalies (0/71) **100 mg/kg body weight:** exencephaly, encephalocele, median spinal cord and abdominal clefts, shortened upper jaw in 2/69 **from 200 mg/kg body weight:** increased foetal weights **200 mg/kg body weight:** defects of all 4 limbs in 1/61 **300 mg/kg body weight:** no anomalies **400 mg/kg body weight:** exencephaly, encephalocele, fissured skull, spinal cord cleft, limb defects in 4/150 **NOAEL < 100 mg/kg body weight**	Willhite 1983

[a] body weight on day 21 minus weight of the gravid uterus minus body weight on day 6

5.6 Genotoxicity

The results of genotoxicity studies are shown in Table 5.

5.6.1 *In vitro*

Tests for the induction of non-replicative DNA synthesis in rat hepatocytes did not suggest that acetonitrile induces DNA excision repair, but details of the concentrations used were not given (Mirsalis *et al.* 1983).

In the *Salmonella* mutagenicity test both with and without an added metabolic activation system, acetonitrile was not mutagenic in the spot test in concentrations up to 123 µg/plate (not toxic; Florin *et al.* 1980), in the preincubation test up to 10 mg/plate (not toxic; Mortelmans *et al.* 1986) and in the plate incorporation test up to 1350 mM (toxic; Schlegelmilch *et al.* 1988). Likewise acetonitrile was not mutagenic either with or without an added metabolic activation system at concentrations up to 5 µg/ml (not toxic) in L5178Y mouse lymphoma cells (Rudd *et al.* 1983) or up to 30 mg/ml (toxic) in the HPRT test in CHO cells (a cell line from Chinese hamster ovary) (Spanish Ministry of Health 2000).

Tests for mutation in a cytochrome P450 competent *Streptomyces griseus* strain (H69) yielded a positive result with the only tested acetonitrile concentration, 200 µg/ml (Buchholz *et al.* 1992). The validity of this test system is questionable because the amino acid glycine also yielded a positive result. The authors ascribed this to effects on the cell wall of *Streptomyces*.

In cytogenetic tests with CHO cells, acetonitrile (5000 µg/ml; 122 mM) induced marginal increases in sister chromatid exchange (21 % increase) only in the absence of metabolic activation, and an increased incidence of chromosomal aberrations only in the presence of metabolic activation. However, the results of the chromosomal aberration tests were considered to be questionable (Galloway *et al.* 1987).

In *Saccharomyces cerevisiae*, strain D7, the authors claimed a weak induction of gene conversion in the presence of a metabolic activation system. Reversions were not induced. The 11 concentrations tested were in the range from 360 to 768 mM. The number of convertants was doubled at 675 and 720 mM with survival of 58 % and 45 %, respectively. Fumaronitrile tested in the same way also induced a doubling of the number of convertants. However, with fumaronitrile the positive results were obtained at 0.51 to 0.765 mM, concentrations three orders of magnitude smaller than the effective acetonitrile concentrations (Schlegelmilch *et al.* 1988). The positive control substance, ethyl methanesulfonate, was used in a concentration of 5 µl/2 ml (about 18 mM). The results demonstrate that *Saccharomyces* evidently takes up ethyl methanesulfonate and fumaronitrile. There is no reason to assume that acetonitrile is not able to enter the cells as the molecule is smaller than the two mentioned above. Thus acetonitrile is only effective at extremely high concentrations which are not relevant for the present assessment.

In the *Saccharomyces cerevisiae* strain D61.M, which was constructed especially for the detection of loss of chromosome 7, acetonitrile induced aneuploidy but neither recombination nor point mutations (Zimmermann *et al.* 1985). This effect of acetonitrile was confirmed in a large inter-laboratory study with the same test system (Whittaker *et al.* 1989). It must be pointed out that acetonitrile is not the only nitrile which induces aneuploidy in the above-mentioned yeast test system; other active nitriles include propionitrile, fumaronitrile, and phenylacetonitrile, but ethyl acetate, acetone, methyl ethyl ketone and 2-methoxyethyl acetate also have this effect. Strain D61.M of *Saccharomyces cerevisiae* reveals the loss of one copy of chromosome 7 accompanied by simultaneous duplication of the remaining homologous chromosome. The genetic process which takes place in this yeast test system corresponds mechanistically to the allelic reduction of tumour suppressor genes to the homozygotic state. This allelic reduction plays an important role in carcinogenesis (loss of heterozygosity). The segregation disorder revealed by the yeast test system can be interpreted as evidence that analogous disorders could be induced in mammalian cells (Liu *et al.* 1997).

It is conceivable that the aneugenic effects result from disturbance of tubulin aggregation (Zimmermann *et al.* 1985). Evidence for this was obtained in *in vitro* tests with tubulin (isolated from *Drosophila melanogaster*, mouse and pig brain) which lost its aggregation capacity in the presence of 0.5 % acetonitrile (Gröschel-Stewart *et al.* 1985, Sehgal *et al.* 1990). Other nitriles also cause inhibition of tubulin aggregation. It is conceivable that this disturbance leads to the formation of an imperfect spindle apparatus which then distributes the chromosomes unequally between the daughter cells.

5.6.2 *In vivo*

A test for unscheduled DNA synthesis (UDS) in rat hepatocytes yielded negative results, but details of the doses administered were not given (Mirsalis *et al.* 1983).

When female larvae or young virgin fruit flies (*Drosophila melanogaster*) were fed with aqueous acetonitrile solutions (2000, 5000, 20000 and 50000 µg/ml), additional chromosomes and also chromosome losses were detected in the oocytes at 20000 and 50000 µg/ml (Osgood *et al.* 1991a).

Exposure of adult *Drosophila melanogaster* females to acetonitrile in a concentration of 131 ml/m^3 for 10, 30, 50 or 70 minutes resulted even after the shortest exposure time in progeny with chromosome losses and duplications significantly more frequently than in the controls. The result was most marked in the first sampled eggs, corresponding to treated mature oocytes. The flies were immobilized after as little as 5 minutes exposure; the recovery time depended on the exposure duration. After the 70-minute exposure, mortality was 42 % and the fertility was reduced (Osgood *et al.* 1991b). Acetonitrile is evidently much more toxic for *Drosophila melanogaster* than for mammals.

Immediately after the 13-week NTP study (NTP 1996) a micronucleus test was carried out with peripheral erythrocytes from the male and female mice. Negative results were obtained for the females and only in the 400 ml/m^3 group males was there a slight increase (2.36 ‰ compared with 1.42 ‰) in the number of micronucleus-containing normochromatic erythrocytes.

Groups of 24 NMRI mice per sex and dose were given single intraperitoneal injections of 40 % and 60 % of the acetonitrile LD_{50} (the value was taken by the authors from the literature and not stated in the publication). After 24, 48 and 72 hours, the bone marrow cells were examined. With 60 % of the LD_{50}, after 24 hours a significant increase in the incidence of micronuclei in the bone marrow was observed. 40 % of the LD_{50} had no effect. The study was also carried out under the same conditions except for a 7-day pretreatment of the animals with low acetonitrile doses to induce the liver monooxygenases (which were not induced by this treatment). In this experiment, treatment with 60 % of the LD_{50} led to a doubling of the control micronucleus incidence in male and female animals 72 hours *post applicationem* (Schlegelmilch *et al.* 1988).

Groups of 5 NMRI mice were given acetonitrile by intraperitoneal injection of doses of 100 mg/kg body weight (2.4 mmol/kg body weight: males) or 125 mg/kg body weight (3.0 mmol/kg body weight: females), the doses being the highest tolerated in a preliminary study. The bone marrow was examined 18, 24 and 36 hours later, the peripheral blood after 24, 48, 72 and 96 hours. The 36-hour examination revealed a slight but statistically significant increase in the incidence of micronucleus-containing polychromatic erythrocytes in the bone marrow of the female animals. The authors considered this increase to be of no biological significance because at 0.7 ‰ it was hardly higher than the 24-hour control value of 0.6 ‰. The micronucleus incidence in peripheral blood was not increased at any sampling time (Jones *et al.* 2001).

Taken as a whole, the published positive results of tests with yeast (Whittaker *et al.* 1989, Zimmermann *et al.* 1985) and the fruit fly (Osgood *et al.* 1991a, 1991b) suggest that high doses of acetonitrile could have aneugenic effects in mammalian cells. This has, however, not been sufficiently investigated to date. There is evidence of other genome-damaging effects, but it does not provide a sufficient basis for the classification of acetonitrile as a clearly DNA-reactive substance. The positive results in the micronucleus test described above are not considered relevant for the present assessment of acetonitrile because the extent of the effects was not biologically relevant (Jones *et al.* 2001) and not concentration-dependent (NTP 1996), or they were observed only sporadically (Schlegelmilch *et al.* 1988).

5.7 Carcinogenicity

In a 2-year study, mice were exposed to acetonitrile in concentrations up to 200 ml/m^3 and rats up to 400 ml/m^3 (NTP 1996; Tables 6 and 7).

In the male rats of the 400 ml/m^3 group, the incidence of liver tumours was slightly increased. A significant positive trend was revealed both with the logistic regression test ($p = 0.045$) and with the Cochran-Armitage test ($p = 0.026$). However, the results of the trend tests were markedly affected by the incidences in the highest concentration group. The tumour incidences were not significantly different when control and exposed groups were compared in pairs.

Table 5. Genotoxicity tests with acetonitrile

Test system	End point	Dose range	Exposure conditions	Results	Experimental details	References
in vitro						
rat hepatocytes	UDS	no data	no data	negative		Mirsalis *et al.* 1983
CHO-W-B1 cells	SCE	160–5000 µg/ml	incubation 12 h	weak positive result at 5000 µg/ml	without rat liver S9	Galloway *et al.* 1987
CHO-W-B1 cells	SCE	500–5000 µg/ml	incubation 2 h	negative	with rat liver S9	
Salmonella typhimurium TA97, TA98, TA100, TA1535, TA1537	point mutations	up to 10 mg/plate	30 minutes preincubation	negative	with and without rat liver S9	Mortelmans *et al.* 1986
Salmonella typhimurium TA98, TA100	point mutations	up to 1350 mM	plate incorporation test	negative	with and without rat liver S9	Schlegelmilch *et al.* 1988
Salmonella typhimurium TA98, TA100, TA1535, TA1537	point mutations	up to 123 µg/plate	spot test	negative	with and without rat liver S9	Florin *et al.* 1980
L5178Y mouse lymphoma cells	mutation/ clastogenicity	up to 5 µg/ml		negative	with and without rat liver S9	Rudd *et al.* 1983
CHO cells	HPRT mutation	up to 30 mg/ml		negative	with and without rat liver S9	Spanish Ministry of Health 2000
CHO-W-B1 cells	chromosomal aberrations	500–5000 µg/ml	incubation 24 h	negative	without rat liver S9	Galloway *et al.* 1987
CHO-W-B1 cells	chromosomal aberrations	500–5000 µg/ml	incubation 2 h	weak positive result at 5000 µg/ml	with rat liver S9	

Table 5. continued

Test system	End point	Dose range	Exposure conditions	Results	Experimental details	References
Streptomyces griseus strain H69 [a]	point mutations	200 µg/ml	24 h at 28°C	highly positive ("3+")	*Streptomyces griseus* strain H69 contains a xenobiotic-oxidizing CYP450 isoenzyme	Buchholz *et al.* 1992
Saccharomyces cerevisiae strain D7 [b]	gene conversion	360–768 mM (14760–31488 µg/ml)	incubation for 3 h at 28°C	positive result at 675 mM negative	with rat liver S9, induction with phenobarbital	Schlegelmilch *et al.* 1988
Saccharomyces cerevisiae strain D7 [b]	reversion	360–768 mM (14760–31488 µg/ml)		negative	with rat liver S9, induction with phenobarbital	
Saccharomyces cerevisiae strain D61.M [c]	point mutations and recombinations	2.91 %, 4.76 % (700, 1160 mM)		negative		Zimmermann *et al.* 1985
Saccharomyces cerevisiae strain D61.M [c]	aneuploidy (mitotic chromosome loss and duplication)	2.91 %, 4.76 % (700, 1160 mM)		positive		
in vivo						
UDS	rat hepatocytes	no data	no data	negative		Mirsalis *et al.* 1983
Drosophila melanogaster larvae or virgin adults	aneuploidy	0.2–5 % (48–1219 mM)		positive, both increased and decreased chromosome numbers	in the FIX and ZESTE systems	Osgood *et al.* 1991a
Drosophila melanogaster adults	aneuploidy	131 ml/m^3	gas phase, 10, 30, 50 and 70 min	positive	in the ZESTE system	Osgood *et al.* 1991b

Table 5. continued

Test system	End point	Dose range	Exposure conditions	Results	Experimental details	References
mouse (NMRI) 4 ♂	micronuclei in bone marrow cells	60 % of the LD_{50} [d]	one i.p. injection	positive result after 24 h		Schlegelmilch *et al.* 1988
		40 % of the LD_{50}		negative		
mouse (NMRI) 4 ♂	micronuclei in bone marrow cells	60 % of the LD_{50} [d]	one i.p. injection	positive result after 72 h	after pretreatment with acetonitrile	
		40 % of the LD_{50}		negative		
mouse (NMRI) 5 ♂, 5 ♀	micronuclei in bone marrow cells and in peripheral blood	♂: 100 mg/kg body weight ♀: 125 mg/kg body weight	one i.p. injection	negative		Jones *et al.* 2001
mouse ($B6C3F_1$) 10 ♂, 10 ♀	micronuclei in peripheral erythrocytes	100, 200, 400, 800 ml/m^3, for 13 weeks	inhalation	♂: significantly increased at 400 ml/m^3 ♀: negative		NTP 1996

[a] reversion to kanamycin resistance; the bacterium has a natural CYP450 isoenzyme which is like the mammalian enzymes and is activated by soya bean meal; the CYP450 in *Streptomyces griseus* makes it unnecessary to use an S9 fraction e.g. from rat liver homogenate

[b] suitable for the detection of reversions, gene conversions and mitotic "crossing-over"

[c] the system detects the expression of recessive markers following loss of the homologous chromosome (chromosome VII), which carries the dominant wild type allele. The same effect is seen with propionitrile and other aprotic polar solvents (Whittaker *et al.* 1989)

[d] from RTECS (Lewis & Sweet, Registry of toxic effects of chemical substances, APECO, Inc., Rockville, microfiche issue of NIOSH (1986)).

CHO cell line from Chinese hamster ovary
UDS unscheduled DNA synthesis
SCE sister chromatid exchange

Table 6. 2-year inhalation study with F-344/N rats (NTP 1996)

Substance:	acetonitrile
Species:	rat, F-344; 48 ♂ and 48 ♀ per group
Administration route:	inhalation (whole animal exposure chamber), 6 h/day, 5 days/week
Concentration:	0, 100, 200, 400 ml/m³ (0, 168, 336, 672 mg/m³)
Duration:	103 weeks
Toxicity:	no difference between males and females in body weight development; in the exposed female rats, the incidence of macroscopic and microscopic liver changes was not higher than in the control group

Effect		Incidence of the effect after exposure to acetonitrile			
		0 ml/m³	100 ml/m³	200 ml/m³	400 ml/m³
survival after 2 years [a]	♂	11/48	13/48	9/48	17/48
	♀	23/48	21/48	26/48	29/48
basophilic liver foci (not atypical in the microscopic examination)	♂	15/48	22/47	25/48 [b]	31/48 [b]
eosinophilic foci	♂	3/48	7/47	5/48	10/48 [c]
clear cell foci	♂	3/48	1/47	2/48	5/48
mixed cell foci	♂	1/48	1/47	1/48	5/48
hepatocellular adenomas	♂	0/48 (0 %)	1/47 (2 %)	1/48 (2 %)	3/48 (6 %)
hepatocellular carcinomas	♂	1/48 (2 %)	0/47 (0 %)	0/48 (0 %)	3/48 (6 %)
hepatocellular adenomas and carcinomas [d]	♂	1/48 (2 %)	1/47 (2 %)	1/48 (2 %)	5/48 (10 %)
hepatocellular adenomas and carcinomas [e]	♂	3.3 %	3.4 %	4.8 %	25.2 %

[a] no difference between control groups and exposed groups
[b] significant in the logistic regression test
[c] $p = 0.038$, Fisher's exact test, one-sided; $p = 0.04$, Cochran-Armitage trend test, one-sided
[d] significant positive trend in the logistic regression test and the Cochran-Armitage trend test; when the control groups and the exposed groups are compared in pairs, no significant differences are found
[e] adjusted according to Kaplan-Meier, not significant in the life-table test

According to the logistic regression test, the number of basophilic liver foci in male rats of the 200 and 400 ml/m³ groups was significantly higher than in the control group. Basophilic liver foci count as preneoplastic changes. The authors of the NTP study doubt the preneoplastic nature of these foci with the argument that they were not atypical. However, basophilic foci do not have to be atypical to be classified as preneoplastic changes (Bannasch and Zerban 1992). In addition, the incidence of eosinophilic foci (also called acidophilic foci), which precede the basophilic foci, was higher in the 400 ml/m³ group males ($p = 0.038$, calculated with Fisher's exact test, one-sided; significant trend with $p = 0.04$, calculated with the Cochran-Armitage test, one-sided) than in the control group.

Table 7. 2-year inhalation study with $B6C3F_1$ mice (NTP 1996)

Substance:	acetonitrile
Species:	mouse, $B6C3F_1$; 50 ♂ and 50 ♀ per group
Administration route:	inhalation (whole animal exposure chamber), 6 h/day, 5 days/week
Concentration:	0, 50, 100, 200 ml/m³ (0, 84, 168, 336 mg/m³)
Duration:	111 weeks
Toxicity:	no treatment-related clinical symptoms; body weights similar in all groups; no effects on liver weights; haematological parameters not determined

Effect		Incidence of the effect after exposure to acetonitrile			
		0 ml/m³	50 ml/m³	100 ml/m³	200 ml/m³
survival [a] after 2 years	♂	32/50	32/50	32/50	43/50
	♀	28/49	33/50	29/50	32/50
epithelial hyperplasia of the forestomach	♂	3/49	3/50	6/48	12/50 [b]
	♀	2/49	7/50	9/50 [c]	19/48 [c]
forestomach papillomas	♂	0/49	0/50	1/48	2/50
	♀	1/49	0/50	1/50	3/48
hepatocellular adenomas	♂	13/50 (26 %)	12/50 (24 %)	18/49 (37 %)	10/50 (20 %)
	♀	4/49 (8 %)	8/50 (16 %)	8/50 (16 %)	6/49 (12 %)
hepatocellular carcinomas	♂	7/50 (14 %)	11/50 (22 %)	13/49 (27 %) [b]	7/50 (14 %)
	♀	7/49 (14 %)	6/50 (12 %)	6/50 (12 %)	5/49 (10 %)
hepatocellular adenomas and carcinomas	♂	19/50 (38 %)	21/50 (42 %)	30/49 (61 %) [b]	15/50 (30 %)
	♀	9/49 (18 %)	13/50 (26 %)	13/50 (26 %)	10/49 (20 %)
hepatocellular adenomas and carcinomas [d]	♂	44.5 %	46.7 %	67.7 %	31.8 %
	♀	28.0 %	33.5 %	36.9 %	28.5 %
alveolar/bronchiolar adenomas	♂	6/50 (12 %)	9/50 (18 %)	8/48 (17 %)	18/50 (36 %) [b]
	♀	7/49 (14 %)	2/50 (4 %)	2/50 (4 %)	0/49 (0 %)
alveolar/bronchiolar carcinomas	♂	4/50 (8 %)	6/50 (12 %)	6/48 (13 %)	4/50 (8 %)
	♀	1/49 (2 %)	1/50 (2 %)	0/50 (0 %)	1/49 (2 %)
alveolar/bronchiolar adenomas and carcinomas [e]	♂	10/50 (20 %)	14/50 (28 %)	14/48 (29 %)	21/50 (42 %) [b]
	♀	8/49 (16 %)	3/50 (6 %)	2/50 (4 %)	1/49 (2 %)
alveolar/bronchiolar adenomas and carcinomas [d]	♂	30.3 %	40.3 %	38.5 %	44.6 %
	♀	27.3 %	9.1 %	6.9 %	3.1 %

[a] no difference between control groups and exposed groups
[b] different from the control, $p < 0.05$, determined with the logistic regression test
[c] different from the control, $p < 0.01$, determined with the logistic regression test
[d] adjusted according to Kaplan-Meier
[e] logistic regression test, trend: $p = 0.038$ for ♂

The historical control incidence for the sum of liver adenomas and carcinomas in male F344 rats in the test laboratory involved was 8/249 (3.2 %). The range was 2 % to 8 %. The arguments of the NTP for the classification of the results as "equivocal" were the absence of a good dose-response relationship, no increase in the incidence of pre-neoplastic lesions or atypical liver foci, and the fact that the incidence of liver tumours was only slightly higher than the maximum (8 %) historical control incidence. The analogous study with female F344/N rats revealed no carcinogenic activity.

Likewise, the 2-year study with male and female B6C3F$_1$ mice (Table 7) yielded no clear evidence of carcinogenicity. The incidence of alveolar/bronchiolar adenomas in the 200 ml/m^3 group males was significantly increased, as was that of hepatocellular adenomas in the 100 ml/m^3 group. No relationship of tumour incidence with exposure concentration was recognizable. Therefore the tumours were considered to be chance findings. The incidence of forestomach papillomas was not significantly increased.

The study must be criticized for the fact that the maximum tolerated concentration was not attained.

6 Manifesto (MAK value/classification)

There are no data for effects in man of long-term exposure to acetonitrile. A MAK value cannot be derived from the short-term study of exposed persons.

In a 2-year inhalation study with F344/N rats and mice, the incidences of hepatocellular adenomas and carcinomas were increased only slightly and only in the male rats of the highest concentration group (400 ml/m^3). As a carcinogenic activity of acetonitrile is considered doubtful both by the authors of the study and by the Commission, and as tumours did not develop at other sites in either rats or mice, and because it is accepted that the substance does not have genotoxic potential, acetonitrile is not classified as carcinogenic.

The 2-year study demonstrated that in male rats exposed to acetonitrile concentrations of 100 ml/m^3 or more, the incidence of basophilic foci in the liver was increased in a dose-dependent manner. At 100 ml/m^3 the incidence was not significantly increased but it was higher than in the control group. Therefore this concentration is to be seen as the LOEL.

The main effects of acetonitrile are ascribed to the release of cyanide in the organism. The values given in the literature for the detoxification capacity of man for cyanide range from 0.1 to 1.0 mg/kg body weight and hour (see MAK documentation for Hydrogen cyanide (2001), this volume). Thus a worker weighing 70 kg can detoxify 56 to 560 mg cyanide within 8 hours. On the assumption that acetonitrile is converted completely to cyanide in the organism, the acetonitrile detoxification capacity calculated from that for cyanide is 88 to 880 mg. If this amount were taken up with 10 m^3 inhaled air, the acetonitrile concentration in the air would be 8.8 to 88 mg/m^3, that is 5 to 50 ml/m^3. The lower limit of this range represents a "worst case" situation because it is based on the assumption that acetonitrile is immediately and quantitatively converted into cyanide. Another "worst case" assumption in this estimation is that 100 % of the acetonitrile is retained in the lungs. However, inhalation studies have shown that even after persons had been exposed to 160 ml for 4 hours, no cyanide was detectable in their blood.

On the basis of the results of animal studies, the MAK value for acetonitrile is established provisionally at 20 ml/m^3; this value is supported by the threshold estimation detailed above.

The half-life of acetonitrile in human serum is 30 hours, that of cyanide 15 hours; the half-life of acetonitrile in mouse serum is 8.5 hours. The kinetic data suggest that a high excursion factor would be appropriate, but because of the uncertainty as to the development and significance of the basophilic liver foci, a low excursion factor has been established: acetonitrile is classified in Peak limitation category II with an excursion factor of 2.

In two valid developmental toxicity studies in which rats inhaled acetonitrile, no prenatal toxic effects were seen in the concentration range between 1200 and 1500 ml/m^3. This exposure level corresponds to a dose of about 500 mg/kg body weight and day. An inhalation study with a second species is not available, but there is a study with oral administration to rabbits in which a maternally toxic dose of 15 mg/kg body weight was also shown to have no prenatal toxic effects. Given observance of a MAK value of 20 ml/m^3, the maximum daily dose would be 5 mg/kg body weight (70 kg body weight, inhaled air volume in 8 hours 10 m^3, absorption 100 %); therefore it is justified to classify acetonitrile in Pregnancy risk group C.

Acetonitrile must be designated with an "H" because the dermal LD_{50} values for undiluted and diluted acetonitrile were shown to be below 1000 mg/kg body weight for the rabbit and because a case of acetonitrile poisoning in man suggests that dermal uptake can be dangerous. Acetonitrile is not designated with an "S".

The database is inadequate for classification of acetonitrile in any of the categories of germ cell mutagens.

7 References

Ahmed AE, Farooqui MYH (1982) Comparative toxicity of aliphatic nitriles. *Toxicol Lett 12*: 157–164

Ahmed AE, Loh JP, Ghanayem B, Hussein G (1992) Studies on the mechanism of acetonitrile toxicity: I. Whole body autoradiographic distribution and macromolecular interaction of 2-^{14}C-acetonitrile in mice. *Pharmacol Toxicol 70*: 322–330

Amdur ML (1959) Accidental exposure to acetonitrile – A clinical study. *J Occup Med 1*: 627–633

Banea-Mayambu JP, Tylleskar T, Tylleskar K, Gebre-Medhin M, Rosling H (2000) Dietary cyanide from insufficiently processed cassava and growth retardation in children in the Democratic Republic of Congo (formerly Zaire). *Ann Trop Paediatr 20*: 34–40

Bannasch P, Zerban H (1992) Predictive value of hepatic preneoplastic lesions as indicators of carcinogenic response. *IARC Sci Publ 116*: 389–427

BP Chemicals (1997) *Delayed contact hypersensitivity study with acetonitrile (HPLC grade) in guinea pigs*. Hill Top Research Inc, Project No. 97-8472-21 (unpublished study report)

Buchholz SE, Omer CA, Viitanen PV, Sariaslani FS, Stahl Jr RG (1992) Activation and detection of (pro)mutagenic chemicals using recombinant strains of *Streptomyces griseus*. *Appl Biochem Biotechnol 32*: 149–158

Caravati EM, Litovitz T (1988) Pediatric cyanide intoxication and death from an acetonitrile-containing cosmetic. *J Am Med Assoc 260*: 3470–3473

Dahl AR, Waruszewski BA (1989) Metabolism of organonitriles to cyanide by rat nasal tissue enzymes. *Xenobiotica 19*: 1201–1205

Dalhamn T, Edfors ML, Rylander R (1968a) Mouth absorption of various compounds in cigarette smoke. *Arch Environ Health 16*: 831–835

Dalhamn T, Edfors ML, Rylander R (1968b) Retention of cigarette smoke components in human lungs. *Arch Environ Health 17*: 746–748

Dequidt J, Furon D, Wattel F, Haguenoer JM, Scherpereel P, Gosselelin B, Ginestet A (1974) Les intoxications par l'acétonitrile a propos d'un cas mortel (Poisoning by acetonitrile. A lethal case) (French). *Eur J Toxicol 7*: 91–97

Feierman DE, Cederbaum AI (1989) Role of cytochrome P-450 IIE1 and catalase in the oxidation of acetonitrile to cyanide. *Chem Res Toxicol 2*: 359–366

Florin I, Rutberg L, Curvall M, Enzell CR (1980) Screening of tobacco smoke constituents for mutagenicity using the Ames' test. *Toxicology 15*: 219–232

Freeman JJ, Hayes EP (1985) Acetone potentiation of acute acetonitrile toxicity in rats. *J Toxicol Environ Health 15*: 609–621

Freeman JJ, Hayes EP (1987) The metabolism of acetonitrile to cyanide by isolated rat hepatocytes. *Fundam Appl Toxicol 8*: 263–271

Freeman JJ, Hayes EP (1988) Microsomal metabolism of acetonitrile to cyanide. *Biochem Pharmacol 37*: 1153–1159

Galloway SM, Armstrong MJ, Reuben C, Colman S, Brown B, Cannon C, Bloom AD, Nakamura F, Ahmed M, Duk S, Rimpo J, Margolin BH, Resnick MA, Anderson B, Zeiger E (1987) Chromosome aberrations and sister chromatid exchanges in Chinese hamster ovary cells: evaluations of 108 chemicals. *Environ Mol Mutagen 10, Suppl 10*: 1–175

Geller RJ, Ekins BR, Iknoian RC (1991) Cyanide toxicity from acetonitrile-containing false nail remover. *Am J Emerg Med 9*: 268–270

Gröschel-Stewart U, Mayer VW, Taylor-Mayer RE, Zimmermann FK (1985) Aprotic polar solvents inducing chromosomal malsegregation in yeast interfere with the assembly of porcine brain tubulin *in vitro*. *Mutat Res 149*: 333–338

Haguenoer JM, Dequidt J, Jacquemont M-C (1975) Intoxication expérimentales par l'acétonitrile. 1ère note: intoxications aguies par voie intraperitoneale (Experimental acetonitrile intoxication I: acute intoxication after intraperitoneal injection) (French). *Eur J Toxicol 8*: 94–101

Houeto P, Hoffman JR, Got P, Dang VB, Baud FJ (1997) Acetonitrile as a possible marker of current cigarette smoking. *Hum Exp Toxicol 16*: 658–661

Huang J, Niknahad H, Khan S, O'Brien PJ (1998) Hepatocyte-catalysed detoxification of cyanide by L- and D-cysteine. *Biochem Pharmacol 55*: 1983–1990

Jaeger A, Tempe JD, Porte A, Stoeckel L, Mantz JM (1977) Acute voluntary intoxication by acetonitrile. *Acta Pharmacol-Toxicol-Copenh 41 (Suppl 2)*: 340

Johannsen FR, Levinskas GJ, Berteau PE, Rodwell DE (1986) Evaluation of the teratogenic potential of three aliphatic nitriles in the rat. *Fundam Appl Toxicol 7*: 33–40

Jones AW, Löfgren A, Eklund A (1992) Two fatalities from ingestion of acetonitrile: limited specificity of analysis by headspace gas chromatography. *J Anal Toxicol 16*: 104–106

Jones E, Fox V, Elliot BM, Moore NP (2001) The mutagenic potential of acetonitrile in the bone marrow and peripheral blood of the mouse. *Mutagenesis 16*: 151–154

Kimura ET, Ebert DM, Dodge PW (1971) Acute toxicity and limits of solvent residue for sixteen organic solvents. *Toxicol Appl Pharmacol 19*: 699–704

Kurt TL, Day LC, Reed WG, Gandy W (1991) Cyanide poisoning from glue-on nail remover. *Am J Emerg Med 9*: 271–272

Liu M, Grant SG, Macina OT, Klopman G, Rosenkranz HS (1997) Structural and mechanistic bases for the induction of mitotic chromosomal loss and duplication ("malsegregation") in the yeast *Saccharomyces cerevisiae*: relevance to human carcinogenesis and developmental toxicology. *Mutat Res 374*: 209–231

Luo G-X, Horowitz PM (1994) The sulfurtransferase activity and structure of rhodanese are affected by site-directed replacement of Arg-186 or Lys-249. *J Biol Chem 269*: 8220–8225

Marine D, Spence AW, Cipra A (1932) Further studies on etiology of goiter with particular reference to the action of cyanides. *Proc Soc Exp Biol Med 29*: 772–775

McKee HC, Rhoades JW, Campbell J, Gross AL (1962) Acetonitrile in body fluids related to smoking. *Public Health Rep 77*: 553–554

McMahon TF, Birnbaum LS (1990) Age-related changes in toxicity and biotransformation of potassium cyanide in male C57BL/6N mice. *Toxicol Appl Pharmacol 105*: 305–314

Michaelis HC, Clemens C, Kijewski H, Neurath H, Eggert A (1991) Acetonitrile serum concentrations and cyanide blood levels in a case of suicidal oral acetonitrile ingestion. *Clin Toxicol 29*: 447–458

Mirsalis J, Tyson K, Beck J, Loh F, Steinmetz K, Contreras C, Austere L, Marin S, Spalding J (1983) Induction of unscheduled DNA synthesis (UDS) in hepatocytes following *in vitro* and *in vivo* treatment. *Environ Mutagen 5*: 482

Morrissey RE, Schwetz BA, Lamb IV JC, Ross MD, Teague JL, Morris RW (1988) Evaluation of rodent sperm, vaginal cytology, and reproductive organ weight data from National Toxicology Program 13-week studies. *Fundam Appl Toxicol 11*: 343–358

Mortelmans K, Haworth S, Lawlor T, Speck W, Tainer B, Zeiger E (1986) *Salmonella* mutagenicity tests: II. Results from the testing of 270 chemicals. *Environ Mutagen 8, Suppl 7*: 1–119

Nagahara N, Ito T, Minami M (1999) Mercaptopyruvate sulfurtransferase as a defense against cyanide toxication: molecular properties and mode of detoxification. *Histol Histopathol 14*: 1277–1286

NL Health Council (1997) *Acetonitrile*. Dutch Expert Committee on Occupational Standards, Health Council of the Netherlands, Den Haag

NTP (National Toxicology Program) (1993) *Technical report on toxicity studies of sodium cyanide, 37*, NIH Publication 94-3386, US Department of Health and Human Services, National Institutes of Health, Bethesda, MD, USA

NTP (1994) *TER91039, Inhalation developmental toxicology studies: acetonitrile in rats*. Battelle, prepared for the National Institute of Environmental Health Sciences, NTP

NTP (1996) *Toxicology and carcinogenesis studies of acetonitrile (CAS No. 75-05-8) in F344/N rats and B6C3F1 mice (inhalation studies)*. National Toxicology Program Technical Report 447, US Department of Health and Human Services, Public Health Service

NTP (2001) communication to the secretariat of the Commission, 21.05.2001

Ohkawa H, Ohkawa R, Yamamoto L, Casida JE (1972) Enzymatic mechanisms and toxicological significance of hydrogen cyanide liberation from various organothiocyanates and organonitriles in mice and houseflies. *Pestic Biochem Physiol* 2: 95–112

Osgood C, Zimmering S, Mason JM (1991a) Aneuploidy in *Drosophila*, II. Further validation of the FIX and ZESTE genetic test systems employing female *Drosophila melanogaster. Mutat Res 259*: 147–163

Osgood C, Bloomfield M, Zimmering S (1991b) Aneuploidy in *Drosophila*, IV. Inhalation studies on the induction of aneuploidy by nitriles. *Mutat Res 259*: 165–176

Ott MG, Teta MJ, Greenberg HL (1989) Lymphatic and hematopoietic tissue cancer in a chemical manufacturing environment. *Am J Ind Med 16*: 631–643

Padmaja G (1995) Cyanide detoxification in cassava for food and feed uses. *Crit Rev Food Sci Nutr 35*: 299–339

Porter DW, Baskin SI (1996) The effect of three alpha-keto acids on 3-mercaptopyruvate sulfurtransferase activity. *J Biochem Toxicol 11*: 45–50

Porter DW, Nealley EW, Baskin SI (1996) *In vivo* detoxification of cyanide by cystathionase γ-lyase. *Biochem Pharmacol 52*: 941–944

Pozzani UC, Carpenter CP, Palm PE, Weil CS, Nair III JH (1959) An investigation of the mammalian toxicity of acetonitrile. *J Occup Med 1*: 634–642

Rudd CJ, Mitchell AD, Spalding J (1983) L8178Y mouse lymphoma cell mutagenesis assay of coded chemicals incorporating analyses of the colony size distributions. *Environ Mutagen 5*: 419

Saillenfait AM, Sabaté JP (2000) Comparative developmental toxicities of aliphatic nitriles: *in vivo* and *in vitro* observations. *Toxicol Appl Pharmacol 163*: 149–163

Saillenfait AM, Bonnet P, Guenier JP, de Ceaurriz J (1993) Relative developmental toxicities of inhaled aliphatic mononitriles in rats. *Fundam Appl Toxicol 20*: 365–375

Schlegelmilch R, Krug A, Wolf HU (1988) Mutagenic activity of acetonitrile and fumaronitrile in three short term assays with special reference to autoinduction. *J Appl Toxicol 8*: 201–209

Sehgal A, Osgood C, Zimmering S (1990) Aneuploidy in *Drosophila*. III. Aneuploidogens inhibit *in vitro* assembly of taxol-purified *Drosophila* microtubules. *Environ Mol Mutagen 16*: 217–224

Silver EH, Kuttab H, Hasan T, Hassan M (1982) Structural considerations in the metabolism of nitriles to cyanide *in vivo*. *Drug Metab Dispos 10*: 495–498

Smith MK, George EL, Zenick H, Manson JM, Stober JA (1987) Developmental toxicity of halogenated acetonitriles: drinking water by-products of chlorine disinfection. *Toxicology 46*: 83–93

Spanish Ministry of Health (2000) *Risk assessment report, acetonitrile*, May 2000

Swenne I, Eriksson UJ, Christoffersson R, Kågedal B, Lundquist P, Nilsson L, Tylleskär T, Rosling H (1996) Cyanide detoxification in rats exposed to acetonitrile and fed a low protein diet. *Fundam Appl Toxicol 32*: 66–71

Tanii H, Hashimoto K (1984) Studies on the mechanism of acute toxicity of nitriles in mice. *Arch Toxicol 55*: 47–54

Taskinen H, Kyyronen P, Hemminki K, Hoikkala M, Lajunen K, Lindbohm ML (1994) Laboratory work and pregnancy outcome. *J Occup Med 36*: 311–319

Tor-Agbidye J, Palmer VS, Sabri MI, Craig AM, Blythe LL, Spencer PS (1998) Dietary deficiency of cystine and methionine in rats alters thiol homeostasis required for cyanide detoxification. *J Toxicol Environ Health 55*: 583–595

Turchen SG, Manoguerra AS, Whitney C (1991) Severe cyanide poisoning from the ingestion of an acetonitrile-containing cosmetic. *Am J Emerg Med 9*: 264–267

Vennesland B, Castric PA, Conn EE, Solomonson LP, Volini M, Westley J. (1982) Cyanide metabolism. *Fed Proc 41*: 2639–2648

Whittaker SG, Zimmermann FK, Dicus B, Piegorsch WW, Resnick MA, Fogel S (1989) Detection of induced mitotic chromosome loss in *Saccharomyces cerevisiae* – an interlaboratory study. *Mutat Res 224*: 31–78

WHO (World Health Organization) (1993) *Acetonitrile. Environmental Health Criteria 154*, International Programme on Chemical Safety, World Health Organization, Geneva

Willhite CC (1981) Inhalation toxicology of acute exposure to aliphatic nitriles. *Clin Toxicol 18*: 991–1003

Willhite CC (1983) Developmental toxicology of acetonitrile in the Syrian golden hamster. *Teratology 27*: 313–325

Willhite CC, Smith RP (1981) The role of cyanide liberation in the acute toxicity of aliphatic nitriles. *Toxicol Appl Pharmacol 59*: 589–602

Zimmermann FK, Mayer VW, Scheel I, Resnick MA (1985) Acetone, methyl ethyl ketone, ethyl acetate, acetonitrile and other polar aprotic solvents are strong inducers of aneuploidy in *Saccharomyces cerevisiae*. *Mutat Res 149*: 339–351

completed 01.03.2001

Bisphenol A diglycidyl ether

MAK value	–
Peak limitation	–
Absorption through the skin (1997)	**H**
Sensitization (1997)	**Sh**
Carcinogenicity (1997)	**Category 3A**
Prenatal toxicity	–
Germ cell mutagenicity	–
BAT value	–
Synonyms	BADGE, DGEBA, DGEBPA 2,2-bis(4-(2,3-epoxypropoxy)phenyl)-propane 2,2-bis(*p*-glycidyloxyphenyl)propane bis(4-glycidyloxyphenyl)dimethylmethane 2,2-bis(4-hydroxyphenyl)propane diglycidyl ether bis(4-hydroxyphenyl)dimethylmethane diglycidyl ether diglycidyl ether of bisphenol A 4,4′-isopropylidenediphenol diglycidyl ether
Chemical name (CAS)	2,2′-[(1-methylethylidene)bis(4,1-phenyl-eneoxymethylene)]bis-oxirane
CAS number	1675-54-3
Structural formula	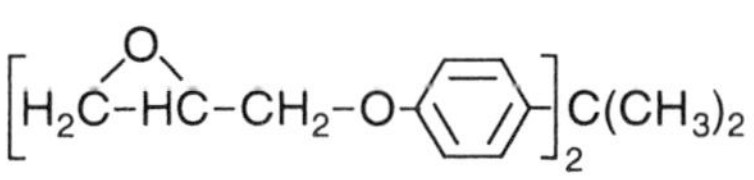
Molecular formula	$C_{21}H_{24}O_4$
Molecular weight	340.42
Melting point	8–12°C
Boiling point	> 200°C
Density at 25°C	1.15–1.2 g/cm^3
Vapour pressure at 25°C	< 0.0001 hPa
$\log P_{ow}$*	about 3.8

* *n*-octanol/water distribution coefficient

Technical grade bisphenol A diglycidyl ether (BADGE) is a viscous liquid at 25°C and contains not only the monomer, which is crystalline in the highly purified form, but also up to about 20 % w/w of oligomers (mainly the dimer and trimer). Low-molecular weight epoxy resins contain up to 95 % bisphenol A diglycidyl ether, higher molecular weight resins contain, as well as a maximum of 10 % bisphenol A diglycidyl ether, various oligomers of the substance. In addition to bisphenol A diglycidyl ether, epoxy resins can contain other glycidyl ethers and epoxides as reactive thinners. Bisphenol A diglycidyl ether is used in combination with polyamines, polyaminoamides, and acid anhydrides or their adducts as cross-linking partners above all in the production of plastics (DECOS 1996). The vapour pressure of bisphenol A diglycidyl ether is very low. Inhalation of the substance is therefore only possible during exposure to powdered polymers, which contain only small amounts of the monomer. During the spraying of paint powder, concentrations of bisphenol A diglycidyl ether of 0.005 to 0.200 mg/m^3 were determined in the breathing zone of the user and of 0.002 to 0.008 mg/m^3 in the air of the room (IARC 1989). During production and use of the substance, however, skin contact with the epoxy resin can occur, in particular if occupational hygiene is inadequate. Natural rubber is not suitable for protection of the skin against bisphenol A diglycidyl ether and formulations with other epoxides and hardeners. Protective gloves should therefore be selected according to the guidelines of the Berufsgenossenschaften der Bauwirtschaft (the professional trade organization of the building industry), and those of the trade organizations Bauchemie und Holzschutzmittel e.V. (chemistry in the building industry) and Klebstoffe e.V. (glues) (APME 1996).

1 Toxic Effects and Mode of Action

Bisphenol A diglycidyl ether applied to the skin is absorbed percutaneously only slowly. Orally administered bisphenol A diglycidyl ether, however, is absorbed rapidly and excreted mainly with the faeces and to a lesser extent with the urine. After oral administration, only small amounts of unchanged bisphenol A diglycidyl ether are systemically available. Independent of the administration route, bisphenol A diglycidyl ether is hydrolysed mainly by epoxide hydrolases to form the bis-diol and is eliminated with the urine in free or conjugated form. In the faeces the main metabolites are α-hydroxycarboxylic acids and carboxylic acids.

The acute dermal and oral toxicity of bisphenol A diglycidyl ether is low. After medium-term oral administration to rats, neither systemic toxicity nor toxic effects on organs were detected.

In vitro, bisphenol A diglycidyl ether causes gene mutation in bacterial and mammalian cells, and gene conversion in yeasts. Positive results are obtained in the SCE (sister chromatid exchange) test, in the test for chromosomal aberrations and in the UDS (unscheduled DNA synthesis) test. However, *in vivo* after peroral administration, no micronuclei or chromosomal aberrations were found in bone marrow or germ cells, and no DNA strand breaks in the liver. Dominant lethal tests with peroral administration and dermal application, and host-mediated assays yielded negative results. It is unclear whether the lack of genotoxicity *in vivo* is the result of the fact that relevant concentrations of bisphenol A diglycidyl ether were not reached in the target organs investigated or because bisphenol A diglycidyl ether is effectively detoxified in these organs by epoxide hydrolases. As studies with mouse skin show, when the epoxide hydrolases are saturated, the glycidyl groups of bisphenol A diglycidyl ether can be oxidatively dealkylated to yield glycidaldehyde. Under such conditions a glycidaldehyde–DNA adduct is formed in mouse skin. The formation of DNA adducts, the direct alkylating effect of bisphenol A diglycidyl ether, and the genotoxic activity *in vitro* suggest in principle the possibility of carcinogenic effects, which, however, have not been confirmed by the available long-term studies with dermal application of bisphenol A diglycidyl ether to mice.

In reproductive toxicological studies, no effects on fertility or on the development of the offspring could be detected in multi-generation or single-generation studies with rats given the substance by gavage. In embryotoxicity and teratogenicity studies of rats and rabbits given oral doses up to the maternally toxic range, no evidence of adverse effects on intrauterine development or of teratogenic effects was found. Nor was evidence of toxic effects on development found with pregnant rabbits treated epicutaneously with irritative doses of bisphenol A diglycidyl ether.

Bisphenol A diglycidyl ether was found to be a contact allergen in man and experimental animals. Commercial bisphenol A diglycidyl ether also produces skin irritation. The data for eye irritation are inconsistent.

2 Mechanism of Action

The saturation of epoxide hydrolases plays a role in the formation of DNA adducts and in the genotoxic activity *in vitro*, as this can result in the formation of glycidaldehyde, the reactive agent (see Sections 3 and 5.5).

3 Toxicokinetics and Metabolism

Bisphenol A diglycidyl ether is absorbed percutaneously only slowly after epicutaneous application. After a single occlusive dermal application of 2-^{14}C-propane-labelled bisphenol A diglycidyl ether (56 mg/kg body weight in acetone) to male CF1 mice, the maximum levels of recovered radioactivity were found on the second and third days and were about 8 % in the faeces and 1.3 % in urine. From days 3 to 8, the amount excreted daily decreased from 5.2 % to 2.6 %; renal excretion decreased from 0.8 % to 0.3 %. On the first day after application, 67 % of the administered radioactivity could be extracted from the skin with methanol–acetone; 97 % of this was the unchanged substance. After 8 days the extractable radioactivity was still as much as 11 % of the dose applied to the skin; 81 % of this was unchanged bisphenol A diglycidyl ether, 19 % consisted of polar metabolites. 1.8 % of the dose could not be extracted at this point, presumably because it was bound to thiol and amino groups in the epidermis (Climie *et al.* 1981a).

In *in vitro* studies, the percutaneous penetration of ^{14}C-ring-labelled bisphenol A diglycidyl ether and its metabolism in dermatomed human skin (mammoplasty), F344 rat skin and in fresh, full-thickness C3H mouse skin were studied by means of a flow-through diffusion chamber (Boogaard *et al.* 2000a). Because the substance has practically no vapour pressure, the recovery of the epidermal dose of bisphenol A diglycidyl ether (5 µmol/cm^2 in acetone) was 100 %. The rate of penetration of bisphenol A diglycidyl ether remained constant during 24 hours independently of the kind of skin tested. The apparent permeability constants for human, rat and mouse skin, respectively, were 0.48 ± 0.22, 5.5 ± 1.5 and 8.6 ± 1.6 (× 10^{-6}cm/h) during 24 hours. The permeability of human skin was less than one tenth of that of rodent skin and it also took longer for the substance to penetrate human skin (Table 1). The authors compare their results obtained with mouse skin (3 % absorption) to those of *in vivo* studies (4 % absorption; Climie *et al.* 1981) and conclude that their model is meaningful. During percutaneous absorption bisphenol A diglycidyl ether is metabolized extensively. Of the substance which penetrated the skin, 79 ± 8 %, 92.9 ± 0.1 % and 96 ± 1% was converted by human, rat and mouse skin, respectively, into the bis-diol (bisphenol A diglycerol ether); 10 ± 5 %, 3.8 ± 0.1 % and 2.0 ± 0.2% were hydrolysed to the mono-diol and only up to 1.1% was unchanged bisphenol A diglycidyl ether. These proportions remained constant during 24 hours (Boogard *et al.* 2000a). When considering the doses involved, it must be remembered that the amount of substance which penetrates human skin is more than an order of magnitude less than that which penetrates rodent skin. Thus only 0.015 % of the applied amount of bisphenol A diglycidyl ether penetrates human skin in the form of the mono-diol and the unchanged substance.

Table 1. Penetration of ^{14}C-bisphenol A diglycidyl ether through dermatomed human and rat skin and through intact mouse skin *in vitro*, measured at 3 hour intervals during a period of 24 hours (Boogaard *et al.* 2000a)

Species	Recovery (% of the dose)	Apparent permeability constant (10^{-6} cm/h)	Percutaneous penetration (% of the dose)	Penetration delay time (h)
man	102 ± 10	0.48 ± 0.22	0.137 ± 0.005	6.1 ± 1.7
rat	98 ± 11	5.5 ± 1.5	1.57 ± 0.41	1.8 ± 0.5
mouse	98 ± 8	8.6 ± 1.6	2.98 ± 0.80	2.3 ± 1.0

Rat skin is much thinner than human skin and pig skin thicker, but when skins are dermatomed there is always a risk of cutting the hair follicles, which are very numerous in such animal skins, and so producing microscopic holes through which the test substances can penetrate in an unphysiological way. Studies with dermatomed animal skin must therefore be interpreted with great caution.

The *in vitro* results described here are, however, in good agreement with the above-mentioned earlier results of *in vivo* studies with rodents (Climie *et al.* 1981).

Unlike after dermal application, the same single dose of ^{14}C-bisphenol A diglycidyl ether administered orally was rapidly excreted by the mice. On average, within as little as 24 hours, 61 % of the dose was detected in the faeces and 9 % in urine. Within 3 days about 80 % of the radioactivity had been recovered in the faeces and 11 % in urine (Climie *et al.* 1981a).

In F344 rats given oral 2-^{14}C-propane-labelled bisphenol A diglycidyl ether doses of 2.7 mg/kg body weight, the maximum level of ^{14}C in plasma was observed 4 hours after administration; less than 10 % was unchanged bisphenol A diglycidyl ether. The level in plasma decreased with a half-time of 4.8 hours. The plasma half-life for an intravenous dose of 0.45 mg/kg body weight was 8.8 hours (Dow Chemical 1981).

In model studies of the sensitivity of bisphenol A diglycidyl ether to hydrolysis during incubation with 0.1 N HCl, a half-life of about 10 minutes was determined. After incubation with 0.01 N HCl (corresponding to the pH value of the gastric juice of the rat), however, no significant hydrolysis was detected within an hour (Ciba Specialty Chemicals 1997). The authors conclude that after oral administration of bisphenol A diglycidyl ether, the reactive epoxy groups are hardly inactivated by acid hydrolysis in the stomach milieu of the animals. *In vitro*, in synthetic stomach fluid, a half-life of 70 minutes was determined for bisphenol A diglycidyl ether (Dow Chemical 1981).

Tissue affinity of systemically available bisphenol A diglycidyl ether could not be detected after either dermal or oral administration to mice: less than 1 % of the radioactivity was found in the organs, in blood or in the body of the animals (Climie *et al.* 1981a). After oral administration to rats, the highest tissue ^{14}C concentrations were found in the liver and intestines. Relative to those found after intravenous administration, the ratio of the concentration in tissue to that in plasma was increased for the liver, intestines and lungs (Dow Chemical 1981).

Within 48 hours after occlusive dermal application of 0.4 mg/animal (about 20 mg/kg body weight) of bisphenol A diglycidyl ether ^{14}C-labelled either on the central C atom (2-^{14}C-propane) or in the glycidyl side-chain, about 36 % of the radioactivity was excreted by C57BL6 mice with the faeces and about 7 % with the urine. In addition, small amounts of $^{14}CO_2$ ($\leq$ 0.5 % of the dose) were exhaled, but only after application of the substance labelled in the side-chain. After higher doses (1.3 or 2 mg/animal or about 65 or 100 mg/kg body weight) excretion with the faeces decreased considerably to levels of 11 % to 16 %, while a higher percentage of the radioactivity remained in the skin at the application site (Bentley *et al.* 1989).

Independent of the administration route (dermal or oral), ^{14}C-bisphenol A diglycidyl ether is metabolized by mice via epoxide hydrolases to form mainly the bis-diol, and is excreted to a small extent (about 8 % of the dose) in free form or as a conjugate. Most of the bis-diol formed is oxidized at one of the diol groups to yield the α-hydroxycarboxylic acid (about 27 % of the dose) and, after decarboxylation, carboxylic acid (about 15 % of the dose). The so-called "phenol diol" is formed in small amounts (about 5 % of the dose) presumably by oxidative dealkylation of the bis-diol and cleavage of glyceraldehyde (see Figure 1). ^{14}C from the glyceraldehyde can enter the normal carbon metabolism of the animals.

In the urine of workers employed in the production of epoxy resin (no further details of exposure), no bisphenol A diglycidyl ether metabolites could be detected (Shell

1983a). Free and conjugated bis-diol were analysed by HPLC and UV-detector; the detection limit was 0.1 to 0.2 μg/ml.

In vitro the alternative metabolic pathway for the formation of the "phenol diol", the oxidative dealkylation of bisphenol A diglycidyl ether after cleavage of the reactive glycidaldehyde (see Figure 1), could be detected to a small extent (1 %) only after inhibition of epoxide hydrolase with 1,1,1-trichloro-2,3-epoxypropane (TCPO) in the presence of liver microsomes from CF1 mice. The authors therefore propose that direct dealkylation of bisphenol A diglycidyl ether to form glycidaldehyde does not take place *in vivo* and the high level of activity of epoxide hydrolase in the liver leads to the more or less complete hydrolysis of bisphenol A diglycidyl ether to the bis-diol (Climie *et al.* 1981b).

On the other hand, it has been shown after epidermal application of ^{14}C-bisphenol A diglycidyl ether labelled in the glycidyl side chain that the alternative metabolic pathway which yields glycidaldehyde plays a role in the skin (Bentley *et al.* 1989). In the skin of C3H mice, small amounts of glycidaldehyde–DNA adducts were detected after epidermal application of 0.8 mg and 2 mg/animal, but not after application of 0.4 mg/animal. Steiner *et al.* (1992) identified the reaction product as a glycidaldehyde–deoxyadenosine adduct. The dose-dependency of adduct formation in the mouse skin, with evidence of a threshold dose, is attributed by the authors to saturable dermal epoxide hydrolase activity (Bentley *et al.* 1989). At higher substrate concentrations the glycidyl residue of bisphenol A diglycidyl ether can be competitively cleaved by the monooxygenase to yield the reactive glycidaldehyde (see Section 5.5.2).

In mice, bisphenol A diglycidyl ether is a substrate for microsomal and cytosolic epoxide hydrolases in the liver and skin. The hydrolytic activity results in metabolic inactivation of this reactive substance (Bentley *et al.* 1989, Ciba Specialty Chemicals 1997; see Section 5.5.2). Efficient metabolic deactivation of bisphenol A diglycidyl ether is to be expected also in man, as it was shown in the 1970s that the epoxide hydrolase activities for polycyclic aromatic hydrocarbons in human liver and skin are higher than in the mouse and rat (Oesch *et al.* 1978, Walker *et al.* 1978). Recent comparative studies of bisphenol A diglycidyl ether metabolism have confirmed that microsomes from human liver have a much higher hydrolytic activity than do rat and mouse liver microsomes (V_{max} 150 compared with 63 and 46.6 nmol/mg/min; K_m 0.63 compared with 0.34 and 0.23 mM). In man and rodents, epoxide hydrolysis is the main and efficient route of detoxification of bisphenol A diglycidyl ether. Glutathione conjugation plays only a subsidiary role. Bisphenol A diglycidyl ether and its mono-diol appear to have similar affinities for epoxide hydrolase (Boogaard *et al.* 2000b).

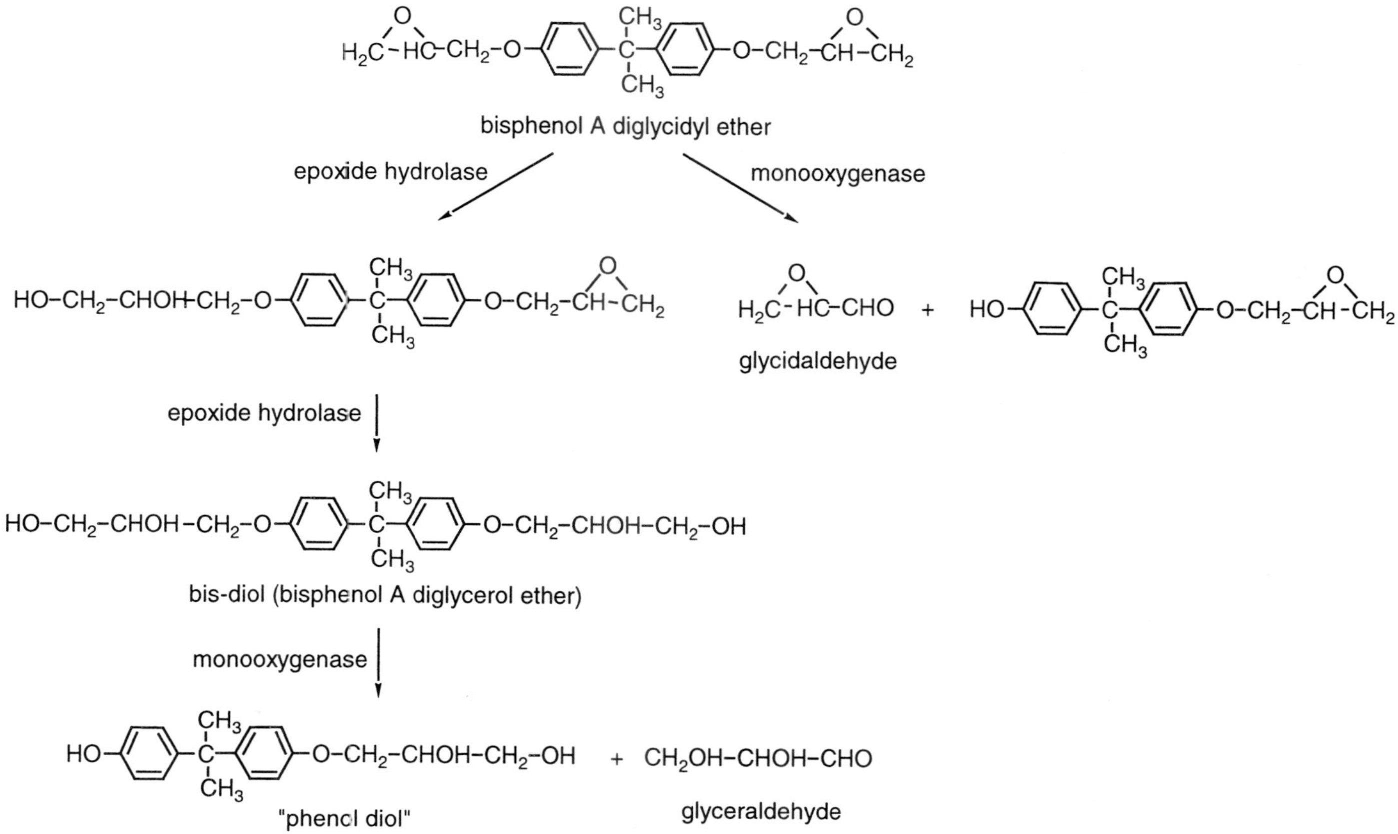

Figure 1. The metabolism of bisphenol A diglycidyl ether to glycidaldehyde and glyceraldehyde

4 Effects in Man

4.1 Allergenic effects

Allergic contact dermatitis resulting from skin contact with epoxy resins is described quite often in the plastics industry (DECOS 1996, Jolanki *et al.* 1987, 1990, Kanerva *et al.* 1988). Data for the incidence of sensitization to bisphenol A diglycidyl ether can be found in Table 2. As the test persons were selected according to different criteria, the incidences vary markedly and cannot be used to estimate the frequency of sensitization within a defined population.

Table 2. Results of patch tests with epoxy resins on a BADGE basis

Patch test*	Persons tested	Positive results	Positive results (%)	References
ICDRG	34	34	100	Fregert and Thorgeirsson 1977
Substance	5	5	100	Niinimäki and Hassi 1983
EU-ST	135	25	19	Van Putten *et al.* 1984
ICDRG	871	3	0.3	Prens *et al.* 1986
EU-ST	542	71	13	Jolanki *et al.* 1987
ICDRG	6	4	67	Van Joost 1988
EU-ST	870	35	4	Jolanki *et al.* 1990
EU-ST	1844	137	7	Jolanki 1991
NACDG	167	30	18	Holness and Nethercott 1993
Substance	39	30	77	Tosti *et al.* 1993
EU-ST	37	6	16	Moura *et al.* 1994
ICDRG	15	14	93	Condé-Salazar *et al.* 1994
EU-ST	6	4	67	Jolanki *et al.* 1995
ICDRG	43	13	30	Tarvainen *et al.* 1995

* BADGE monomer (molecular weight 340) contained in:
EU-ST European Standard Patch Test Series (see Flyvholm, 1991)
ICDRG International Contact Dermatitis Research Group routine test series
NACDG North American Contact Dermatitis Group routine test series

Monomeric bisphenol A diglycidyl ether (molecular weight 340) is thought to be mainly responsible for sensitization (Thorgeirsson *et al.* 1978, Tosti *et al.* 1993). Higher molecular weight oligomers were found to be of low allergenicity or not allergenic. In a patch test with bisphenol A diglycidyl ether (molecular weight 340) purified by gel chromatography, 34 persons with contact dermatitis after occupational exposure to epoxy resins were tested; all produced a reaction. Patch tests with oligomers of molecular weight 624 and 908 were carried out in 23 of these persons, and patch tests with the oligomer of molecular weight 1192 in 7 persons; all the tests yielded negative results. Positive results in patch tests with oligomers of molecular weight 1280 and 1850 were due to contamination with monomer. Bisphenol A diglycidyl ether is therefore regarded as mainly responsible for the allergenic effects (Fregert and Thorgeirsson 1977; see also Section 5.3).

It must, however, be borne in mind that the bisphenol A diglycidyl ether epoxy resin of technical grade quality which is usually used for the patch tests contains an often undeclared level of oligomers (n = 1, n = 2). The commercially available epoxy resins can also contain reactive thinners (glycidyl ethers), epichlorohydrin and hardeners, usually aliphatic polyamines and acid anhydrides, and thus components which have allergenic potential for the skin or respiratory tract. Exposure to a mixture of substances is therefore frequent, and it is often difficult to identify the causal relationship between the symptoms and a specific allergen.

Two patients (from the group of Jolanki *et al.* 1990 mentioned in Table 2) were found to have respiratory symptoms such as rhinitis and asthma as well as allergic contact dermatitis. In both cases prick tests with commercial bisphenol A diglycidyl ether yielded positive results and specific IgE titres against bisphenol A diglycidyl ether-human serum albumin conjugate were detected. In one of these patients, the IgE test for phthalic anhydrides also yielded positive results. In five other patients who complained of respiratory symptoms after similar use of epoxy resins, however, no evidence of IgE-mediated reactivity to bisphenol A diglycidyl ether or epoxy resins was found (Kanerva *et al.* 1991).

A cement worker who had developed recurring contact dermatitis induced by bisphenol A diglycidyl ether and 4,4′-diaminodiphenylmethane and produced a reaction in a patch test with an epoxyacrylate was examined for suspected occupational asthma 6 years after the skin disorder was first diagnosed. Spirometric data and total IgE levels were normal but mild unspecific bronchial hyperreactivity was detected. Provocation by application without occlusion of a resin said to contain bisphenol A diglycidyl ether resulted first in severe acute erythroderma (described as an allergic reaction) of the exposed skin areas (face, ears, neck and hands) followed by a decrease in the peak expiratory flow (PEF) by 22 % to 36 % (Kanerva *et al.* 2000). The product used for provocation and the method of application are not documented, nor are details given of the course of changes in the PEF. This case report is not sufficient to demonstrate that bisphenol A diglycidyl ether causes airway sensitization.

In a sensitization study with 118 men and 24 women, two technical grade bisphenol A diglycidyl ether products each produced symptoms of contact dermatitis in 9 persons (incidence about 0.8 %). Bisphenol A diglycidyl ether was applied 3 times per week as a patch to the skin of the test persons, 9 times for 24 hours, and provocation was carried out after a further 2 weeks (Dow Chemical 1958).

4.2 Genotoxicity

There are three studies of chromosomal changes in the peripheral lymphocytes of persons exposed to epoxy resins on a bisphenol A diglycidyl ether basis.

In one study (Mitelman *et al.* 1980), 9 workers aged between 30 and 66 years had had intermittent skin contact with an epoxy resin used for glueing during periods of 5 to 16 years (mean 6.5 years). The epoxy resin had a mean molecular weight of about 900 with an unspecified content of bisphenol A diglycidyl ether monomer and a hardener of aliphatic amine type. As controls served 9 workers of the same age group who were employed in the same factory but did not use epoxy resins. Cultures of lymphocytes from these workers were examined for chromosomal aberrations (200 metaphases per person) and after PHA (phytohaemagglutinin) stimulation for sister chromatid exchange (20 cells with 46 chromosomes/person). The same cytogenetic investigations were carried out in a further 9 workers aged between 18 and 59 years who had carried out spraying jobs with powdered epoxy resins (mean molecular weight about 2000) in various factories during 3 to 10 years and were subjected to high-level dermal exposure and probably also exposure by inhalation. In the persons exposed to low or high molecular weight epoxy resins and in the controls the incidences of chromosomal aberrations were similar (exposed persons: 2.3 %, controls: 2.9 %), as were the incidences of sister chromatid exchange (exposed persons: 9.0, controls: 8.7 %).

On the other hand, a significant increase in the incidence of chromosomal aberrations in peripheral lymphocytes was reported in 146 workers exposed to "epoxy resin ED-20": 5.5 % compared with 2.4 % in 74 control persons (Suskov and Sazonova 1982). The duration of exposure was given as 4 months to 30 years. It is unclear whether the persons were exposed to the precursors epichlorohydrin and bisphenol A or to epoxy resins with different degrees of polymerization. According to the authors the mean epichlorohydrin concentration was 1 mg/m^3. The second communication (Sazonova and Suskov 1985) describes a mean incidence of chromosomal aberrations of 5.88 % in 90 men exposed for 8 months to 29 years to "epoxy resin" (controls: 2.67 %) and in 66 similarly exposed women an incidence of 4.31 % (controls: 2.0 %); higher age groups were found to have higher values independent of the duration of exposure. Unlike the control groups, the exposed groups were made up of mainly older employees. In this publication, there are insufficient data for the substances handled by the persons and the type and extent of exposure, so that these studies do not permit conclusions as to the genotoxic potential of bisphenol A diglycidyl ether.

5 Animal Studies

5.1 Acute toxicity

Results are available from numerous early studies carried out with the pure substance or technical grade bisphenol A diglycidyl ether with an oligomer content of up to 20 %. The data come mostly from internal company sources and cannot be individually validated; they are quoted below from the reviews by Gardiner *et al.* (1992) and DECOS (1996).

The acute oral toxicity of bisphenol A diglycidyl ether is low. The oral LD_{50} for rats is given as > 1000 to > 4000 mg/kg and for mice as > 500 mg/kg body weight. Manifestations of intoxication were reduced motor activity, respiration rate, and weight loss.

After dermal application the systemic effects are also slight. Without details of the vehicle and mode of application, the dermal LD_{50} values given for mice are > 800 mg/kg and for rats > 1200 to > 1600 mg/kg body weight. Rabbits tolerated single dermal doses of 2000 mg/kg body weight. It is therefore not possible to draw conclusions about the uptake of toxic amounts of bisphenol A diglycidyl ether through the skin.

There are no useful data for inhalation exposure to bisphenol A diglycidyl ether. Exposure of persons by inhalation of the substance, which has a very low vapour pressure and generally does not occur as an aerosol, is only expected during work with powdered polymers (see Section 1). Attempts to obtain a sufficient concentration in the air at room temperature by heating bisphenol A diglycidyl ether were not successful (Dow Chemical 1981, Gardiner *et al.* 1992).

5.2 Subacute, subchronic and chronic toxicity

Organ-specific toxicity was not detected in several studies in which bisphenol A diglycidyl ether was administered orally to rats. Daily gavage doses of technical grade bisphenol A diglycidyl ether of 50 to 1000 mg/kg body weight (Araldite® GY 250 in 0.5 % carboxymethylcellulose with 0.1 % Tween® 80) were tolerated by rats for 28 days without symptoms of toxicity (Ciba-Geigy 1984a). In a 90-day study with rats, administration of 3 % bisphenol A diglycidyl ether in the diet (about 1500 mg/kg body weight and day) led to weight loss in the animals as a result of food refusal. In a 26-week study with rats, all the animals died after being fed a diet containing 5 % bisphenol A diglycidyl ether (about 2500 mg/kg body weight and day) for 20 weeks; concentrations of 0.2 % and 1 % (about 100 and 500 mg/kg body weight and day) were tolerated. Gross pathological and histopathological examinations yielded no evidence of systemic toxicity in any dose group (Gardiner *et al.* 1992).

There are no inhalation studies available with bisphenol A diglycidyl ether (see Section 5.1).

5.3 Local effects on skin and mucous membranes

Epicutaneous application of a 20 % solution of bisphenol A diglycidyl ether in acetone was tolerated without reaction, and application of the pure substance or a 20 % solution in toluene caused only slight or no skin irritation in rabbits within 24 hours, even under occlusion or after scarification of the skin. Repeated dermal application intensified the skin irritation (Gardiner *et al.* 1992).

The data for primary irritative effects of bisphenol A diglycidyl ether in the rabbit eye are inconsistent. Marked irritation was reported after application of 2 mg bisphenol A diglycidyl ether for 24 hours (Lewis 1992). On the other hand, there are other unpublished reports in which no or slight eye irritation is described after administration of undiluted monomeric bisphenol A diglycidyl ether, or monomeric bisphenol A diglycidyl ether in a formulation with solvent, or oligomers (DECOS 1996, Gardiner *et al.* 1992).

5.4 Allergenic effects

Commercial bisphenol A diglycidyl ether was found in the guinea pig maximization test of Magnusson and Kligman to be a powerful contact allergen after intradermal induction with a 5 % solution in acetone and Freund's adjuvant (Thorgeirsson and Fregert 1977). After induction with bisphenol A diglycidyl ether (molecular weight 340) or the dimer (molecular weight 624), 100 % and about 60 % of the animals were sensitized, respectively; with increasing degree of polymerization of the epoxy resin the number of animals which produced reactions decreased. A resin of molecular weight 1850 did not cause sensitization. Furthermore, by topical application of 20 % bisphenol A diglycidyl ether (molecular weight 340) in acetone, alone and without lauryl sulfate, that is, without simultaneous skin irritation, no sensitization was achieved (Thorgeirsson *et al.* 1978). As in the maximization test no cross-reactions between bisphenol A diglycidyl ether and the higher oligomers were observed, the authors conclude that sensitization to epoxy resins is mainly an effect of the residual bisphenol A diglycidyl ether monomer that they contain.

5.5 Reproductive and developmental toxicity

5.5.1 Fertility

In a one-generation study (Ciba-Geigy 1989), male CD rats (24 ♂ and 24 ♀/group) were given daily gavage doses of commercial bisphenol A diglycidyl ether (Araldite® CY250) of 0, 20, 60, 180 or 540 mg/kg body weight (in 0.5 % aqueous carboxymethylcellulose with 0.1 % Tween® 80) for 10 weeks before mating. Female rats received the same dose, but treatment was started 2 weeks before mating. Both sexes were exposed for up to 21 days *post partum*; male animals for a total of 19 weeks, female animals for 11 weeks. The highest dose led to significantly reduced weight gains in the F_0 males in spite of normal food consumption and to slight delays in growth of the pups. From 180 mg/kg body weight, development of the "air righting reflex" was delayed. No effects on the fertility index, litter size or postnatal survival were detected. The gross pathological and microscopic examinations of the sexual organs and the intestinal tracts of the parent animals in the highest dose group yielded no evidence of treatment-related changes.

A two-generation study with rats was carried out with purified bisphenol A diglycidyl ether (purity 99.7 %) (SPI 1996). Groups of 30 male and 30 female Sprague-Dawley rats were given gavage doses of 0, 50, 540 or 750 mg/kg body weight and day (in aqueous 0.5 % Methocel® with 0.1 % Tween®) for 14 weeks (P_1 parent animals) or 12 weeks (P_2 from the F_1 generation) before mating and during pregnancy and rearing of the pups. For reasons of practicability, this form of administration was preferred to dermal application.

The body weight gains of the P_1 and P_2 parent animals were reduced relative to those of the controls by 8 % to 11 % from doses of 540 mg/kg body weight in the males and in the females of the 750 mg/kg group. There was no evidence of impairment of fertility or postnatal development as a result of the prolonged exposure to bisphenol A diglycidyl ether.

5.5.2 Developmental toxicity

Studies of foetotoxicity and teratogenicity after oral administration were carried out with rats and rabbits given low-molecular weight bisphenol A diglycidyl ether (Araldite® CY250, in 0.5 % aqueous carboxymethylcellulose with 0.1 % Tween® 80). The rats were given doses of 0, 60, 180 and 540 mg/kg body weight and day (gestation days 6–15), the rabbits 0, 20, 60 and 180 mg/kg body weight and day (gestation days 7–19). The highest doses caused slight maternal toxicity in both species but no signs of impairment of embryonal or foetal development, or of teratogenic effects (Ciba-Geigy 1988a, 1988b).

Occlusive epidermal application of bisphenol A diglycidyl ether in doses of 0, 30, 100 or 300 mg/kg body weight in polyethylene glycol 400 for about 6 hours daily from days 6 to 18 of gestation led in New Zealand White rabbits to cutaneous irritation at the application site in the middle and high dose groups. Systemic maternal toxicity was not seen, however, and no treatment-related changes in the implantation rate, litter size or prenatal development were observed (Breslin *et al.* 1988, Dow Chemical 1987).

5.6 Genotoxicity

5.6.1 *In vitro*

The numerous studies of the *in vitro* genotoxicity of bisphenol A diglycidyl ether and its technical grade products have yielded positive results in practically all test systems (see Table 3). The substance was found to be mutagenic in the *Salmonella* mutagenicity test with strains TA100 and TA1535 with and without metabolic activation; more base-pair substitutions than frame shift mutations seem to be produced. In yeast cells, gene conversion was induced by bisphenol A diglycidyl ether with and without metabolic activation. The substance had mutagenic effects in mouse lymphoma cells without metabolic activation and in CHO cells (a cell line derived from Chinese hamster ovary) after metabolic activation. In CHO cells, bisphenol A diglycidyl ether caused sister chromatid exchange directly and in primary rat hepatocytes DNA repair synthesis and chromosomal aberrations. In human lymphocytes DNA repair synthesis could not be detected.

Table 3. Studies of the *in vitro* genotoxicity of BADGE

Test system	End-point	Concentrations, test substance	Results without/with S9		Authors
Salmonella typhimurium TA98, TA1535	gene mutation	0.5–2.0 µmol/plate, BADGE and distilled BADGE	TA98: TA1535:	–/– +/+	Dow Chemical 1977
S. typhimurium TA100, TA1535	gene mutation	0–54 µmol/plate, technical grade BADGE		+/+	Andersen *et al.* 1978
S. typhimurium TA100, TA1535	gene mutation	0.05–10.0 mg/plate, BADGE and technical grade BADGE		–/–	Wade *et al.* 1979
S. typhimurium TA98, TA100	gene mutation	not specified, technical grade BADGE	TA98: TA100:	–/– +/+	Murray and Cummings 1979
S. typhimurium TA100, TA1535	gene mutation	not specified		+/+	Ringo *et al.* 1982
S. typhimurium TA98, TA100, TA1535, TA1537	gene mutation	0–10 mg/plate, not specified	TA98, TA1537: TA100, TA1535:	–/– +/+	Canter *et al.* 1986
S. typhimurium TA98, TA100, TA1535, TA1537, TA1538 *Escherichia coli* WP2, WP2 uvrA	gene mutation	0.2–2000 µg/plate, BADGE and technical grade BADGE	TA 98, TA100, TA1538; *E. coli* WP2, WP2 uvrA: TA1535: TA1537:	 –/– +/+ –/+	Shell 1981a
S. typhimurium TA98, TA100, TA1535, TA1537	gene mutation	100–1000 µg/plate, BADGE	TA98, TA1537: TA100, TA1535:	–/– –/+	Kier *et al.* 1986, Union Carbide 1978
E. coli WP2 uvrA	gene mutation	20–10000 µM, technical grade BADGE		not determined/+	Hemminki *et al.* 1980

Table 3. continued

Test system	End-point	Concentrations, test substance	Results without/with S9	Authors
Saccharomyces cerevisiae JD1	gene conversion	0.01–5 mg/ml, BADGE and technical grade BADGE	+/+	Shell 1981a
mouse lymphoma cells L5178Y	gene mutation	0.027 µl/ml, technical grade BADGE	+	Ciba-Geigy 1978a
CHO cells	gene mutation	0.03 µl/ml, technical grade BADGE	–/+	Union Carbide 1981a
human lymphocytes	UDS	1–500 ppm, BADGE and distilled BADGE	–	Dow Chemical 1977
rat liver cells	UDS	0.003–0.01 µl/ml, technical grade BADGE	+	Union Carbide 1981a
CHO cells	SCE	0.0075–0.12 µl/ml, technical grade BADGE	+	Union Carbide 1981a
rat liver cells	chromosomal aberrations	3.75–20 µg/ml, BADGE and technical grade BADGE	+ (increased incidence of gaps, chromatid breaks, chromosome breaks)	Shell 1981a

CHO cells: a cell line from Chinese hamster ovary

Table 4. Studies of the genotoxicity of BADGE *in vivo*

Test system	End-point/test method	Administration, dose, test substance	Result	Comments	Authors
host mediated assay in mice with *S. typhimurium*	gene mutations in i.p. inoculated bacteria	peroral, 1000 mg/kg body weight, BADGE and distilled BADGE	–	method inadequately documented	Dow Chemical 1977
host mediated assay in mice (DBA/2f/Bom) with mouse lymphoma cells	gene mutations in i.p. inoculated cells	peroral, 2600 mg/kg body weight, technical grade BADGE	–	administration 3 days after inoculation; exposure for 3 days	Ciba-Geigy 1978b
rat (Wistar) 2 ♂, 2 ♀	DNA strand breaks in the liver alkaline elution	peroral, 500 mg/kg body weight, BADGE	–	6 hours *post applicationem*	Shell 1981b
mouse (B6D2F$_1$) groups of 10 ♀	micronucleus test bone marrow	peroral, 1000 mg/kg body weight and day, 5 days BADGE and distilled BADGE	–		Dow Chemical 1977
Chinese hamster groups of 6 ♂, 6 ♀	micronucleus test bone marrow	peroral, 0, 825, 1650, 3300 mg/kg body weight and day, 2 days, technical grade BADGE	–	24 hours after 2nd dose	Ciba-Geigy 1978c
Chinese hamster groups of 4 ♂, 4 ♀	chromosomal aberrations bone marrow	peroral, 0, 825, 1650, 3300 mg/kg body weight and day, 2 days, technical grade BADGE	–	6 hours after 2nd dose	Ciba-Geigy 1982a
mouse (NMRI) ♂: 12, 15, 18 animals	chromosomal aberrations germ cells	peroral, 0, 1000, 3000 mg/kg body weight and day, 5 days (day 0, 2, 3, 5, 9), technical grade BADGE	–	3 days after last dose	Ciba-Geigy 1982c
mouse (Tif.MAGf) groups of 24 ♂	chromosomal aberrations germ cells	peroral, 0, 375, 750, 1500, 3000 mg/kg body weight and day, 5 days, technical grade BADGE	–	1 day after last dose	Ciba-Geigy 1984b
mouse (NMRI) groups of 20 ♂	dominant lethal test	peroral, 0, 3333, 10000 mg/kg body weight, technical grade BADGE	–	mating for 6 weeks	Ciba-Geigy 1982b
mouse (B6D2F$_1$) ♂	dominant lethal test	dermal, about 3000 mg/kg body weight and application, 3 × weekly, 8 weeks, BADGE and distilled BADGE	–	after end of exposure mating for 2 weeks	Dow Chemical 1977

In the model reaction for alkylating substances with *p*-nitrobenzyl pyridine, the alkylating activity of bisphenol A diglycidyl ether was found to be three times that of styrene oxide (Ciba Specialty Chemicals 1997, Hemminki *et al.* 1980). Because of this alkylating activity, bisphenol A diglycidyl ether is a direct mutagen in the Ames test. The mutagenic effects can be prevented by S9, but only in the absence of NADPH. When NADPH is added to the S9 system, mutagenic activity is observed (Andersen *et al.* 1978). The metabolism of bisphenol A diglycidyl ether offers an explanation for these results (see Section 3). Bisphenol A diglycidyl ether is inactivated by the epoxide hydrolases of S9, but with increasing substrate concentration the glycidyl side chain can also be oxidatively dealkylated by monooxygenases (Figure 1). NADPH is a necessary cofactor for the monooxygenases and so, in the presence of the complete S9 activation system *in vitro,* the dealkylation of bisphenol A diglycidyl ether to the mutagenic glycidaldehyde takes precedence over the hydrolysis to the bis-diol.

5.6.2 *In vivo*

Unlike the *in vitro* studies, the *in vivo* tests summarized in Table 4 yielded no evidence of a genotoxic potential of bisphenol A diglycidyl ether or its technical grade products. After peroral administration, clastogenic effects could not be found in either the micronucleus test with bone marrow cells or in cytogenetic studies of bone marrow or the male seminiferous epithelium. Dominant lethal tests in the mouse after dermal application and peroral administration of the substance yielded no evidence of induction of germ cell mutations. Host-mediated assays with *Salmonella typhimurium* or mouse lymphoma cells also yielded negative results after oral administration of bisphenol A diglycidyl ether. Orally administered bisphenol A diglycidyl ether did not cause any single strand breaks in rat liver DNA that could be detected by means of alkaline elution. The *in vivo* genotoxicity studies were, however, almost all carried out with oral administration. It is not clear from the data available for the distribution and toxicokinetics of bisphenol A diglycidyl ether (see Section 3) whether this mode of administration produces significant concentrations of the test substance in the target organs investigated. Effective inactivation by epoxide hydrolase could also account for the negative results.

However, as shown in studies of the mouse epidermis after occlusive dermal application of bisphenol A diglycidyl ether ^{14}C-labelled in the glycidyl side chain, this metabolic detoxification is not always complete. Covalent DNA adduct formation was detected chromatographically in DNA hydrolysates from the skin of C3H mice after application of 0.8 and 2 mg/animal (270 and 667 μg/cm^2, in acetone) (2.6 and 6.5 pmol/mg DNA), but not after application of 0.4 mg/animal (133 μg/cm^2). On this basis, the dose of 0.4 mg/animal with a detection limit of 0.4 pmol/mg DNA can be regarded as the NOEL (no observed effect level) for adduct formation. At this dose, when an inhibitor of microsomal epoxide hydrolase (2,3-epoxycyclopentyl ether) was administered simultaneously, DNA adducts were also detected (1.3–2.0 pmol/mg DNA). The formation of DNA adducts therefore apparently results from the saturation of epoxide hydrolase, which is possible at higher bisphenol A diglycidyl ether doses, or at lower bisphenol A diglycidyl ether doses up to 133 μg/cm^2 from the effective inhibition of epoxide hydrolase. In both cases, *O*-dealkylation of the intact glycidyl residue by monooxygenases and the formation of glycidaldehyde as the reactive agent takes place as the alternative metabolic pathway (see Figure 1). With bisphenol A diglycidyl ether which was ^{14}C-labelled at the central C atom, the formation of DNA adducts could not be detected. DNA-associated radioactivity was the result of contamination of the DNA fraction with test substance or ^{14}C-labelled proteins (Bentley *et al.* 1989).

The hypothesis that adducts are formed indirectly was confirmed by the detection of the epidermal glycidaldehyde–DNA adduct. Epicutaneous application of 2 mg bisphenol A diglycidyl ether (333 µg/cm^2) or 2 mg glycidaldehyde led to the formation of highly fluorescent hydroxymethylethenodeoxyadenosine-3′-monophosphate (HMEdAp) in C3H mice. Determination of this cyclic deoxyadenosine adduct by HPLC with fluorescence detection in mice treated with glycidaldehyde yielded on average 166 adducts per 10^6 nucleotides after 24 hours, while 48 hours after exposure to bisphenol A diglycidyl ether 0.14 to 0.80 adducts per 10^6 nucleotides were detected in epidermal DNA. The yield of glycidaldehyde adducts induced by a bisphenol A diglycidyl ether dose of 2 mg per animal corresponds to a glycidaldehyde dose of up to 10 µg per animal, assuming a linear relationship between the glycidaldehyde dose and adduct formation. As 2 mg bisphenol A diglycidyl ether contains 0.85 mg glycidaldehyde equivalents, it can be deduced from the amount of HMEdAp formed that, after dermal application of a bisphenol A diglycidyl ether dose of 333 µg/cm^2, at most 1.1 % of the glycidaldehyde fraction was available for DNA binding (Steiner *et al.* 1992). An increase in DNA adduct formation after dermal application of higher bisphenol A diglycidyl ether doses was observed in another study (see above; Bentley *et al.* 1989). It is unclear whether this adduct formation has detectable biological consequences. The dose of 2 mg per animal administered to determine adduct formation was much lower than the nominal epicutaneous dose of about 20 mg per animal applied 2 to 3 times a week in carcinogenicity studies (see Section 5.7 and Table 5).

There are no studies of bisphenol A diglycidyl ether-induced DNA adduct formation in man or of the metabolism of bisphenol A diglycidyl ether by human epoxide hydrolases. Metabolic deactivation of bisphenol A diglycidyl ether is, however, to be expected also in man, as the epoxide hydrolase activity measured in human skin samples with epoxides of polycyclic aromatic hydrocarbons was higher by a factor of 1.6 to 5.4 than that in mouse skin. Significant interindividual variation was, however, reported. The effect of known inhibitors (1,1,1-trichloropropene oxide, cyclohexene oxide, transstilbene oxide) on the activity of epoxide hydrolases in human skin was comparable to that on microsomal epoxide hydrolase from human and rat liver. On the basis of these studies it was concluded that the specificity of dermal and hepatic epoxide hydrolases for epoxides of polycyclic aromatic hydrocarbons are similar, although there are considerable quantitative differences in activity in the two tissues (Oesch *et al.* 1978). That the hydrolytic activity of hepatic epoxide hydrolase is greater than that in the skin was confirmed also with bisphenol A diglycidyl ether as a substrate in studies with 3 strains of mouse (Bentley *et al.* 1989). The rates of hydrolysis with the mouse hepatic enzyme are comparable for bisphenol A diglycidyl ether and the model substrate styrene oxide. Bisphenol A diglycidyl ether, however, has a lower affinity for the hepatic epoxide hydrolase (microsomal enzyme). In equimolar concentrations, bisphenol A diglycidyl ether inhibits the hydrolysis of styrene oxide by the microsomal enzyme by merely 20 %. A potent inhibitor such as 2,3-epoxycyclopentyl ether inhibits this reaction under the same conditions by 65 % (Ciba Specialty Chemicals 1997). The transformation of bisphenol A diglycidyl ether by mouse dermal epoxide hydrolase is inhibited by almost equimolar concentrations of 2,3-epoxycyclopentyl ether by 50 % (Bentley *et al.* 1989). Cyclic epoxides of this kind are hardly encountered in combination with bisphenol A diglycidyl ether at the workplace (Gardiner *et al.* 1992). It is difficult to draw conclusions about the genotoxic activity of bisphenol A diglycidyl ether *in vivo* on the basis of these studies, as the comparatively high alkylating activity (see Section 5.6.1) is counteracted by relatively efficient detoxification (Ciba Specialty Chemicals 1997).

5.7 Carcinogenicity

5.7.1 Short-term studies

In the cell transformation test with BHK cells (a cell line from baby hamster kidney), both pure and technical grade bisphenol A diglycidyl ether (test concentrations 25–200 µg/ml) yielded positive results (Shell 1981a).

5.7.2 Long-term studies

In more recent studies, the carcinogenic effects of bisphenol A diglycidyl ether have been investigated only after dermal application of the substance to the rat and mouse (see Table 5). These studies were mainly carried out with commercially available technical grade bisphenol A diglycidyl ether. Details of the bisphenol A diglycidyl ether content and impurities, where available, are given in Table 5.

Some earlier studies with dermal application of the substance (Hine *et al.* 1958, Union Carbide 1964) do not meet present-day requirements for carcinogenicity studies with regard to the characterization of the test substance, the methods used and the documentation of the results, and cannot be included in the evaluation of the carcinogenic potential of bisphenol A diglycidyl ether. In a review (Gardiner *et al.* 1992; see Table 5), a study is described in which resin mixtures containing 27 % to 76 % bisphenol A diglycidyl ether were applied dermally; this study can be included in the evaluation only with reservations, as specification of the bisphenol A diglycidyl ether products investigated is inadequate. In addition it is unclear whether and for how long the animals were observed after the 94-week exposure period. An early study with oral administration of bisphenol A diglycidyl ether (Hine *et al.* 1958) does not meet present-day requirements for a carcinogenicity study and cannot be included in the evaluation of systemic carcinogenic effects.

Rat

Pure bisphenol A diglycidyl ether was applied epicutaneously in concentrations of 0, 0.06 %, 6.0 % or 60 % (w/v) in acetone (bisphenol A diglycidyl ether doses about 1, 100 or 1000 mg/kg body weight and application) without occlusion to female F344 rats on 5 days weekly for a period of 2 years (Table 5). With a view to detecting any systemic effects, female F344 rats were used and not males because of the known high incidence of spontaneous endocrine tumours in males. The main effects ascribed to the long-term treatment were reduced body weight gains in the animals of the 60 % group during the first months and mild, histologically verified chronic dermatitis at the application sites of the 6 % and 60 % solutions. The incidence and size of foci of altered liver cells was increased above the concurrent control value in the middle and high dose groups and in the high dose group also the incidence of hepatocellular hypertrophy. In high dose group animals, total bilirubin and alkaline phosphatase and aspartate aminotransferase activities showed a tendency to increase and there was a statistically significant increase in the alanine aminotransferase activity in the serum. The authors consider that the clinico-chemical and morphological findings suggest slight hepatotoxicity after long-term dermal application of 60 % bisphenol A diglycidyl ether (Dow Chemical Company 1998a).

In this 2-year study with dermal application to female F344 rats, foci of altered cells were found frequently but their incidence corresponds to the high incidences in control animals known from National Toxicology Program (NTP) studies. Both in the highest dose group and in the acetone control, only 2 primary liver adenomas were detected. This low incidence lies within the range for spontaneous tumours known from NTP studies (Haseman *et al.* 1998). Particularly the frequent appearance of foci or islets of altered cells in the rat liver and the hepatocellular hypertrophy were seen as signs of systemic effects. In young rats, early induction of liver foci suggests the induction of potentially preneoplastic hyperplasia and has therefore been used in models for testing hepatocarcinogens (Kayser 1986). In long-term studies, however, foci of altered hepatocytes can often appear spontaneously with increasing age of the experimental animals without an associated increase in the incidence of hepatocellular tumours. This applies particularly to basophilic foci (both homogenous foci and the 'tigroid' kind with marginalized basophilic cytoplasm) which have a special significance in short-term tests for tumour development (Squire 1989). In ageing F344 rats, the absence of an association between the occurrence of foci of altered cells and liver tumours is particularly clear: after 24 months basophilic foci were detected in 100 % of the female control rats and the incidence of spontaneous foci of other histological phenotypes varied between 27.5 % and 68.8 % (Harada *et al.* 1989). The general occurrence of liver foci in old rats, especially F344 rats, together with an unchanged low incidence of spontaneous hepatomas in long-term studies was reason enough for the NTP to conclude that an increase in the incidence of foci in the absence of tumour induction should not be interpreted as a sign of carcinogenic potential (Maronpot *et al.* 1989). This opinion is shared by many authorities (Mahon 1989, Richter-Reichhelm 1989, Moch *et al.* 1996).

The incidences of systemic tumours were not different in the exposed groups and controls apart from a statistically significant trend towards a dose-dependent reduction in the incidence of all mammary tumours.

In some exposed animals (1/50, 3/50, 1/50 in the low, middle and high dose groups, respectively) carcinomas of the oral cavity and the tongue were detected. In all cases these tumours had developed in areas affected by foreign body reactions to hairs which had penetrated the mucosa. The authors expressed the opinion that the tumours were secondary mucosal tumours which are known to develop in the rat as a result of unspecific irritation by foreign bodies (Kociba and Keyes 1985). In the present case, such reactions could have been induced by the penetration of the mucosa by hair shafts stiffened with resin, probably after the animals had licked the freshly shaved application sites. A papilloma on the muzzle developed in one of the animals in the acetone control group. Compared with control data from other studies of the same laboratory, the incidence of oral tumours is not increased significantly in this study (Dow Chemical Company 1998a).

However, specific properties of the test substance produce difficulties in non-occlusive application and so the extent of direct skin contact with bisphenol A diglycidyl ether achieved at the application sites is unclear. It is documented in the test records of this rat study that the test substance spread onto the skin and the fur area adjacent to the application site, either because of its specific flow properties or because of the method used to clean the fur. At the low concentration after evaporation of the acetone vehicle, a thin film of solid formed on the skin. At the highest concentration the viscous, sticky test material took all night to dry to a thick layer; then when the application site was shaved again, hairs stiff with the substance were spread over the whole fur of the animal. As a consequence, the animals ingested the substance while grooming and so were exposed systemically, as described in the study documentation.

An independent peer review considered the tumours detected in this rat study to be chance findings and found no evidence of a carcinogenic effect of bisphenol A diglycidyl ether. In addition, this report mentions that in control groups from NTP studies up to 3/50 female F344 rats develop spontaneous squamous cell tumours (papillomas, carcinomas) of the oral cavity (Dow Chemical Company 1998c).

Mouse

The epicutaneous application of a commercial bisphenol A diglycidyl ether product in doses of 15 or 75 mg/animal and week for 2 years did not lead to the formation of tumours in C3Hf/Bd mice. When inbred mice of the C57BL/6Bd strain were treated in the same way, however, non-significant increases in the incidence of skin tumours were found; they were interpreted as evidence of a weak carcinogenic effect (Holland *et al.* 1979, Union Carbide 1978). These skin tumours cannot be definitively attributed to bisphenol A diglycidyl ether, as subsequent analysis (Union Carbide 1981b) of the product used revealed that in addition to 85 % bisphenol A diglycidyl ether it contained comparatively high levels of epichlorohydrin, phenyl glycidyl ether and 2,3-dichlorohydrin. The testing according to the same protocol of 3 other commercial products containing 78 % to 96 % bisphenol A diglycidyl ether and fewer impurities confirmed the lack of tumorigenic effect in C3H mice. No studies were carried out with these products in C57BL mice (Union Carbide 1981b).

In two laboratories (Peristianis *et al.* 1988, Zakova *et al.* 1985) parallel studies were carried out using the same protocol with mice of the same provenance and dermal application of bisphenol A diglycidyl ether; the aim was to compare the effects of the pure substance with those of three commercially available technical grade products (Shell 1980). The outbred strain CF1 was selected for its known sensitivity to directly alkylating, carcinogenic substances (propane sultone, β-propiolactone) applied epicutaneously (Doak *et al.* 1976). Technical grade bisphenol A diglycidyl ether (Araldite® GY 250) containing 4.3 ppm epichlorohydrin did not affect the lifespan of the animals, nor did it cause skin tumours or influence the incidence of systemic tumours in CF1 mice. The reference carcinogen β-propiolactone (not listed in Table 5), however, caused a large number of malignant skin tumours (Zakova *et al.* 1985). In the other study, mice of the same strain were given pure bisphenol A diglycidyl ether (Peristianis *et al.* 1988, Shell 1983b) or the technical grade substance containing ≤ 29 ppm epichlorohydrin (Epon® 828) or ≤ 3 ppm epichlorohydrin (Epikote® 828) (Peristianis *et al.* 1988). After application of Epikote® 828, a few epithelial tumours (7/200) of various origins (see Table 5) were observed around the application site. The authors regard this as possible evidence of low carcinogenic potential, as in historical controls fibrosarcomas occurred only rarely on their own. After application of pure bisphenol A diglycidyl ether, local sarcomas were found in 4/200 animals (1/200 in the vehicle controls). Other tumours were observed on distant, untreated skin areas. The type of skin tumours observed after application of Epon® 828 (2/200) were occasionally also found in the historical controls, so that the authors see no connection with the exposure to bisphenol A diglycidyl ether. The incidence of local skin tumours after application of the various products was low and not significantly increased above that in the acetone controls.

For systemic tumours, a statistically significant trend towards a dose–response relationship was found in some groups only for kidney tumours and lympho-reticular tumours (Peristianis *et al.* 1988). For the kidney tumours in the male animals treated with Epon® 828, this result was attributed to the unusual lack of kidney tumours in the males of the low dose group (1 % group: 0/50; 10% group: 8/50; controls: 6/100). In the parallel studies there were no such differences. A statistically significant trend was observed for the incidence of lympho-reticular tumours in female mice only after application of the pure substance and Epikote® 828. The systemic tumour findings were interpreted by the authors (Peristianis *et al.* 1988) not as evidence of a carcinogenic effect of bisphenol A diglycidyl ether, but as coincidental changes in the relatively high spontaneous incidence in the strain of mouse used.

Male B6C3F$_1$ mice were also treated with pure bisphenol A diglycidyl ether applied epicutaneously in concentrations of 0, 0.005 %, 0.5 % or 5.0% (w/v) in acetone (bisphenol A diglycidyl ether doses about 0.1, 10 or 100 mg/kg body weight and application) without occlusion on 5 days weekly for a period of 2 years (Table 5). In the middle and high dose groups the treatment caused very severe local irritation and histological examination revealed severe chronic and recurring dermatitis. These results suggest a persistent dermatotoxic effect of the test substance remaining on the skin of the animals after evaporation of the vehicle. The 0.005 % concentration was tolerated without such effects, as was the vehicle alone (pure acetone).

Table 5. Dermal carcinogenicity studies with bisphenol A diglycidyl ether (BADGE)

Author:	Peristianis *et al.* 1988, Shell 1983b		
Substance:	EPON® 828: commercially available technical grade BADGE, epichlorohydrin content < 29 ppm Epikote® 828: commercially available technical grade BADGE, epichlorohydrin content < 3 ppm BADGE: pure BADGE		
Species:	CF1 mouse		
	exposure groups:		groups of 50 ♂ and 50 ♀
	control group (vehicle: acetone):		99 ♂, 100 ♀
	positive controls:		50 ♂, 50 ♀
	historical controls:	A (untreated animals)	50 ♀
		B (ethanol/Nonidet)	100 ♂, 100 ♀
		C (liquid paraffin)	50 ♀
Administration:	dermal, shaved dorsal skin; 2 years, twice a week		
Dose:	200 µl 1 % (v/v) BADGE in acetone: about 4.8 mg/animal and week or about 96 mg/kg body weight and week 200 µl 10 % (v/v) BADGE in acetone: about 48 mg/animal and week or about 960 mg/kg body weight and week positive controls: 200 µl 2 % (v/v) β-propiolactone in acetone		
Toxicity:	mortality not affected		
Comments:	as a result of the low spontaneous incidence of animals with skin tumours (2/199 ♂, 2/300 ♀; only fibrosarcomas) in the laboratory, the authors regarded the slightly increased incidence of skin tumours in animals exposed to BADGE (EPON 828: 2/100; Epikote 828: 7/200; BADGE: 7/200; different tumour types and cellular origin) as evidence of a low carcinogenic potential for BADGE		

Group		EPON 828		Epikote 828		BADGE		Controls	Historical controls		
Concentration		1 %	10 %	1 %	10 %	1 %	10 %	Vehicle	A	B	C
animals with skin tumours on treated skin (local)	♂	**0/50**	**1/50**	**3/50**	**2/50**	**0/50**	**3/50**	**1/99**	–	**1/100**	–
		0 %	**2 %**	**6 %**	**4 %**	**0 %**	**6 %**	**1 %**		**1 %**	
	♀	**0/50**	**1/50**	**1/50**	**1/50**	**1/50**	**0/50**	**0/100**	**0/50**	**1/100**	**0/50**
		0 %	**2 %**	**2 %**	**2 %**	**2 %**	**0 %**	**0 %**	**0 %**	**1 %**	**0 %**

Table 5. continued

Group		EPON 828		Epikote 828		BADGE		Controls	Historical controls		
Concentration		1 %	10 %	1 %	10 %	1 %	10 %	Vehicle	A	B	C
squamous cell carcinoma – local	♂	0/50	1/50	0/50	0/50	0/50	0/50	0/99	–	0/100	–
		0 %	2 %	0 %	0 %	0 %	0 %	0 %		0 %	
squamous cell papilloma – local	♂	0/50	0/50	1/50	0/50	0/50	0/50	0/99	–	0/100	–
		0 %	0 %	2 %	0 %	0 %	0 %	0 %		0 %	
basal cell carcinoma – local	♂	0/50	0/50	1/50	1/50	0/50	0/50	0/99	–	0/100	–
		0 %	0 %	2 %	2 %	0 %	0 %	0 %		0 %	
	♀	0/50	0/50	0/50	1/50	0/50	0/50	0/100	0/50	0/100	0/50
		0 %	0 %	0 %	2 %	0 %	0 %	0 %	0 %	0 %	0 %
sebaceous adenoma – local	♂	0/50	0/50	0/50	1/50	0/50	0/50	0/99	–	0/100	–
		0 %	0 %	0 %	2 %	0 %	0 %	0 %		0 %	
fibrosarcoma – local	♂	0/50	0/50	0/50	0/50	0/50	0/50	1/99	–	0/100	–
		0 %	0 %	0 %	0 %	0 %	0 %	1 %		0 %	
	♀	0/50	0/50	0/50	0/50	0/50	0/50	0/100	0/50	1/100	0/50
		0 %	0 %	0 %	0 %	0 %	0 %	0 %	0 %	1 %	0 %
haemangiosarcoma – local	♂	0/50	0/50	0/50	0/50	0/50	2/50	0/99	–	0/100	–
		0 %	0 %	0 %	0 %	0 %	4 %	0 %		0 %	
haemangioendothelioma – local	♀	0/50	0/50	1/50	0/50	0/50	0/50	0/100	0/50	0/100	0/50
		0 %	0 %	2 %	0 %	0 %	0 %	0 %	0 %	0 %	0 %
fibroma – local	♂	0/50	0/50	1/50	0/50	0/50	0/50	0/99	–	0/100	–
		0 %	0 %	2 %	0 %	0 %	0 %	0 %		0 %	
subcutis: fibrosarcoma – local	♂	0/50	0/50	0/50	0/50	0/50	1/50	0/99	–	1/100	–
		0 %	0 %	0 %	0 %	0 %	2 %	0 %		1 %	
	♀	0/50	1/50	0/50	0/50	0/50	0/50	0/100	0/50	0/100	0/50
		0 %	2 %	0 %	0 %	0 %	0 %	0 %	0 %	0 %	0 %
subcutis: anaplastic sarcoma – local	♀	0/50	0/50	0/50	0/50	1/50	0/50	0/100	0/50	0/100	0/50
		0 %	0 %	0 %	0 %	2 %	0 %	0 %	0 %	0 %	0 %

Table 5. continued

Group		EPON 828		Epikote 828		BADGE		Controls	Historical controls		
Concentration		1 %	10 %	1 %	10 %	1 %	10 %	Vehicle	A	B	C
animals with skin tumours	♂	**0/50**	**0/50**	**0/50**	**1/50**	**0/50**	**0/50**	**0/99**	–	**0/100**	–
on untreated skin (distal)		**0 %**	**0 %**	**0 %**	**2 %**	**0 %**	**0 %**	**0 %**		**0 %**	
	♀	**0/50**	**0/50**	**0/50**	**0/50**	**3/50**	**0/50**	**0/100**	**0/50**	**0/100**	**1/50**
		0 %	**0 %**	**0 %**	**0 %**	**6 %**	**0 %**	**0 %**	**0 %**	**0 %**	**2 %**
squamous cell carcinoma – distal	♀	0/50	0/50	0/50	0/50	1/50	0/50	0/100	0/50	0/100	0/50
		0 %	0 %	0 %	0 %	2 %	0 %	0 %	0 %	0 %	0 %
basal cell carcinoma – distal	♂	0/50	0/50	0/50	1/50	0/50	0/50	0/99	–	0/100	–
		0 %	0 %	0 %	2 %	0 %	0 %	0 %		0 %	
haemangioendothelioma – distal	♀	0/50	0/50	0/50	0/50	1/50	0/50	0/100	0/50	0/100	0/50
		0 %	0 %	0 %	0 %	2 %	0 %	0 %	0 %	0 %	0 %
fibrosarcoma – distal	♀	0/50	0/50	0/50	0/50	0/50	0/50	0/100	0/50	0/100	1/50
		0 %	0 %	0 %	0 %	0 %	0 %	0 %	0 %	0 %	2 %
subcutis: fibrosarcoma – distal	♀	0/50	0/50	0/50	0/50	1/50	0/50	0/100	0/50	0/100	0/50
		0 %	0 %	0 %	0 %	2 %	0 %	0 %	0 %	0 %	0 %
kidney tumours	♂	0/50	8/50*	5/50	2/50	5/50	3/50	6/99	–	–	–
(mainly carcinomas)		0 %	16 %	10 %	4 %	10 %	6 %	6 %			
all tumours of the lymphoreticular/	♀	12/50	18/50	14/50	24/50**	22/50	15/50	27/100	–	–	–
haematopoietic system		24 %	36 %	28 %	48 %	44 %	30 %	27 %			
reticulosarcomas	♀	4/50	8/50	11/50	15/50*	12/50	7/50	16/100	–	–	–
		8 %	16 %	22 %	30 %	24 %	14 %	16 %			
lymphoblastic sarcomas	♀	5/50	5/50	2/50	4/50*	9/50	3/50	2/100	–	–	–
		10 %	10 %	4 %	8 %	18 %	6 %	2 %			
lymphosarcomas of the thymus	♀	2/50	4/50	–	–	2/50	5/50*	3/100	–	–	–
		4 %	8 %			4 %	10 %	3 %			

statistically significant linear trend for the dose–response relationship: * $p \leq 0.05$; ** $p \leq 0.01$

Table 5. continued

Author:	Zakova *et al.* 1985
Substance:	Araldite® GY 250: commercially available technical grade BADGE, epichlorohydrin content 4.3 ppm
Species:	CF1 mouse
	exposed groups: 1 %: 50 ♂ and 50 ♀; 10 %: 50 ♂ and 49 ♀
	control group (vehicle: acetone): 50 ♂, 50 ♀
	positive controls: 50 ♂, 50 ♀
	controls from another experiment (vehicle: acetone): 100 ♂, 100 ♀
Administration:	dermal, shaved dorsal skin; 2 years, twice a week
Dose:	200 µl 1 % (v/v) solution: about 4.8 mg/animal and week or about 96 mg/kg body weight and week
	200 µl 10 % (v/v) solution: about 48 mg/animal and week or about 960 mg/kg body weight and week
	positive controls: 200 µl 2 % β-propiolactone in acetone
Toxicity:	mortality and body weight gains not affected
Comments:	no evidence of carcinogenic effects

Group Concentration		Araldite GY 250 1 %		10 %		Controls: vehicle		Controls from another experiment	
animals with skin tumours	♂	**0/50**	**0 %**	**0/50**	**0 %**	**1/50**	**2 %**	**0/100**	**0 %**
on treated skin (local)	♀	**0/50**	**0 %**	**0/49**	**0 %**	**0/50**	**0 %**	**0/100**	**0 %**
fibrosarcoma – local	♂	0/50	0 %	0/50	0 %	1/50		0/100	0 %
animals with skin tumours	♂	**0/50**	**0 %**	**1/50**	**2 %**	**1/50**	**2 %**	**0/100**	**0 %**
on untreated skin (distal)	♀	**2/50**	**4 %**	**1/49**	**2 %**	**1/50**	**2 %**	**6/100**	**6 %**
squamous cell carcinoma – distal	♂	0/50	0 %	1/50	2 %	0/50	0 %	0/100	0 %
	♀	0/50	0 %	0/49	0 %	0/50	0 %	1/100	1 %
basal-cell carcinoma – distal	♀	0/50	0 %	0/49	0 %	0/50	0 %	1/100	1 %
fibrosarcoma – distal	♂	0/50	0 %	0/50	0 %	1/50	2 %	0/100	0 %
	♀	2/50	4 %	0/49	0 %	0/50	0 %	4/100	4 %
undifferentiated sarcoma	♀	0/50	0 %	1/49	2 %	1/50	2 %	0/100	0 %
kidney tumours (adenomas and carcinomas)	♂	6/50	12 %	4/50	8 %	4/50	8 %	11/100	11 %
lymphoreticular tissue (malignant lymphomas)	♂	17/50	34 %	12/50	24 %	16/50	32 %	24/100	24 %
	♀	18/50	36 %	15/49	39 %	12/50	24 %	21/100	21 %

Table 5. continued

Author:	Holland *et al.* 1979, Union Carbide 1978
Substance:	commercially available technical grade BADGE composition (Union Carbide 1981b): 85 % BADGE (monomer), 7 % BADGE dimer, 1 % BADGE trimer impurities: 1 % diol of BADGE, epichlorohydrin 1476 ppm, phenylglycidyl ether 369 ppm, 2,3-dichlorohydrin 180 ppm
Species:	C3Hf/Bd mouse — groups of 40 ♂, 40 ♀ C57BL/6Bd mouse — groups of 20 ♂, 20 ♀
Administration:	dermal, shaved dorsal skin; 2 years, 3 times a week
Dose:	50 µl 10 % (v/v) solution: about 15 mg/animal and week or about 300 mg/kg body weight and week 50 µl 50 % (v/v) solution: about 75 mg/animal and week or about 1500 mg/kg body weight and week control group: vehicle acetone
Toxicity:	C3H mouse: body weight slightly decreased at high dose; mortality: ♂ slightly increased at high dose, ♀ no effects C57BL mouse: body weight: ♂ significantly decreased at high dose ($p < 0.05$), ♀ no effects; mortality: ♂ significantly increased at high dose ($p < 0.05$), ♀ no effects
Comments:	not significantly increased incidence of skin tumours in C57BL mouse interpreted as evidence of weak carcinogenic effects (Holland *et al.* 1979); influence of impurities (Union Carbide 1981b) cannot be excluded

Strain Concentration		C3H mouse Controls		10 %		50 %		C57BL mouse Controls		10 %		50 %	
animals with skin tumours	♂	**0/40**	**0 %**	**0/40**	**0 %**	**0/40**	**0 %**	**0/20**	**0 %**	**1/20**	**5 %**	**6/20**	**30 %**
	♀	**1/40**	**2.5 %**	**0/40**	**0 %**	**0/40**	**0 %**	**0/20**	**0 %**	**0/20**	**0 %**	**2/20**	**10 %**
skin: carcinoma – local	♂	0/40	0 %	0/40	0 %	0/40	0 %	0/20	0 %	0/20	0 %	6/20	30 %
	♀	0/40	0 %	0/40	0 %	0/40	0 %	0/20	0 %	0/20	0 %	1/20	5 %
skin: papilloma	♂	0/40	0 %	0/40	0 %	0/40	0 %	0/20	0 %	1/20	5 %	0/20	0 %
	♀	1/40	2.5 %	0/40	0 %	0/40	0 %	0/20	0 %	0/20	0 %	1/20	5 %
reticulosarcomas	♂	4/40	10 %	11/40	27.5 %	3/40	7.5 %	3/20	15 %	7/20	35 %	8/20	40 %
	♀	6/40	15 %	5/40	12.5 %	3/40	7.5 %	6/20	30 %	7/20	35 %	10/20	50 %

Table 5. continued

Author:	Union Carbide 1981b
Substance:	Epi-Rez 508: commercially available technical grade BADGE; 96 % BADGE (monomer); 1 % BADGE dimer impurities: 2 % diol of BADGE; epichlorohydrin 4 ppm; phenylglycidyl ether 220 ppm EPON® 828: commercially available technical grade BADGE; 78 % BADGE (monomer); 9 % BADGE dimer; 2 % BADGE trimer impurities: 4 % diol of BADGE; epichlorohydrin 29 ppm; 1,3-dichlorohydrin 15 ppm Araldite® 6010: commercially available technical grade BADGE; 78 % BADGE; 7 % BADGE dimer; 2 % BADGE trimer impurities: 4 % diol of BADGE; epichlorohydrin 3 ppm; 1,3-dichlorohydrin 10 ppm
Species:	C3Hf/Bd mouse; 2 years, 3 times a week groups of 40 ♂ and 40 ♀
Administration:	dermal, shaved dorsal skin
Dose:	50 µl 50 % (w/v) solution: about 75 mg/animal and week or 1500 mg/kg body weight and week; vehicle control group: acetone
Toxicity:	body weight not affected; mortality: ♂ significantly increased for all BADGE products tested ($p < 0.01$); ♀ no effects; haematological and clinicochemical parameters not affected
Tumours:	no skin tumours in any group; no other tumours investigated
Author:	Gardiner *et al.* 1992
Substance:	A: commercially available epoxy resin with a BADGE content of 42 % (no details of other components and impurities) B: commercially available epoxy resin with a BADGE content of 76 % (no details of other components and impurities) C: commercially available epoxy resin with a BADGE content of 27 % (no details of other components and impurities)
Species:	C3H/HeJ mouse groups of 50 ♂
Administration:	dermal, dorsal skin; 94 weeks, twice a week
Dose:	A: 50 µl 50 % in acetone, BADGE concentration 21 %; about 25 mg/animal and week or about 500 mg/kg body weight and week B: 50 µl 25 % in acetone, BADGE concentration 19 %; about 22 mg/animal and week or about 440 mg/kg body weight and week C: 50 µl 100 % in acetone, BADGE concentration 27 %; about 32 mg/animal and week or about 640 mg/kg body weight and week control group: vehicle acetone; control group: untreated animals; positive controls: 0.025 % benz(*a*)pyrene in acetone
Toxicity:	no data available
Comments:	insufficient data in the review to be able to characterize the test substances and duration of the experiment and follow-up period
Tumours:	A: dermal papillomas in 3/50 animals no skin tumours in groups B or C or the control groups positive controls: in 48/50 animals skin tumours after a mean latency period of 32.4 weeks

Table 5. continued

Author:	Dow Chemical Company 1998a
Substance:	Bisphenol A diglycidyl ether, purity 99.32 ± 0.11 %
Species:	rat, Fischer 344, 70 ♀ per group; interim killing of 20 animals per group after 12 months
Administration:	2 years, 5 days weekly, dermal, shaved dorsal skin, about 10 % of the body surface area
Dose:	300 µl 0.06 %, 6.0 % or 60 % bisphenol A diglycidyl ether in acetone per application, that is, doses of about 1, 100 or 1000 mg/kg body weight per application
Toxicity:	0.06%: no toxic effects from 6.0 %: mild chronic dermatitis, 60 %: reduced body weight gains

Tumours:		Controls (historical controls)		Bisphenol A diglycidyl ether concentration					
				0.06 %		6.0 %		60 %	
oral cavity	papillomas	1/50 (2/≅600)	2 % (0.33 %)	0/50	0 %	1/50	2 %	0/50	0 %
	squamous cell carcinomas	0/50 (2/≅600)	0 % (0.33 %)	1/50	2 %	2/50	4 %	0/50	0 %
	undifferentiated carcinomas [1]	0/50 (0/≅600)	0 % (0 %)	0/50	0 %	1/50[1)]	2 %	0/50	0 %
tongue	squamous cell carcinomas	0/50 (1/≅600)	0 % (0.16 %)	0/50	0 %	0/50	0 %	1/50	2 %
mammary gland	adenomas	1/50	2 %	0/50	0 %	0/50	0 %	1/50	2 %
	fibroadenomas	7/50	14 %	13/50	26 %	4/50	8 %	0/50	0 %
	adenocarcinomas	0/50	0 %	0/50	0 %	2/50	4 %	0/50	0 %
liver	centrilobular hypertrophy	0/49	0 %	0/50	0 %	1/50	2 %	14/50	28 %
	basophilic foci	4/49	8 %	4/50	8 %	3/50	6 %	9/50	18 %
	tigroid foci	39/49	80 %	40/50	80 %	42/50	84 %	43/50	86 %
	eosinophilic foci	17/49	35 %	16/50	32 %	26/50	52 %	39/50	78 %
	pale cell foci	1/49	2 %	0/50	0 %	0/50	0 %	1/50	2 %
	hepatocellular adenomas	2/49	4 %	0/50	0 %	1/50	2 %	2/50	4 %
	bile duct adenomas	0/50	0 %	0/50	0 %	0/50	0 %	1/50	2 %

[1] questionable whether this was a primary tumour

Table 5. continued

Author:	Dow Chemical Company 1998b
Substance:	Bisphenol A diglycidyl ether, purity 99.32 ± 0.11 %
Species:	mouse, $B6C3F_1$, 70 ♂ per group, interim killing of 20 animals per group after 12 months
Administration:	2 years, 3 days weekly, dermal, shaved dorsal skin, about 10 % of the body surface area
Dose:	50 µl 0.005 %, 0.5 % or 5.0 % bisphenol A diglycidyl ether in acetone per application, that is, about 0.1, 10 or 100 mg/kg body weight per application
Toxicity:	0,005 %: no toxic effects from 0.5 %: chronic dermatitis, histological examination revealed epidermal hyperplasia, scab formation and occasional ulceration

Tumours		Controls		Bisphenol A diglycidyl ether concentration					
				0.005 %		0.5 %		5.0 %	
skin	scaling and fissuring	0/42	0 %	1/33	3 %	23/38	61 %	18/36	50 %
	epidermal hyperplasia	44/50	88 %	33/49*	67 %	47/47*	100 %	48/49	98 %
	chronic recurring dermatitis	16/50	32 %	15/49	31 %	47/47*	100 %	48/49*	98 %
	squamous cell carcinomas [1]	0/50	0 %	0/49	0 %	1/47	2 %	0/49	0 %

* $p < 0.05$

[1] historical control incidences ♂: 0–1/50 carcinomas
♀: 0–2/50 carcinomas, 0–1/50 papillomas

The incidence of epidermal hyperplasia after application of 0.005 % bisphenol A diglycidyl ether was significantly lower than in the acetone control group, but higher after application of 0.5 %. In one mouse from the 0.5 % group, a squamous cell carcinoma was detected at the application site. This tumour was considered by the authors to be spontaneous because such tumours were also observed in animals of this strain in the historical control groups (see Table 2). No evidence of systemic effects was found nor of carcinogenic potential of the dermally applied bisphenol A diglycidyl ether (Dow Chemical Company 1998b).

In the long-term study described above, male mice were chosen to complement the female rats in the parallel study (Haseman and Huff 1987). Earlier studies with C57BL mice found an increased (but not significantly) incidence of skin tumours, which suggested that bisphenol A diglycidyl ether could have weak carcinogenic effects (Holland *et al.* 1979). It may be assumed that the $B6C3F_1$ and C57BL mice react similarly because the former is the product of a cross between the C57BL and the C3H mouse. In addition, the sensitivity of the $B6C3F_1$ mouse has been investigated in comparative NTP studies (NTP 1996). The strain proved to be similar to the CD-1 and SENCAR mouse in its reaction to the repeated application of known skin carcinogens (dimethylbenzanthracene, methylnitrosonitroguanidine).

Taken together, the results of the long-term studies provide no clear evidence of a carcinogenic potential of bisphenol A diglycidyl ether.

It can be concluded from the negative results or only low tumour incidences after long-term exposure of mouse skin to bisphenol A diglycidyl ether alone (highest tolerable concentration) that the diglycidyl ether was effectively hydrolysed under these conditions. Carcinogenicity studies with the mouse seem to be particularly relevant, as the epoxide hydrolase activity in mouse skin is lower than that in human skin (Oesch *et al.* 1978; see also Section 5.6). Long-term epicutaneous application of bisphenol A diglycidyl ether together with a cyclic epoxide (2,3-epoxycyclopentyl ether) in the ratio 1:1 did, however, result in increased tumour formation in mouse skin (Holland *et al.* 1979, Union Carbide 1978, 1981b). This cycloaliphatic epoxide was found to be an effective inhibitor of microsomal epoxide hydrolase (Bentley *et al.* 1989, Ciba Specialty Chemicals 1997; see also Section 5.6.2).

6 Manifesto (MAK value/classification)

The recent dermal carcinogenicity studies are not sufficient to waive the suspicion of carcinogenic effects of bisphenol A diglycidyl ether, raised by the clear genotoxicity of the substance *in vitro* and the formation of DNA adducts *in vivo* as well as by effects seen with other structurally related glycidyl ethers. After topical application of bisphenol A diglycidyl ether to mouse skin, glycidaldehyde–DNA adducts were detected. Glycidaldehyde can be formed in the oxidative dealkylation of glycidyl residues of bisphenol A diglycidyl ether. This metabolic pathway can be activated on saturation of epoxide hydrolase or direct inhibition of this enzyme. The dose-dependency of DNA adduct formation in mouse skin allows the deduction of a NOEL (0.4 mg/animal, 133 µg/cm^2) for this metabolic pathway. If such a NOEL were demonstrated in human skin, it would be possible to define an exposure level at which no significant contribution to cancer risk would be expected. That would fulfill the conditions for the classification of bisphenol A diglycidyl ether in Carcinogen category 5. Further studies are required to establish whether DNA adducts are formed in human skin. A MAK value cannot be established at present. Therefore bisphenol A diglycidyl ether is classified in Carcinogen category 3A.

Designation with "Sh" has been retained.

The *in vivo* studies demonstrate that glycidaldehyde formed from bisphenol A diglycidyl ether can bind to DNA in the skin. Because the DNA is contained in structures under the skin barrier, this DNA binding must be seen as a systemic effect, especially as it cannot be assumed that a substance which binds to dermal DNA is captured entirely by this reaction. Therefore and because bisphenol A diglycidyl ether is a genotoxic substance, the designation with an "H" is retained.

Reproductive toxicity was not observed in rabbits exposed either dermally or orally or in rats exposed orally to bisphenol A diglycidyl ether doses up to the maternally toxic range. A risk of prenatal toxicity for man is not to be expected.

7 References

Andersen M, Kiel P, Larsen H, Maxild J (1978) Mutagenic action of aromatic epoxy resins. *Nature 276*: 391–392

APME (Association of Plastics Manufacturers in Europe) (1996) *Protective gloves for epoxy resin systems.* Technical paper produced by Association of Plastics Manufacturers in Europe. Brussels, Belgium, March 1996

Bentley P, Bieri F, Kuster H, Muakkassah-Kelly S, Sagelsdorff P, Stäubli W, Waechter F (1989) Hydrolysis of bisphenol A diglycidylether by epoxide hydrolases in cytosolic and microsomal fractions of mouse liver and skin: inhibition by epoxycyclopentylether and the effects upon the covalent binding to mouse skin DNA. *Carcinogenesis 10*: 321–327

Breslin WJ, Kirk HD, Johnson KA (1988) Teratogenic evaluation of diglycidyl ether of bisphenol A (DGEBPA) in New Zealand White rabbits following dermal exposure. *Fundam Appl Toxicol 10*: 736–743

Boogaard PJ, Denneman MA, Van Sittert NJ (2000a) Dermal penetration and metabolism of five glycidyl ethers in human, rat and mouse skin. *Xenobiotica 30*: 469–483

Boogaard PJ, De Kloe KP, Bierau J, Kuiken G, Borkulo PED, Van Sittert NJ (2000b) Metabolic inactivation of five glycidyl ethers in lung and liver of humans, rats and mice *in vitro*. *Xenobiotica 30*: 485–502

Canter DA, Zeiger E, Haworth S, Lawlor T, Mortelmans K, Speck W (1986) Comparative mutagenicity of aliphatic epoxides in *Salmonella. Mutat Res 172*: 105–138

Ciba-Geigy (1978a) *Point mutation assay with mouse lymphoma cells. I. In vitro test.* Internal report 78-2314, unpublished

Ciba-Geigy (1978b) *Point mutation assay with mouse lymphoma cells. II. Host-mediated assay with TK 12386.* Internal report 7823-18, unpublished

Ciba-Geigy (1978c) *Nucleus anomaly test in somatic interphase nuclei. TK 12386. Chinese hamster.* Internal report 78-3011, unpublished

Ciba-Geigy (1982a) *Chromosome studies in somatic cells. TK 12386. Chinese hamster.* Internal report 78-3110, unpublished

Ciba-Geigy (1982b) *Dominant lethal study. TK12386. Mouse.* Internal report 790529, unpublished

Ciba-Geigy (1982c) *Chromosome studies in male germinal epithelium. TK 12386. Mouse.* Internal report 782932, unpublished

Ciba-Geigy (1984a) *Final report. TK10490. 28 days subacute, oral toxicity study in rats.* GU Project 830199, unpublished

Ciba-Geigy (1984b) *Chromosome studies on male germinal epithelium of mouse spermatogonia. Test material: TK 10490 (Araldite GY 250).* Internal report 830336, unpublished

Ciba-Geigy (1988a) *A study of the effect of TK 10490 on pregnancy of the rat.* HRC Report No. CBG 442/871640, Huntingdon Research Centre, Huntingdon, England, unpublished

Ciba-Geigy (1988b) *A study of the effect of TK 10490 on pregnancy of the rabbit.* HRC Report No. CBG 440/871639, Huntingdon Research Centre, Huntingdon, England, unpublished

Ciba-Geigy (1989) *A study of the effect of TK 10490 on reproductive function of one generation in the rat.* HRC Report No. CBG 451/881061, Huntingdon Research Centre, Huntingdon, England, unpublished

Ciba Specialty Chemicals (1997) *CY 179, CY 184 and MY 790. Investigation of alkylating potency and epoxide hydrolysis.* Report CB93/25-A, Novartis Crop Protection, Toxicology Services/ Cell Biology, Basle, unpublished

Climie IJG, Hutson DH, Stoydin G (1981a) Metabolism of the epoxy resin component 2,2-bis[4-(2,3-epoxypropoxy)phenyl]propane, the diglycidyl ether of bisphenol A (DGEBPA) in the mouse. Part I. A comparison of the fate of a single dermal application and of a single oral dose of ^{14}C-DGEBPA. *Xenobiotica 11*: 391–399

Climie IJG, Hutson DH, Stoydin G (1981b) Metabolism of the epoxy resin component 2,2-bis[4-(2,3-epoxypropoxy)phenyl]propane, the diglycidyl ether of bisphenol A (DGEBPA) in the mouse. Part II. Identification of metabolites in urine and feces following a single oral dose of ^{14}C-DGEBPA. *Xenobiotica 11*: 401–424

Condé-Salazar L, Gonzalez de Domingo MA, Guimaraens D (1994) Sensitization to epoxy resin systems in special flooring workers. *Contact Dermatitis 31*: 157–160

DECOS (Dutch Expert Committee on Occupational Standards) (1996) *Bisphenol A and its diglycidyl ether. Health-based recommended occupational exposure limit.* Publication No. 1996/02WGD, Health Council of the Netherlands, Den Haag

Doak SMA, Simpson BJE, Hunt PF, Stevenson DE (1976) The carcinogenic response in mice to the topical application of propane sultone to the skin. *Toxicology 6*: 139–154

Dow Chemical (1958) *Results of skin irritation and sensitization studies on: epoxy resin X2633.1 (DOW ET-257), epoxy resin X2633.2 (DOW ET-256).* NTIS/OTS 0206671, EPA/OTS Doc ID 878214854

Dow Chemical (1977) *Integrated mutagenicity testing program on several epoxy compounds.* Report to the Dow Chemical Company, unpublished

Dow Chemical (1981) *Diglycidyl ether of bisphenol A (DGEBA): Fate in male Fischer 344 rats (probc).* NTIS/OTS 0206138, EPA/OTS Doc ID 818211589

Dow Chemical (1987) *Diglycidyl ether of bisphenol A (DGEBA): Dermal teratology study in rabbits.* NTIS/OTS 0523698, EPA/OTS new Doc ID 40-8740513

Dow Chemical Company (1998a) *DGEBPA: two-year dermal chronic toxicity/oncogenicity study in female Fischer 344 rats.* The Dow Chemical Company, Health & Environmental Research Laboratories, Midland, Michigan, USA, unpublished

Dow Chemical Company (1998b) *DGEBPA: two-year dermal chronic toxicity/oncogenicity study in the male B6C3F$_1$ mouse.* Dow Chemical Company, Health & Environmental Research Laboratories, Midland, Michigan, USA, unpublished

Dow Chemical Company (1998c) *DGEBPA: A two-year dermal chronic toxicity/oncogenicity study in female Fischer 344 rats.* Pathology Peer Review Report, Dow Chemical Company, Experimental Pathology Laboratories, Inc., Herndon, USA, unpublished

Flyvholm M-A (1991) Contact allergens in registered chemical products. *Contact Dermatitis 25*: 49–56

Fregert S, Thorgeirsson A (1977) Patch testing with low molecular oligomers of epoxy resins in humans. *Contact Dermatitis 3*: 301–303

Gardiner TH, Waechter Jr JM,Wiedow MA, Solomon WT (1992) Glycidyloxy compounds used in epoxy resin systems: A toxicology review. *Reg Toxicol Pharmacol 15*: S34–S77

Harada T, Maronpot RR, Morris RW, Stitzel KA, Boorman GA (1989) Morphological and stereological characterization of hepatic foci of cellular alteration in control Fischer 344 rats. *Toxicol Pathol 17*: 579–593

Haseman JK, Huff JE (1987) Species correlation in long-term carcinogenicity studies. *Cancer Lett 37*: 125–132

Haseman JK, Hailey JR, Morris RW (1998) Spontaneous neoplasm incidences in Fischer 344 rats and B6C3F$_1$ mice in two-year carcinogenicity studies: a National Toxicology Program update. *Toxicol Pathol 26*: 428–441

Hemminki K, Falck K, Vainio H (1980) Comparison of alkylation rates and mutagenicity of directly acting industrial and laboratory chemicals. Epoxides, glycidyl ethers, methylating and ethylating agents, halogenated hydrocarbons, hydrazine derivatives, aldehydes, thiuram and dithiocarbamate derivatives. *Arch Toxicol 46*: 277–285

Hine CH, Guzman RJ, Coursey NM, Wellington WS, Anderson HH (1958) An investigation on the oncogenic activity of two representative epoxy resins. *Cancer Res 18*: 20–26

Holland JM, Gosslee DG, Williams NJ (1979) Epidermal carcinogenicity of bis(2,3-epoxycyclopentyl)ether, 2,2-bis(*p*-glycidyloxyphenyl)propane and *m*-phenylenediamine in male and female C3H and C57BL/6 mice. *Cancer Res 39*: 1718–1725

Holness DL, Nethercott JR (1993) The performance of specialized collections of bisphenol A epoxy resin system components in the evaluation of workers in an occupational health clinic population. *Contact Dermatitis 28*: 216–219

IARC (International Agency for Research on Cancer) (1989). *IARC monographs on the evaluation of carcinogenic risks to humans. Some organic solvents, resin monomers and related compounds, pigments and occupational exposures in paint manufacture and painting. Vol. 47*: 237–261

Jolanki R (1991) Occupational skin diseases from epoxy compounds. Epoxy resin compounds, epoxy acrylates and 2,3-epoxypropyl trimethyl ammonium chloride. *Acta Derm Venereol (Stockh) Suppl 159*: 1–80

Jolanki R, Estlander T, Kanerva L (1987) Occupational contact dermatitis and contact urticaria caused by epoxy resins. *Acta Derm Venereol (Stockh) Suppl 134*: 90–94

Jolanki R, Kanerva L, Estlander T, Tarvainen K, Keskinen H, Henriks-Eckerman ML (1990) Occupational dermatoses from epoxy resin compounds. *Contact Dermatitis 23*: 172–183

Jolanki R, Kanerva L, Estlander T (1995) Occupational allergic contact dermatitis caused by epoxy diacrylate in ultraviolet-light-cured paint, and bisphenol A in dental composite resin. *Contact Dermatitis 33*: 94–99

Kanerva L, Estlander T, Jolanki R (1988) Occupational skin disease in Finland: An analysis of 10 years of statistics from an occupational dermatology clinic. *Int Arch Occup Environ Health 60*: 89–94

Kanerva L, Jolanki R, Tupasela O, Halmepuro L, Keskinen H, Estlander T, Sysilampi M-L (1991) Immediate and delayed allergy from epoxy resins based on diglycidyl ether of bisphenol A. *Scand J Work Environ Health 17*: 208–215

Kanerva L, Estlander T, Keskinen H, Jolanki R (2000) Occupational allergic airborne contact dermatitis and delayed bronchial asthma from epoxy resin revealed by bronchial provocation test. *Eur J Dermatol 10*: 475–477

Kayser D (Ed.) (1986) *Kurzzeittests zum Nachweis von Kanzerogenen* (Short-term tests to detect carcinogens) (German). *bga-Schriften 1/86*, MMV Medizin Verlag München

Kier LE, Brusick DJ , Auletta AE, Von Halle ES, Brown MM, SimmonVF, Dunkel V, McCann J, Mortelmans K, Prival M, Rao TK, Ray V (1986) The *Salmonella typhimurium*/mammalian microsomal assay. A report of the US Environmental Protection Agency Gene-Tox Program. *Mutat Res 168*: 69–204

Kociba RJ, Keyes DG (1985) Squamous cell carcinoma, tongue, rat. in: Jones TC, Mohr U, Hunt RD (Eds) *Digestive System*, ILSI Monographs of Pathology of Laboratory Animals, Springer Verlag, Berlin-Heidelberg

Lewis Sr RJ (1992) *Sax's Dangerous Properties of Industrial Chemicals*, 8th ed, Vol. II, Van Nostrand Reinhold, New York, 511–512

Mahon DC (1989) Altered hepatic foci in rat liver as weight of evidence of carcinogenicity: the Canadian perspective. *Toxicol Pathol 17*: 709–715

Maronpot RR, Harada T, Murthy ASK, Boorman GA (1989) Documenting foci of hepatocellular alteration in two-year carcinogenicity studies: current practices of the National Toxicology Program. *Toxicol Pathol 17*: 675–684

Mitelman F, Fregert S, Hedner K, Hillbertz-Nilsson K (1980) Occupational exposure to epoxy resins has no cytogenetic effect. *Mutat Res 77*: 345–348

Moch RW, Dua PM, Hines FA (1996) Problems in consideration of rodent hepatocarcinogenesis for regulatory purposes. *Toxicol Pathol 24*: 138–146

Moura C, Dias M, Vale T (1994) Contact dermatitis in painters, polishers and varnishers. *Contact Dermatitis 31*: 51–53

Murray MP, Cummings JE (1979) Mutagenic activity of epoxy embedding reagents employed in electron microscopy. *Environ Mutagen 1*: 307–313

Niinimäki A, Hassi J (1983) An outbreak of epoxy dermatitis in insulation workers at an electric power station. *Dermatosen 31*: 23–25

NTP (1996) *TR-441. Comparative initiation/promotion skin paint studies of B6C3F$_1$ mice, Swiss (CD-1®) mice, and SENCAR mice.* National Toxicology Program, Bethesda, USA

Oesch F, Schmassmann H-U, Bentley P (1978) Specificity of human, rat and mouse skin epoxide hydratase towards K-region epoxides of polycyclic hydrocarbons. *Biochem Pharmacol 27*: 17–20

Peristianis GC, Doak SMA, Cole PN, Hend RW (1988) Two-year carcinogenicity study on three aromatic epoxy resins applied cutaneously to CF1 mice. *Food Chem Toxicol 26*: 611–624

Prens EP, de Jong G, van Joost T (1986) Sensitization to epichlorohydrin and epoxy system components. *Contact Dermatitis 15*: 85–90

Ringo DL, Brennan EF, Cota-Robles EH (1982) Epoxy resins are mutagenic. Implications for electron microscopists. *J Ultrastruct Res 80*: 280–287

Richter-Reichhelm H-B (1989) View of the regulatory use of rat liver foci data in the Federal Republic of Germany. *Toxicol Pathol 17*: 721–723

Sazonova LA, Suskov II (1985) Cytogenetic effects of epoxy resin in man: dependence on sex, age and period of exposure. *Mutat Res 155*: 127–129

Shell (1980) *Skin cancer studies on bisphenol A based epoxy resins: overview of two recently completed studies.* Shell International Petroleum, Group Toxicology Division, Bericht vom 9.3.80

Shell (1981a) *Toxicity studies with epoxy resins: in vitro genotoxicity studies with diglycidyl ether of bisphenol A, Epikote 828, Epikote 1007.* Sittingbourne Research Centre, Shell Toxicology Laboratory, group research report TLGR. 80.123, unpublished

Shell (1981b) *Studies on the effect of diglycidyl ether of bisphenol A, on the integrity of rat liver DNA in vivo.* Sittingbourne Research Centre, Shell Toxicology Laboratory, group research report TLGR. 80.102; 80.152, unpublished

Shell (1983a) *Diglycidylether of bisphenol A (DGEBPA). Development of a method to determine a urinary metabolite.* Sittingbourne Research Centre, Shell Toxicology Laboratory, group research report SBGR 83.170, unpublished

Shell (1983b) *A cutaneous carcinogenicity study with mice on diglycidyl ether of bisphenol A. Systemic pathology report.* Shell Toxicology Laboratory, addendum to group research report TLGR 80.076. NTIS/OTS 0523689, new Doc ID 40-8340501

SPI (Society of the Plastics Industry, Epoxy Resin Systems, DGEBPA Task Force) (1996) *DGEBPA. Two-generation oral gavage reproductive study in Sprague-Dawley rats.* Dow Chemical, Toxicology Research Laboratory Health and Environmental Sciences, Midland, USA, unpublished

Squire R (1989) Evaluation and grading of rat liver foci in carcinogenicity tests. *Toxicol Pathol 17*: 685–688

Steiner S, Hönger G, Sagelsdorff P (1992) Molecular dosimetry of DNA adducts in C3H mice treated with bisphenol A diglycidylether. *Carcinogenesis 13*: 969–972

Suskov II, Sazonova LA (1982) Cytogenetic effects of epoxy, phenolformaldehyde and polyvinylchloride resins in man. *Mutat Res 104*: 137–140

Tarvainen K, Kanerva L, Jolanki R, Estlander T (1995) Occupational dermatoses from the manufacture of plastic composite products. *Am J Contact Dermatitis 6*: 95–104

Thorgeirsson A, Fregert S (1977) Allergenicity of epoxy resins in the guinea pig. *Acta Derm Venereol (Stockh) 57*: 253–256

Thorgeirsson A, Fregert S, Ramnäs O (1978) Sensitization capacity of epoxy resin oligomers in the guinea pig. *Acta Derm Venereol (Stockh) 58*: 17–21

Tosti A, Guerra L, Vincenci C, Peluso A-M (1993) Occupational skin hazards from synthetic plastics. *Toxicol Ind Health 9*: 493–502

Union Carbide (1964) *Results of long-term tests for mouse skin carcinogenicity of eight epoxides.* Mellon Institute Special Report 27-152, unpublished

Union Carbide (1978) *Epidermal carcinogenicity of bis(2,3-epoxycyclopentyl)ether, 2,2-bis(p-glycidyloxyphenyl)propane and m-phenylenediamine in male and female C3H and C57BL/6 mice.* Oak Ridge National Laboratory Report No. ORNL 5375, NTIS, EPA/OTS Doc ID 40-7840392

Union Carbide (1981a) *BAKELITE® epoxy resin ERL-2774. In vitro mutagenesis studies: 3-test battery.* Bushy Run Research Center, project report 43-129, unpublished

Union Carbide (1981b) *Chronic dermal toxicity of epoxy resins. I. Skin carcinogenic potency and general toxicity.* Oak Ridge National Laboratory Report No. ORNL 5762. NTIS/OTS 0204933, new Doc ID 88-8100262

USEPA (1994) *Testing consent order for bisphenol A diglycidyl ether.* US Environmental Protection Agency, Federal Register 59: 38917–38920

Van Joost T (1988) Occupational sensitization to epichlorohydrin and epoxy resin. *Contact Dermatitis 19*: 278–280

Van Putten PB, Coenraads PJ, Nater JP (1984) Hand dermatoses and contact allergic reactions in construction workers exposed to epoxy resin. *Contact Dermatitis 10*: 146–150

Wade MJ, Moyer JW, Hine CH (1979) Mutagenic action of a series of epoxides. *Mutat Res 66*: 367–371

Walker CH, Bentley P, Oesch F (1978) Phylogenetic distribution of epoxyde hydratase in different vertebrate species, strains and tissues measured using three substrates. *Biochim Biophys Acta 539*: 427–434

Zakova N, Zak F, Froehlich E, Hess R (1985) Evaluation of skin carcinogenicity of technical 2,2-bis-(*p*-glycidyloxyphenyl)-propane in CF1 mice. *Food Chem Toxicol 23*: 1081–1089

completed 01.03.2001

n-Butyl acetate

MAK value (1997)	**100 ml/m³ (ppm) ≙ 480 mg/m³**
Peak limitation (2000)	**Category I, excursion factor 2**
Absorption through the skin	**–**
Sensitization	**–**
Carcinogenicity	**–**
Prenatal toxicity (1999)	**Pregnancy risk group C**
Germ cell mutagenicity	**–**
BAT value	**–**
Synonyms	–
Chemical name (CAS)	acetic acid butyl ester
CAS number	123-86-4
Structural formula	$CH_3COO–CH_2–(CH_2)_2–CH_3$
Molecular formula	$C_6H_{12}O_2$
Molecular weight	116.16
Melting point	–77°C
Boiling point	126°C
Density at 20°C	0.88 g/cm³
Vapour pressure at 20°C	13.3 hPa
log P_{ow}*	1.82
1 ml/m³ (ppm) ≙ 4.82 mg/m³	**1 mg/m³ ≙ 0.207 ml/m³ (ppm)**

The present documentation is based on reviews of the toxicological data for *n*-butyl acetate (ACGIH 1997, ECB 1995).

* *n*-octanol/water distribution coefficient

1 Toxic Effects and Mode of Action

Once absorbed by the organism or in contact with mucous membranes, *n*-butyl acetate is hydrolysed to *n*-butyl alcohol and acetic acid. The acute toxicity of *n*-butyl acetate is low after oral and dermal absorption and inhalation exposure to the vapour. After high concentrations, irritation of the eyes and respiratory passages and central nervous and narcotic effects are observed. The substance does not affect rabbit skin, but has weak irritative effects on the rabbit eye. *n*-Butyl acetate does not cause sensitization in man or animals.

The critical effect for man is irritation of the eyes and upper respiratory tract which begins from concentrations of about 200 ml/m^3. The few data available from experiments with animals suggest that the toxicity of the substance after repeated exposure is also low; again, irritation is the main effect. *n*-Butyl acetate is not mutagenic in prokaryotes and does not induce chromosomal aberrations in mammalian cells. Studies of the carcinogenic effects of the substance are not available.

Exposure of pregnant rats to the maternally toxic concentration of 1500 ml/m^3 leads in the offspring to reduced body sizes and in some cases to skeletal variations. Malformations do not occur. In rabbits this treatment has no effects on the parent animals, but in the offspring the incidence of morphological variations is increased.

2 Mechanism of Action

There are no data available for the mechanism of action of *n*-butyl acetate. The irritative effects may be caused by the acetic acid formed during hydrolysis.

3 Toxicokinetics and Metabolism

The rate of absorption of ingested or inhaled *n*-butyl acetate is not known. *n*-Butyl acetate does not readily penetrate the skin. With isolated human skin, a permeability constant of 1.6 ± 0.1 g/m^2 and hour was determined (Ursin *et al.* 1995). From the distribution coefficients liver/air, muscle/air and fat/air for *n*-butyl acetate of 281, 157 and 900 determined at 37°C with rat tissue (Poulin and Krishnan 1996) it can be concluded that the absorbed substance is distributed in the whole body with the greatest accumulation in adipose tissue. In persons exposed to *n*-butyl acetate concentrations of 200 mg/m^3, the *n*-butyl acetate concentration in the exhaled air was found to be 100 mg/m^3 (ACGIH 1997).

After contact with mucous membranes and absorption by the organism, butyl acetates may be assumed to be hydrolysed to the corresponding butyl alcohols and acetic acid in reactions catalysed by various esterases; this was demonstrated with isolated nasal tissue from the rat. The rate of hydrolysis decreased with increasing branching of the butyl residue and was between 42 and 77 nmol/mg protein and minute. This can be explained by the increasing steric hindrance (Dahl *et al.* 1987). Exposure of rats for 5 hours to *n*-butyl acetate concentrations of 1000 ml/m^3 via tracheal cannula led to rapid equilibration at a blood level of 25 µM. During the first 30 minutes of exposure the level of *n*-butyl alcohol in the blood increased to 52 µM and then remained constant (Groth and Freundt 1991, 1994). In male Sprague-Dawley rats given intravenous injections of ^{14}C-labelled *n*-butyl acetate (about 30 mg/kg body weight) the elimination half-time from the blood was 0.4 minutes. During the first 2.5 minutes after administration the substance could also be detected in the brain. The maximum level of *n*-butyl alcohol could be detected in the blood and brain 2.5 minutes after administration. Somewhat later, but only in the blood, also butyraldehyde and polar metabolites (probably conjugates of *n*-butyl alcohol with sulfuric acid or glucuronic acid and intermediates from the citric acid cycle) were found. The elimination half-time for *n*-butyl alcohol (1–1.2 minutes) was also short. 20 minutes after the administration of *n*-butyl acetate no more *n*-butyl alcohol could be detected (CMA 1997). In an earlier study, 4-hydroxy-3-methoxymandelic acid was shown to be the main metabolite in the urine of mammals (no other details). It was shown later, however, that this substance is excreted with the urine also under normal or pathological conditions, which is the reason it is not clear whether 4-hydroxy-3-methoxymandelic acid is a true butyl acetate metabolite (ACGIH 1997, ECB 1995).

4 Effects in Man

Workers repeatedly exposed to *n*-butyl acetate and other substances complained of irritation of the eyes, nose and throat, and tightness of the chest. Very high concentrations (no other details) led also to weakness, dizziness and unconsciousness. As the exposure was always to a mixture of solvents, the symptoms cannot be attributed exclusively to *n*-butyl acetate. Repeated contact with the liquid substance led to defatting and slight irritation of the affected area of skin (ACGIH 1997, ECB 1995).

In a controlled study, 10 volunteers were exposed to *n*-butyl acetate for 2 to 5 minutes. At concentrations of 200 ml/m^3 most of the persons complained of throat irritation, at 300 ml/m^3 the symptoms given were irritation of the eyes and nose and severe throat irritation (Nelson *et al.* 1943). Exposure for 5 minutes in a room in which 3 minutes previously *n*-butyl acetate had been sprayed and then distributed evenly, led at the low concentration of 1000 mg/m^3 (210 ml/m^3) to slight eye, nose and throat irritation, and to very mild irritation of the airways. At the high concentration of 10000 mg/m^3 (2100 ml/m^3), a moderate irritation score was given for all symptoms (Flury and Wirth 1933). Persons who were exposed for a short time (no other details) to concentrations of 3300, 7000 and 14000 ml/m^3, reported a strong, unpleasant odour and irritation of the eyes and nasal mucosa which was more severe with increasing concentration. The odour threshold for *n*-butyl acetate was given as 0.39 ml/m^3 (ACGIH 1997). In another inhalation study with controlled exposure in a chamber, 24 healthy non-smokers were exposed to *n*-butyl acetate according to 3 different exposure patterns. The irritative effects were evaluated on the basis of the subjective feeling of irritation, clinical examination of eye irritation and the lung function. In the first experiment the persons were exposed for 20 minutes to concentrations of 74, 147, 221 and 295 ml/m^3. Evaluation of the irritative effects by the exposed group and the not exposed control group revealed no significant differences. In the second experiment one group was exposed for 20 minutes to *n*-butyl acetate concentrations of 15 ml/m^3, the other group to 295 ml/m^3. The group exposed to the low concentration was regarded as the control group. The authors later judged this concentration, however, to be too high for a control exposure. The persons from the high concentration group gave their symptoms of irritation a significantly higher score than did the control group, although the irritation was not felt to be severe. An unpleasant odour was described only by the control group. In the third experiment the persons were exposed for 4 hours to concentrations of 15 ml/m^3 (controls) and 147 ml/m^3. In the high concentration group, throat irritation, breathing difficulties and the perception of an unpleasant odour were significantly more marked than in the control group. Red eyes were observed in 50 % of the persons exposed to the high concentration and in 17 % of the control persons. Nevertheless, the authors concluded from these investigations that *n*-butyl acetate has only a low irritative potential (Iregren *et al.* 1993). For exposure to *n*-butyl acetate for about 2 seconds a threshold value of 3650 ml/m^3 was given for nasal irritation in anosmic persons (Abraham *et al.* 1996).

Unexposed persons and persons with solvent-induced toxic encephalopathy were exposed for 2 hours to increasing *n*-butyl acetate concentrations (3–50 ml/m^3). The persons of the second collective were mostly significantly more sensitive to the odour, taste, unpleasant effects, irritation and tiredness associated with the exposure than were the healthy persons (Ørbæk *et al.* 1998). As there was no group exposed only to air, it is not possible to make a statement about the irritation threshold of *n*-butyl acetate.

A 4 % *n*-butyl acetate preparation in petrolatum did not cause skin irritation after occlusive application for 48 hours. Nail varnish containing 25.5 % *n*-butyl acetate was not found to have irritative effects after repeated occlusive application to the skin. Also after repeated application of the nail varnish to the nails, no irritative effects were seen on the nails, cuticle or skin. In an accident at work, butyl acetate caused corneal corrosion. After removal of the damaged epithelium down to Bowman's membrane, the cornea healed within 48 hours (no other details) (ACGIH 1997, ECB 1995).

Several maximization tests with *n*-butyl acetate did not lead to sensitization. Induction was carried out by repeated occlusive application of 0.5 ml of the pure substance, of 4 % *n*-butyl acetate in petrolatum or of nail varnish containing 25.5 % *n*-butyl acetate. Provocation took place 10 days after the last induction treatment. A patch test with 5 % *n*-butyl acetate in olive oil produced positive results in one worker who had occupational contact with *n*-butyl acetate and other substances and suffered from dermatitis on the hands, arms and face. In 36 control persons the results were negative. Of 149 patients with dermatitis, one produced a positive reaction to *n*-butyl acetate (ACGIH 1997, ECB 1995).

5 Animal Experiments and *in vitro* Studies

5.1 Acute toxicity

5.1.1 Inhalation

The symptoms observed after single exposures to *n*-butyl acetate were irritation of the eyes and respiratory tract; with high concentrations severe lung damage was observed (haemorrhage, congestion, oedema) and was regarded as the cause of death. In addition, there were central nervous and narcotic effects.

In two studies with rats, inhaled *n*-butyl acetate was found to be of unexpectedly high toxicity with 4-hour LC_{50} values of 160 and 391 ml/m^3. This was attributed to the high fraction of the substance present as an aerosol in the test atmosphere, but could not be reproduced in other studies with *n*-butyl acetate aerosol. The acute inhalation toxicity of *n*-butyl acetate vapour is low. For rats exposed for 4 hours, an LC_{50} value of 2000 ml/m^3 was determined. In other studies carried out according to OECD guideline 403 the 4-hour LC_{50} values were all over 4000 ml/m^3. All rats survived exposure for 4 hours to an atmosphere more or less saturated with *n*-butyl acetate (about 10000 ml/m^3) and exposure for 6 hours to 1850 ml/m^3. All animals died after exposure for 8 hours to an atmosphere saturated with *n*-butyl acetate and exposure for 6 hours to 14050 ml/m^3 (ACGIH 1997, ECB 1995). After exposure of rats for 6 hours to *n*-butyl acetate concentrations of 1500, 3000 and 6000 ml/m^3 no deaths occurred. The clinical and gross pathological examinations after autopsy did not yield unusual findings. Particular consideration was given to possible neurotoxic effects. In the two high concentration groups the average motor activity was reduced and some behavioural parameters changed on the day of exposure. On days 1, 7 and 14 after the end of exposure such anomalies could no longer be detected. The animals of all concentration groups seemed to be slightly sedated during the exposure and reacted more slowly to external stimuli (CMA 1994).

Mice survived exposure to 5900 ml/m^3 for 6 hours and to 7350 ml/m^3 for 3.6 hours. In an earlier investigation an LC_{50} value of 1260 ml/m^3 was found for mice (ACGIH 1997, ECB 1995). Exposure of mice for 20 minutes to *n*-butyl acetate led from concentrations of 8000 ml/m^3 to reduced motor activity, and numerous parameters investigated in a behavioural test battery were changed relative to the control values. All symptoms were reversible within a few minutes after the end of exposure (Bowen and Blaster 1997). In Swiss-OF1 mice, the concentration which reduced the respiration rate by 50 % (RD_{50} value) after exposure for several minutes was found to be 730 ml/m^3. The concentration–effect relationship was not given (Muller and Greff 1984).

The lowest lethal concentration for guinea pigs after exposure for 4 hours was 14000 ml/m^3 and for cats after exposure for 1.2 hours was 14300 ml/m^3 (ACGIH 1997, ECB 1995).

5.1.2 Ingestion and dermal absorption

The acute oral toxicity of *n*-butyl acetate is also low. For the rat, the LD_{50} values were in the range from 10700 to 14130 mg/kg body weight. A value of 7060 mg/kg body weight was determined for the mouse. For the rabbit, in two different studies LD_{50} values of 7440 and 3200 mg/kg body weight were obtained. The dermal LD_{50} value for the rabbit was found to be over 17600 mg/kg body weight. In guinea pigs, occlusive application of *n*-butyl acetate doses of up to 10 ml/kg body weight (8800 mg/kg body weight) for 24 hours was not lethal (ACGIH 1997, ECB 1995).

5.2 Subacute, subchronic and chronic toxicity

Groups of 10 male and 10 female Sprague-Dawley rats were exposed to *n*-butyl acetate concentrations of 500, 1500 or 3000 ml/m^3 for 6 hours a day over 14 weeks for a total of 65 days. Specific investigations to evaluate the neurotoxic potential of the substance were carried out. The results of tests for functional behaviour, motor activity and operant behaviour were not unusual at the lowest concentration. At concentrations of 1500 and 3000 ml/m^3 there were transient signs of sedation and hypoactivity. Motor activity was significantly increased in the male animals of the high concentration group only in week 4 of the study, but not in weeks 8 or 13. Histological examination of the central and peripheral nervous systems did not reveal any differences between the control group and the 3000 ml/m^3 group. Other organs were not investigated. The only sign of systemic toxicity was reduced body weights from concentrations of 1500 ml/m^3 in the female animals and from 3000 ml/m^3 in the male animals (David *et al.* 1998). From this study, a NOEL (no observed effect level) of 500 ml/m^3 can be deduced for rats, at least for neurotoxic effects.

The following data for toxicity after repeated exposure are from studies carried out over 60 years ago.

Mice tolerated without symptoms exposure for 4 days (17 hours/day) to an atmosphere saturated with *n*-butyl acetate. Cats were exposed to concentrations of 1600, 3100 or 4200 ml/m^3, 6 hours/day for 6 days. At 1600 ml/m^3 increased salivation and slight eye irritation were observed. At 3100 ml/m^3 morphological changes in the blood cells were detected, and after 4200 ml/m^3 the animals appeared weak, body weights were reduced and slight irritation of the airways was observed. In guinea pigs exposed to *n*-butyl acetate concentrations of 1000 ml/m^3, 6 hours daily on 6 days weekly for 5 weeks, neither urine and blood analyses nor autopsy revealed pathological findings (ACGIH 1997, ECB 1995).

5.3 Local effects on skin and mucous membranes

0.01 ml *n*-butyl acetate applied for 24 hours to the rabbit skin produced effects which were given an irritation index of 1 (maximum value 10), and thus the substance was classified as not irritative. Also after occlusive application of 0.5 ml *n*-butyl acetate to rabbit skin for 4 hours no signs of irritation were observed. Occlusive application of *n*-butyl acetate doses of up to 10 ml/kg body weight for 24 hours to the depilated abdominal skin of guinea pigs was not found to have irritative effects. The observation period was 2 weeks (ACGIH 1997, ECB 1995).

In an earlier study of the irritative effects of *n*-butyl acetate on the rabbit eye, severe, yet reversible corrosion of the cornea was described after the instillation of 0.005 ml into the conjunctival sac. The irritative effects were given a score of 5 on a scale with a maximum value of 10. On the other hand, in several more recent studies the substance had only weak irritative effects or was not irritating to the eye when applied for 24 hours. The effects of 0.1 ml *n*-butyl acetate were given an irritation index of 8 on a scale with a maximum value of 110. With a 30 % solution an irritation index of 11 was obtained, with a 10 % solution an irritation index of 19 (ACGIH 1997, ECB 1995).

5.4 Allergenic effects

In the maximization test with guinea pigs and the mouse ear swelling test, *n*-butyl acetate was not found to have sensitizing effects (ACGIH 1997, ECB 1995).

5.5 Reproductive and developmental toxicity

The reproductive toxicity of *n*-butyl acetate was investigated with Sprague-Dawley rats (37–43 animals/group) exposed to 1500 ml/m^3 for 7 hours daily. The animals from group 1 served as controls and were exposed to filtered air according to the exposure pattern of group 4 (see below). The animals from group 2 were exposed from days 7 to 16 of gestation, those from group 3 from days 1 to 16. The animals from group 4 were exposed during the 3 weeks before mating on 5 days per week, then mated with untreated male animals and exposed from days 1 to 16 of gestation. In all groups food consumption, and as a result body weights, were reduced at the end of the study, an indication of maternal toxicity. The histological examination of the dams did not reveal any pathological findings. The course and success of gestation were not affected in any group. The length of the foetuses was reduced in all groups. Whether this is the result of effects of *n*-butyl acetate on the dams or a direct effect on the foetuses cannot be determined. The incidence of abnormally formed ribs (wavy, fused, branched) was significantly increased in the foetuses of groups 2 and 3, but not in the foetuses of group 4. Extended urethras were found in an increased number only in groups 3 and 4, but not in group 2. As these findings are generally regarded as variations and not as malformations and they did not occur equally in all groups, they are not to be seen as teratogenic effects (NIOSH 1982).

In another test, artificially inseminated New Zealand White rabbits (21–25 animals per group) were investigated. The rabbits were exposed to *n*-butyl acetate concentrations of 1500 ml/m^3 from days 7 to 19 of gestation (group 2) or from days 1 to 19 (group 3) for 7 hours daily. The control animals (group 1) were exposed to filtered air according to the exposure pattern of group 3. Food consumption was reduced in the treated groups; body weights, however, corresponded with those of the controls up to the end of the study. The course and success of gestation were not affected in any group. The histological examination of the dams did not reveal any pathological findings. In the offspring of the two treated groups no malformations were detected. The increase in the incidence of morphological variations of the gall bladder and of other anomalies (retinal folds, unsymmetrically fused breastbone) was statistically significant only in the offspring of group 3. In the opinion of the authors *n*-butyl acetate had no significant influence on the development of the foetuses (NIOSH 1982).

In rats, the *n*-butyl acetate metabolite *n*-butyl alcohol had no effects on the offspring and dams up to concentrations of 3500 ml/m^3 (see "*n*-Butyl alcohol" in this volume). For the other metabolite, acetic acid, the effective principle of any prenatal toxicity could be acidosis. However, exposure of rabbits for over 2 hours even to very high concentrations of various alkyl acetate esters (e.g. *n*-propyl acetate concentrations of 54000 mg/m^3, corresponding to 13000 ml/m^3) did not produce such effects (Flury and Wirth 1933), so that in the range of the MAK value for *n*-butyl acetate germ cell mutagenicity resulting from acetic acid is not to be expected. This is confirmed by a study with rabbits given cider vinegar in doses of 1600 mg /kg body weight from days 6 to 18 of gestation. Assuming an acetic acid content of 5 %, this corresponds to acetic acid doses of 80 mg/kg body weight. Neither maternal toxicity nor foetal toxicity, nor an increased incidence of malformations in the foetuses were observed (FDA 1977).

5.6 Genotoxicity

n-Butyl acetate was not found to be mutagenic in the *Salmonella typhimurium* strains TA92, TA94, TA98, TA100, TA1535, TA1537 or TA1538, or in *Echerichia coli* WP2 uvr either with or without an added metabolic activation system. The highest tested concentration was 10000 µg/plate. With *Saccharomyces cerevisiae* D61.M, the substance did not induce mitotic aneuploidy up to the highest concentration tested of 4 mg/ml.

Tests for chromosomal aberrations in CHL cells (a cell line derived from Chinese hamster liver) yielded negative results up to the highest concentration tested of 2 mg/ml (ACGIH 1997, ECB 1995).

There are no data available for the carcinogenicity of *n*-butyl acetate.

6 Manifesto (MAK value/classification)

For the workplace situation, the critical effects of *n*-butyl acetate are the irritative effects on the eyes, nasal mucosa and throat. The studies of the threshold concentration in man do not give consistent results. Symptoms are described by 2 authors after a maximum of 5 minutes exposure to 200 ml/m^3. At 300 ml/m^3 throat irritation was described as severe. On the other hand, the symptoms were not evaluated differently by the persons exposed for 20 minutes to concentrations of 74 to 295 ml/m^3 than by the control persons. Yet in the same study, volunteers exposed for 4 hours to 147 ml/m^3 reported symptoms of irritation, in particular throat irritation, significantly more often.

It was concluded from these findings that the MAK value for *n*-butyl acetate which applied until 1999 (200 ml/m^3), and in particular the peak exposure value of 400 ml/m^3, did not provide adequate protection against the irritative effects of the substance. The MAK value was therefore reduced to 100 ml/m^3. In the study of Iregren *et al.* (1993), important for determining the excursion factor, no irritative effects occurred at concentrations of 221 ml/m^3 for 20 minutes. *n*-Butyl acetate is therefore classified in Peak limitation category I with an excursion factor of 2.

Systemic-toxic effects are not to be expected if the MAK value is observed. A medium-term inhalation study with rats yielded a NOEL of 500 ml/m^3. It must, however, be pointed out that mainly neurotoxic parameters were investigated. The systemic toxicity of *n*-butyl acetate can also be estimated by evaluating the systemic toxicity of the metabolites. For *n*-butyl alcohol, the MAK value is also 100 ml/m^3, a concentration at which no systemic effects occur (see "*n*-Butyl alcohol" in this volume). The amount of acetic acid taken up daily during exposure at the level of the MAK value, at most 2.5 g, is of the order of magnitude of the average amount taken up with food (Guest *et al.* 1982). In addition, acetic acid is formed endogenously and is produced in numerous pathways of intermediary metabolism, so that no adverse effects are expected as a result of the additional occupational exposure.

The maternally toxic concentration of 1500 ml/m^3 was also foetotoxic in rats, but not teratogenic. Concentrations not toxic for the dams were not tested; it is therefore not possible to use this study to evaluate the potential reproductive toxicity of *n*-butyl acetate. However, the metabolite *n*-butyl alcohol also did not have effects on the offspring of rats exposed at the concentration 3500 ml/m^3, which was not toxic for the dams. Given observation of the MAK value for *n*-butyl acetate, germ cell mutagenicity is not to be expected from acetic acid, the other metabolite (see Section 5.5). In rabbits, *n*-butyl acetate concentrations of 1500 ml/m^3 were not maternally toxic and led in the offspring merely to an increase in the incidences of various variations not regarded as adverse. There is thus no reason to fear a risk of damage to the embryo or foetus when the MAK value is observed, and *n*-butyl acetate has been classified in Pregnancy risk group C.

Because of the high dermal LD_{50} value in rabbits and the poor skin penetration of *n*-butyl acetate, the substance is not designated with an "H".

n-Butyl acetate was not found to have sensitizing effects in controlled investigations with man and animals, so that despite the two case reports with positive results the substance has not been designated with an "S".

7 References

Abraham MH, Andonian-Haftvan J, Cometto-Muñiz JE, Cain WS (1996) An analysis of nasal irritation thresholds using a new solvation equation. *Fundam Appl Toxicol 31*: 71–76

ACGIH (American Conference of Governmental Industrial Hygienists) (1997) *n*-Butyl acetate. in: *Documentation of the threshold limit values and biological exposure indices*, ACGIH, Cinncinnati, OH, USA

Bowen SE, Blaster RL (1997) A comparison of the acute behavioral effects of inhaled amyl-, ethyl-, and butyl-acetate in mice. *Fundam Appl Toxicol 35*: 189–196

CMA (Chemical Manufacturers Association) (1994) *n-Butyl acetate. An acute inhalation neurotoxicity study in the rat.* Eastman Kodak Company, Toxicological Sciences Laboratory, Corporate Health and Environment Laboratories, Project No. 93-0305I2, unpublished

CMA (Chemical Manufacturers Association) (1997) *The in vivo pharmacokinetics of n-butyl acetate in rats after intravenous administration.* Eastman Kodak Company, Toxicological Sciences Laboratory, Health and Environment Laboratories, Project No. 94-0306BT01, unpublished

Dahl AR, Miller SC, Petridou-Fischer J (1987) Carboxylesterases in the respiratory tracts of rabbits, rats and Syrian hamsters. *Toxicol Lett 36*: 129–136

David RM, Tyler TR, Ouellette R, Faber WD, Banton MI, Garman RH, Gill MW, O'Donoghue JL (1998) Evaluation of subchronic neurotoxicity of *n*-butyl acetate vapor. *Neurotoxicology 19*: 809–822

ECB (European Chemicals Bureau) (1995) *IUCLID data sheet, n-butyl acetate*, 23.10.95, Ispra, Italy

FDA (Food and Drug Administration) (1977) *Evaluation of the health aspects of acetic acid, sodium acetate, and sodium diacetate as food ingredients.* Report No. FDA/BF-78/16; NTIS Doc No. PB-274670, NTIS, Springfield, VA, USA

Flury F, Wirth W (1933) Zur Toxikologie der Lösungsmittel (Toxicology of solvents) (German). *Arch Gewerbepathol Gewerbehyg 5*: 1–90

Groth G, Freundt KJ (1991) Blood alcohol under the presence of *n*-butyl acetate. *Blutalkohol 28*: 166–173

Groth G, Freundt KJ (1994) Inhaled *tert*-butyl acetate and its metabolite *tert*-butyl alcohol accumulate in the blood during exposure. *Hum Exp Toxicol 13*: 478–480

Guest D, Katz GV, Astill BD (1982) Aliphatic carboxylic acids. in: Clayton GD, Clayton FE (Eds) Patty's Industrial Hygiene and Toxiology, Vol 2C, John Wiley & Sons, New York, 4910

Iregren A, Löf A, Toomingas A, Wang Z (1993) Irritation effects from experimental exposure to *n*-butyl acetate. *Am J Ind Med 24*: 727–742

Muller J, Greff G (1984) Recherche de relations entre toxicité de molécules d'intérêt industriel et propriétés physico-chimiques: test d'irritation des voies aériennes supérieures appliqué à quatre familles chimiques (Relation between the toxicity of molecules of industrial value and their physico-chemical properties: test of upper airway irritation applied to 4 chemical groups) (French). *Food Chem Toxicol* 22: 661–664

Nelson KW, Ege Jr JF, Ross M, Woodman LE, Silverman L (1943) Sensory response to certain industrial solvent vapors. *J Ind Hyg Toxicol 25*: 282–285

NIOSH (National Institute for Occupational Safety and Health) (1982) *Teratogenic study of ethylene and propylene oxide and n-butyl acetate.* Battelle, Pacific Northwest Laboratories, PB83-258038, NTIS, Springfield, VA, USA

Ørbæk P, Österberg K, Akesson B, Bergendorf U, Karlson B, Seger L (1998) Suprathreshold intensity and annoyance reactions in experimental challenge to toluene and *n*-butyl acetate among subjects with long-term solvent exposure. *Scand J Work Environ Health 24*: 432–438

Poulin P, Krishnan K (1996) A tissue composition-based algorithm for predicting tissue:air partition coefficients of organic chemicals. *Toxicol Appl Pharmacol 136*: 126–130

Ursin C, Hansen CM, Van Dyk JW, Jenson PO, Christensen IJ, Ebbehoej J (1995) Permeability of commercial solvents through living human skin. *Am Ind Hyg Assoc J 56*: 651–660

completed 21.04.1999, peak limitation factor completed 04.05.2000

sec-Butyl acetate

MAK value	**not yet established, see Section IIb of the *List of MAK and BAT Values***
Peak limitation	–
Absorption through the skin	–
Sensitization	–
Carcinogenicity	–
Prenatal toxicity	–
Germ cell mutagenicity	–
BAT value	–
Synonyms	acetic acid *sec*-butyl ester 2-butyl acetate
Chemical name (CAS)	acetic acid 1-methylpropyl ester
CAS number	105-46-4
Structural formula	$CH_3COO–CH(CH_3)–CH_2–CH_3$
Molecular formula	$C_6H_{12}O_2$
Molecular weight	116.16
Melting point	–99°C
Boiling point	112.2°C
Density at 20°C	0.87 g/cm^3
Vapour pressure at 20°C	25.3 hPa
log P_{ow}*	1.51 (calculated)
1 ml/m^3 (ppm) ≙ 4.82 mg/m^3	**1 mg/m^3 ≙ 0.207 ml/m^3 (ppm)**

* *n*-octanol/water distribution coefficient

1 Toxic Effects and Mode of Action

Once absorbed by the organism or in contact with mucous membranes, *sec*-butyl acetate is probably hydrolysed to *sec*-butyl alcohol and acetic acid. Quantitative data for the reaction are not available. The acute toxicity of *sec*-butyl acetate is low after oral and inhalation exposure. According to unpublished results, the irritative effects in man are weaker than those of *n*-butyl acetate. Other effects were not investigated.

2 Mechanism of Action

There are no data available for the mechanism of action of *sec*-butyl acetate. The irritative effects may be caused by the acetic acid formed during hydrolysis.

3 Toxicokinetics and Metabolism

After contact with the mucous membranes and absorption by the organism, butyl acetates may be assumed to be hydrolysed to the corresponding butyl alcohols and acetic acid in reactions catalysed by various esterases; this was demonstrated with isolated nasal tissue from the rat. The rate of hydrolysis decreased with increasing branching of the butyl residue and was between 42 and 77 nmol/mg protein and minute. This can be explained by the increasing steric hindrance (Dahl *et al.* 1987). There is no other information available.

4 Effects in Man

Unpublished results indicate that the irritative effects of *sec*-butyl acetate are weaker than those of *n*-butyl acetate (ACGIH 1991). A threshold value of 3950 ml/m^3 was given for nasal irritation in anosmic persons exposed to *sec*-butyl acetate for about 2 seconds (Abraham *et al.* 1996).

5 Animal Experiments and *in vitro* Studies

There are no data available for the subacute, subchronic and chronic toxicity of *sec*-butyl acetate, nor the local effects of the substance on the skin and mucous membranes, allergenic effects, or toxic effects on reproduction, or its genotoxicity or carcinogenicity.

5.1 Acute toxicity

LC_{50} values are not available. Exposure of rats to concentrations of 3500 ml/m^3 for 6 hours was not lethal. At concentrations of 24000 ml/m^3 all rats died within 4 hours. The oral LD_{50} for the rat was found to be between 3200 and 6400 mg/kg body weight (Eastman Kodak 1970).

6 Manifesto (MAK value/classification)

For the workplace situation, the critical effects of *sec*-butyl acetate are the irritative effects of the substance. There are no data for the irritation threshold of *sec*-butyl acetate. It can be assumed, however, that this is not very different from the irritation threshold of the better investigated *n*-butyl acetate; this has a MAK value of 100 ml/m^3 (see *n*-Butyl acetate" in this volume). As, however, there are no data available for the systemic toxicity of either *sec*-butyl acetate or its metabolite *sec*-butyl alcohol (see "*sec*-Butyl alcohol" in this volume), a scientifically grounded MAK value cannot be evaluated. The previous MAK value for *sec*-butyl acetate has therefore been withdrawn and the substance is listed in Section IIb of the *List of MAK and BAT Values*.

Because there are no data for the systemic toxicity of *sec*-butyl acetate, it cannot be decided whether or not designation with an "H" is necessary. Likewise because of lack of data, the substance is not designated with an "S" (for substances which cause sensitization) and is not classified in one of the categories for germ cell mutagens.

7 References

Abraham MH, Andonian-Haftvan J, Cometto-Muñiz JE, Cain WS (1996) An analysis of nasal irritation thresholds using a new solvation equation. *Fundam Appl Toxicol 31*: 71–76

ACGIH (American Conference of Governmental Industrial Hygienists) (1991) Isobutyl acetate, *sec*-butyl acetate, *tert*-butyl acetate. in: *Documentation of the threshold limit values and biological exposure indices*, ACGIH, Cincinnati, OH, USA

Dahl AR, Miller SC, Petridou-Fischer J (1987) Carboxylesterases in the respiratory tracts of rabbits, rats and Syrian hamsters. *Toxicol Lett 36*: 129–136

Eastman Kodak (1970) *Toxicity and health hazard summary of sec-butyl acetate with cover letter dated 02/15/94*. NTIS/OTS 0556683, EPA/OTS 86940000087, NTIS, Springfield, VA, USA

completed 06.04.1999

tert-Butyl acetate

MAK value (1999)	**20 ml/m^3 (ppm) ≙ 96 mg/m^3**
Peak limitation (2002)	**Category II, excursion factor 4**
Absorption through the skin	**–**
Sensitization	**–**
Carcinogenicity	**–**
Prenatal toxicity (1999)	**Pregnancy risk group D**
Germ cell mutagenicity	**–**
BAT value	**–**
Synonyms	acetic acid *tert*-butyl ester
Chemical name (CAS)	acetic acid 1,1-dimethylethyl ester
CAS number	540-88-5
Structural formula	$CH_3COO–C(CH_3)_3$
Molecular formula	$C_6H_{12}O_2$
Molecular weight	116.16
Melting point	not stated
Boiling point	97°C
Density at 20°C	0.87 g/cm^3
Vapour pressure at 20°C	not stated
log P_{ow}*	1.38, 1.79
1 ml/m^3 (ppm) ≙ 4.82 mg/m^3	**1 mg/m^3 ≙ 0.207 ml/m^3 (ppm)**

* *n*-octanol/water distribution coefficient

1 Toxic Effects and Mode of Action

Once absorbed by the organism or in contact with mucous membranes, *tert*-butyl acetate is hydrolysed to *tert*-butyl alcohol and acetic acid; quantitative data are not available. Longer exposures can be expected to result in accumulation of the metabolite *tert*-butyl alcohol in the blood. According to unpublished results, the irritative effects on the eyes and mucous membranes of the upper respiratory tract are weaker than those of *n*-butyl acetate. The acute toxicity of *tert*-butyl acetate is low after inhalation, oral and dermal exposure. The substance has only weak irritative effects on rabbit skin.

2 Mechanism of Action

There are no data available. The irritative effects may be caused by the acetic acid formed during hydrolysis.

3 Toxicokinetics and Metabolism

After contact with the mucous membranes and absorption by the organism, butyl acetates may be assumed to be hydrolysed to the corresponding butyl alcohols and acetic acid in reactions catalysed by various esterases; this has been demonstrated *in vitro* with nasal tissue from the rat for all butyl acetate isomers. The rate of hydrolysis decreased with increasing branching of the butyl residue and was between 42 and 77 nmol/mg protein and minute. This can be explained by the increasing steric hindrance (Dahl *et al.* 1987).

During exposure of rats to *tert*-butyl acetate in a concentration of 440 ml/m^3 via a tracheal cannula for 5 hours or to 900 ml/m^3 for 4.25 hours, the concentrations of *tert*-butyl acetate and *tert*-butyl alcohol in blood increased continuously until the end of exposure. At 440 ml/m^3 a maximum *tert*-butyl acetate concentration of about 280 µM and a maximum *tert*-butyl alcohol concentration of about 340 µM was achieved. At 900 ml/m^3 the corresponding values were about 550 and 450 µM; 45 minutes after the end of the exposure to 900 ml/m^3 the concentration in blood of *tert*-butyl acetate had decreased to about 300 µM, while the *tert*-butyl alcohol concentration remained constant. This is because tertiary alcohols are poor substrates for alcohol dehydrogenase and accordingly are metabolized slowly (Groth and Freundt 1994).

4 Effects in Man

Unpublished results indicate that the irritative effects of *tert*-butyl acetate are weaker than those of *n*-butyl acetate (ACGIH 1991). For exposure to *tert*-butyl acetate for about 2 seconds a threshold value of 9460 ml/m^3 was given for nasal irritation in anosmic persons (Abraham *et al.* 1996).

5 Animal Experiments and *in vitro* Studies

There are no data available for the subacute, subchronic and chronic toxicity of *tert*-butyl acetate, or the allergenic effects, genotoxicity or carcinogenicity of the substance.

5.1 Acute toxicity

An LC_{50} value of 13300 mg/m^3 (2750 ml/m^3) was determined for Sprague-Dawley rats exposed for 4 hours. The oral LD_{50} for this strain of rat was 3300 mg/kg body weight. All 4 rabbits survived occlusive application to the dorsal skin for 24 hours of doses up to 23 ml/kg body weight (20200 mg/kg body weight). The only symptom was transient erythema (Albemarle Corporation 1958).

The RD_{50} for mice was 75850 mg/m^3 (about 15200 ml/m^3). The concentration–effect relationship was not described (Muller and Greff 1984).

5.2 Local effects on skin and mucous membranes

In the rabbit eye 0.1 ml of the undiluted test substance caused only minimal irritation. The highest irritation index of 4.8 (maximum value of 110) was obtained one hour after application (Albemarle Corporation 1958).

5.3 Reproductive and developmental toxicity

There are no data available. There is, however, information on the toxic effects on reproduction of the metabolites. *tert*-Butyl alcohol has been given a MAK value of 20 ml/m^3 and is classified in Pregnancy risk group D, as embryotoxic and foetotoxic effects were detected in 2 studies, in some cases at concentrations or doses which did not have toxic effects on the dams. On the other hand, the tested concentrations and doses were so high that metabolism or elimination may have been saturated (see "*tert*-Butyl alcohol" in this volume). It is therefore only possible to make a conclusive evaluation when studies have been carried out in lower concentration ranges. The other metabolite, acetic acid, is at present not classified in any Pregnancy risk group. The effective principle of a potential prenatal toxicity in this case could be acidosis. However, even exposure of rabbits for over 2 hours to very high concentrations of various alkyl acetate esters (e.g. *n*-propyl acetate concentrations of 54000 mg/m^3, corresponding to 13000 ml/m^3) did not produce such effects (Flury and Wirth 1933), so that in the range of the MAK value prenatal toxicity resulting from acetic acid is not to be expected for *tert*-butyl acetate. This is confirmed by a study with rabbits given cider vinegar in doses of 1600 mg /kg body weight from days 6 to 18 of gestation. Assuming an acetic acid content of 5 %, this corresponds to an acetic acid dose of 80 mg/kg body weight. Neither maternal toxicity nor foetal toxicity, nor increased malformations in the foetuses were observed (FDA 1977).

6 Manifesto (MAK value/classification)

For the situation at the workplace, the irritative effects of *tert*-butyl acetate are among the critical effects, but also systemic-toxic effects, as a result of the accumulation of *tert*-butyl alcohol after the inhalation of *tert*-butyl acetate, cannot be excluded. There is no information available about the irritation threshold. It can be assumed, however, that this is not very different from the irritation threshold of the better investigated *n*-butyl acetate; this has a MAK value of 100 ml/m^3 (see "*n*-Butyl acetate" in this volume). Studies of the systemic toxicity of *tert*-butyl acetate itself are not available, but this can be estimated by evaluating the systemic toxicity of the metabolites. For *tert*-butyl alcohol, the MAK value was established at 20 ml/m^3 on the basis of the dose-response relationship for systemic effects in male rats, the most sensitive species (see "*tert*-Butyl alcohol" in this volume). The MAK value for *tert*-butyl acetate must therefore also be lowered to 20 ml/m^3. The amount of acetic acid taken up daily during exposure at the level of the MAK value, at most 0.5 g, is less than the average amount taken up with food (Guest *et al.* 1982). In addition, acetic acid is formed endogenously and is produced in numerous pathways of intermediary metabolism, so that no adverse effects are expected as a result of the additional occupational exposure.

As *tert*-butyl acetate is metabolized to *tert*-butyl alcohol and because an excursion factor of 4 has been established for the latter substance (see "*tert*-Butyl alcohol" in this volume), *tert*-butyl acetate is also classified in Peak limitation category II with an excursion factor of 4.

Studies of the reproductive toxicity of *tert*-butyl acetate itself are not available. Evaluation of the toxic effects on reproduction of the metabolites can, however, provide information about the substance itself. Given observation of the MAK value for *tert*-butyl acetate, prenatal toxicity is not to be expected from acetic acid (see Section 5.3). *tert*-Butyl alcohol itself, with a MAK value of 20 ml/m^3, is classified in Pregnancy risk group D (see "*tert*-Butyl alcohol" in this volume). In analogy, *tert*-butyl acetate is also classified in Pregnancy risk group D.

As a result of the lack of data, the substance is not designated with "H" or "S" or classified as a germ cell mutagen.

7 References

Abraham MH, Andonian-Haftvan J, Cometto-Muñiz JE, Cain WS (1996) An analysis of nasal irritation thresholds using a new solvation equation. *Fundam Appl Toxicol 31*: 71–76

ACGIH (American Conference of Governmental Industrial Hygienists) (1991) Isobutyl acetate, *sec*-butyl acetate, *tert*-butyl acetate. in: *Documentation of the threshold limit values and biological exposure indices*, ACGIH, Cinncinnati, OH, USA

Albemarle Corporation (1958) *Toxicity studies on TFA-168, tert-butyl acetate, with cover letter dated 03/23/94.* Indus Bio-Test Laboratories. NTIS/OTS 0556824, EPA/OTS 86940000229, NTIS, Springfield, VA, USA

Dahl AR, Miller SC, Petridou-Fischer J (1987) Carboxylesterases in the respiratory tracts of rabbits, rats and Syrian hamsters. *Toxicol Lett 36*: 129–136

FDA (Food and Drug Administration) (1977) *Evaluation of the health aspects of acetic acid, sodium acetate, and sodium diacetate as food ingredients.* Report No. FDA/BF-78/16; NTIS Doc No. PB-274670, NTIS, Springfield, VA, USA

Flury F, Wirth W (1933) Zur Toxikologie der Lösungsmittel (Toxicology of solvents) (German). *Arch Gewerbepathol Gewerbehyg 5*: 1–90

Groth G, Freundt KJ (1994) Inhaled *tert*-butyl acetate and its metabolite *tert*-butyl alcohol accumulate in the blood during exposure. *Hum Exp Toxicol 13*: 478–480

Guest D, Katz GV, Astill BD (1982) Aliphatic carboxylic acids. in: Clayton GD, Clayton FE (Eds) *Patty's Industrial Hygiene and Toxiology, Vol. 2C*, John Wiley & Sons, New York, 4910

Muller J, Greff G (1984) Recherche de relations entre toxicité de molécules d'intérêt industriel et propriétés physico-chimiques: test d'irritation des voies aeriennes supérieures appliqué a quatre familles chimiques (Relation between the toxicity of molecules of industrial value and their physico-chemical properties: test of upper airway irritation applied to 4 chemical groups) (French). *Food Chem Toxicol 22*: 661–664

completed 06.04.1999, peak limitation category 28.02.2002

n-Butyl alcohol

MAK value (1958)	**100 ml/m³ (ppm) ≙ 300 mg/m³**
Peak limitation (2000)	**Category I, excursion factor 1**
Absorption through the skin	**–**
Sensitization	**–**
Carcinogenicity	**–**
Prenatal toxicity (1999)	**Pregnancy risk group C**
Germ cell mutagenicity	**–**
BAT value	**10 mg/g creatinine (urine: end of shift)** **2 mg/g creatinine (beginning of next shift)**
Synonyms	butyl alcohol propyl carbinol
Chemical name (CAS)	1-butanol
CAS number	71-36-3
Structural formula	$CH_3–(CH_2)_2–CH_2OH$
Molecular formula	$C_4H_{10}O$
Molecular weight	74.12
Melting point	–90°C
Boiling point	117–118°C
Density at 20°C	0.81 g/cm³
Vapour pressure at 20°C	6.3 hPa
log P_{ow}*	0.88
1 ml/m³ (ppm) ≙ 3.03 mg/m³	**1 mg/m³ ≙ 0.330 ml/m³ (ppm)**

A MAK value of 100 ml/m³ was set in 1958 for *n*-butyl alcohol and *tert*-butyl alcohol in analogy to the TLV value at the time. In 1969 this was extended to all butyl alcohol isomers. The present documentation was based on reviews of toxicological data for *n*-butyl alcohol (DECOS 1994, ECB 1995, WHO 1987) and supplemented with more recent data.

---

* *n*-octanol/water distribution coefficient

1 Toxic Effects and Mode of Action

n-Butyl alcohol is absorbed well via all routes of exposure and is distributed rapidly in the organism. The substance is metabolized mainly by alcohol dehydrogenase and aldehyde dehydrogenase with subsequent degradation to CO_2.

Exposure of man to *n*-butyl alcohol vapour from concentrations of about 200 ml/m^3 causes central nervous disorders and irritation of the mucous membranes of the upper respiratory tract and eyes which is accompanied by characteristic corneal changes, vacuolar keratopathy. Allergic reactions on the skin to *n*-butyl alcohol are very rare and usually interpreted as cross-reactions. The acute toxicity of the substance in experimental animals was found to be low to moderate depending on the route of exposure. In rats given repeated oral doses of the substance, no adverse effects were observed other than central nervous effects after doses of 500 mg/kg body weight or more. On rabbit skin *n*-butyl alcohol was found to be moderately irritative, in the rabbit eye highly irritative. The damage to the rabbit eye regressed within 72 hours. *n*-Butyl alcohol is not mutagenic in *Salmonella typhimurium* nor clastogenic in various mammalian cell lines. In female rats, even doses as high as 5000 mg/kg body weight administered for 8 weeks did not lead to impaired fertility. Studies of the toxic effects of *n*-butyl alcohol on reproduction are only available for rats; evidence was found of foetotoxic effects even in the dose range not toxic for the dams.

2 Mechanism of Action

The irritative effects of the butyl alcohol isomers are probably the result of non-covalent interaction with the receptors of the sensory nerve endings in the mucous membranes of the respiratory tract and are a function of the physico-chemical properties (e.g. vapour pressure, lipophilia) of the substances (Abraham *et al.* 1996, Alarie *et al.* 1998, Hansen and Nielsen 1994). The central nervous effects which occur at higher concentrations are probably the result of the interaction of the relatively hydrophobic substances with neuronal membranes (Fang *et al.* 1997, Frantik *et al.* 1994).

3 Toxicokinetics and Metabolism

n-Butyl alcohol is absorbed readily via the lungs. In 12 volunteers exposed to 200 ml/m^3 for 30 minutes at rest or for 90 minutes during light physical exercise (50 watts), or to 100 ml/m^3 for 30 minutes at rest and 30-minute periods with increasing physical stress (50, 100, 150 watts), pulmonary retention rates of about 47 % were determined during the rest periods. During physical activity the value decreased to about 40 % (Astrand *et al.* 1976). In another study with persons exposed to 65 ml/m^3 for 2 hours, retention of about 60 % was determined under both resting conditions and during physical activity (25–75 watts). In dogs exposed to 50 ml/m^3 for 6 hours, about 55 % of the inhaled *n*-butyl alcohol was absorbed. *n*-Butyl alcohol is absorbed almost completely from the gastrointestinal tract. Rats given single oral doses of 4.5, 45 or 450 mg/kg body weight excreted only 1 % of the dose with the faeces (DECOS 1994, WHO 1987).

There are no data available for dermal absorption of *n*-butyl alcohol in man. Liquid *n*-butyl alcohol was applied to about 3 cm^2 of the shaved dorsal skin of guinea pigs in a depot for 4 hours. The concentration of *n*-butyl alcohol in the blood increased continuously and at the end of exposure had reached a value of about 1.8 mg/l blood (Boman *et al.* 1995). In dogs, the dermal penetration rate for *n*-butyl alcohol was 0.53 mg/cm^2 and hour. With isolated human skin a penetration rate of 0.048 mg/cm^2 and hour through the epidermis and of 0.82 mg/cm^2 and hour through the dermis was determined for undiluted *n*-butyl alcohol; distilled water was used as the absorbing medium on the other side (DECOS 1994, WHO 1987). With the isolated skin of minipigs, dermal flux of *n*-butyl alcohol from the liquid and gaseous phases was determined with the ^{14}C-labelled substance. The flux was found to be 0.3 mg/cm^2 and hour for an *n*-butyl alcohol solution (concentration not stated) and 0.085 mg/cm^2 and hour for an atmosphere with 26000 mg *n*-butyl alcohol per m^3 (Jacobs and Phanprasit 1993). On the basis of the physico-chemical properties of *n*-butyl alcohol, a penetration rate through human skin of 1.88 mg/cm^2 and hour (Fiserova-Bergerova *et al.* 1990) or 0.21 mg/cm^2 and hour (Guy and Potts 1993) was calculated for a saturated aqueous *n*-butyl alcohol solution. Assuming both hands and forearms were wetted for an hour (skin area of about 2000 cm^2), with a flux of about 1 mg/cm^2 and hour the amount of *n*-butyl alcohol absorbed would be 2000 mg or a dose of about 30 mg/kg body weight.

n-Butyl alcohol in the bloodstream is rapidly distributed in the organism or metabolized. In volunteers exposed to 50 ml/m^3 for 30 minutes, no *n*-butyl alcohol was detected in the blood (detection limit 0.08 mg/l). When the exposure concentrations were increased to up to 200 ml/m^3 and the volunteers carried out physical exercise, the maximum levels of *n*-butyl alcohol determined in blood were 1.3 mg/l, which decreased to below the detection limit within 30 minutes after the end of exposure (Astrand *et al.* 1976, DECOS 1994, WHO 1987). The *n*-butyl alcohol concentration in the blood of Wistar rats exposed for 7 hours to 100 ml/m^3 increased during the first hour of exposure to about 0.16 mg/l and until the end of exposure ranged between 0.16 and 0.22 mg/l (Swiercz *et al.* 1995). In volunteers given intravenous bolus doses of ^{15}O-labelled butyl alcohol, the distribution of and decrease in radioactivity was monitored using positron-emission tomography. Three minutes after administration, the amount of radioactivity in the liver reached its maximum of 11.5 % of the administered dose; 10 minutes later it had decreased to 5.6 %. The maximum level in the brain was reached after only 30 seconds and was about 4 % of the administered dose. After 13 minutes only 0.9 % was still detectable. It can be concluded from these data that *n*-butyl alcohol passes the blood-brain barrier rapidly and to a high degree. During the period investigated (0–13 minutes after dosing), about 1 % of the administered activity was found in the kidneys and in the spleen; one minute after administration 19.5 % was found in blood where the level had decreased to 8 % in the 13th minute (Herzog *et al.* 1994). It can be concluded also from the distribution coefficients liver/air, muscle/air and fat/air for *n*-butyl alcohol of 1250, 900 and 1520 (Poulin and Krishnan 1996) and of 1250, 900 and 900 (Kaneko *et al.* 1994) determined at 37°C with rat tissue, that the inhaled substance can be distributed evenly in the whole body. In rats given single oral doses of ^{14}C-labelled *n*-butyl alcohol of 450 mg/kg body weight, 2.64 % of the administered activity was found in the liver 4 hours after administration, 0.51 % in the blood and 0.24 % in the kidneys. After a further 4 hours, the values in the liver and blood had increased to 3.88 % and 0.74 %, while in the kidneys the level decreased to 0.18 %. In the lungs, heart, adipose tissue, brain and adrenal glands the levels were all below 0.1 %. 24 hours after administration, 2.64 % of the administered radioactivity was still detectable in the liver and 0.38 % in the blood. In another study with rats, radioactivity was detected in the liver, kidneys, small intestine and lungs one hour after oral administration of ^{14}C-labelled *n*-butyl alcohol. After another 4 hours, the levels in all organs were decreased (DECOS 1994, WHO 1987). *In vitro*, binding to plasma proteins of rat blood was found to be about 10 % (ECB 1995).

For the metabolism and excretion of *n*-butyl alcohol the only data available are from experiments with animals and *in vitro* studies. These results show that the substance can be oxidized, mainly by alcohol dehydrogenase, but also by cytochrome P450 isoenzymes, to form the corresponding aldehyde, which is then transformed by aldehyde dehydrogenase to the corresponding acid, with subsequent degradation to the main metabolite CO_2. The affinity of alcohol dehydrogenase for *n*-butyl alcohol is higher than its affinity for ethanol. Some of the *n*-butyl alcohol can be conjugated with glucuronic acid or sulfate (DECOS 1994, WHO 1987) or esterified in the liver with endogenous fatty acids such as palmitic acid, stearic acid and oleic acid (Carlson 1994).

n-Butyl alcohol and its metabolites are eliminated mainly via the lungs. In rats given single oral doses of ^{14}C-labelled *n*-butyl alcohol, 80 % of the radioactivity was exhaled in the form of CO_2 within 24 hours. The amount of unchanged *n*-butyl alcohol in the exhaled air was less than 1 % of the dose. About 1 % was eliminated with the faeces, 3 % to 5 % with the urine, mainly in conjugated form. Dogs given ^{14}C-labelled *n*-butyl alcohol in intravenous doses of 1 mg/kg body weight did not exhale any *n*-butyl alcohol during the first 8 hours after administration. In all, 12 % to 16 % of the administered activity was exhaled as CO_2 and 2 % to 3 % was eliminated with the urine. What happened to the rest of the activity is not stated (DECOS 1994, WHO 1987). The renal excretion of *n*-butyl alcohol and its conjugates was investigated in female Wistar rats exposed to concentrations of 50, 100 or 200 ml/m^3 for 8 hours. The level of *n*-butyl alcohol—determined after acid hydrolysis—in the urine collected over 16 hours after the end of exposure increased linearly with the exposure concentration. The proportion of free *n*-butyl alcohol at the low exposure concentration was about 8 %, at both of the high concentrations 35 % to 40 % (Kawai *et al.* 1997), possibly the result of saturation of the conjugation reaction. In the post-shift urine of 11 workers exposed to a number of substances including *n*-butyl alcohol in concentrations of 0.6 to 2.7 ml/m^3, after acid hydrolysis *n*-butyl alcohol concentrations of 76 to 446 µg/l were detected; in control persons the corresponding values were between 20 and 47 µg/l. Free *n*-butyl alcohol was not detected in any sample (Kawai *et al.* 1997).

4 Effects in Man

In workers occupationally exposed to *n*-butyl alcohol in 6 factories, the most frequent finding was irritation of the eyes, which led to characteristic corneal inflammation. In addition, an intolerable odour, slight headaches or dizziness, and irritation of the mucous membranes of the nose and throat were mentioned. Contact with liquid butyl alcohol led in some cases to dermatitis around the bed of the nail and on the fingers, which regressed when gloves were used. Only the workers from the factory with the lowest *n*-butyl alcohol concentrations (5–14 ml/m^3) were free of symptoms. The *n*-butyl alcohol concentrations at the other factories were in the range of 20 to 115 ml/m^3. Reduction of the *n*-butyl alcohol concentrations at the workplace led also to regression of the symptoms. Some of the workers were exposed also to other solvents (2-butanone, ethanol, naphtha or diacetone alcohol). The authors concluded that eye irritation is to be expected with *n*-butyl concentrations at the workplace of over 50 ml/m^3, while systemic-toxic effects do not occur below 100 ml/m^3 (Tabershaw *et al.* 1944). The described correlation between the exposure data and the effects does not, however, confirm this. The levels of exposure in these 6 factories were determined only once and probably retrospectively; it is, therefore, uncertain which levels of exposure caused the symptoms and conclusions cannot be drawn about the irritation threshold of *n*-butyl alcohol.

Damage to the eyes observed in one of these factories was analysed more closely in another study. The eye damage occurred first of all after the solvent 2-butanone, which had been used for a long time, was replaced by *n*-butyl alcohol to which varying amounts of diacetone alcohol and ethanol had been added. The incidence of complaints was higher in areas with higher concentrations. 34 exposed female workers and 1 exposed male worker underwent ophthalmological examination. Although the male worker was exposed to high concentrations, he was free of symptoms. Of the 34 female workers, 28 were found to have between 10 and 1000 symmetrically distributed vacuoles in the corneal epithelium. 19 women complained also of subjective symptoms of eye irritation, such as burning and itching of the eyes or the feeling there was a foreign body in the eye. In some cases the lids were swollen and the eyes red. In the morning the symptoms were most severe; on absence from work for a longer period they were completely reversible (Cogan and Grant 1945). The characteristic changes to the cornea caused by *n*-butyl alcohol vapour have also been described by other authors and defined as vacuolar keratopathy (Jaeger 1955, Rowe and McCollister 1982). The vacuoles are thought to be inclusions of liquid in the corneal epithelium and can be formed as a result of penetration of *n*-butyl alcohol into and through the epithelial cell layers. As these are flushed out after the end of exposure and the corneal epithelium has a high regeneration capability, episodic exposure is not expected to cause permanent damage.

Other authors determined, over a period of 10 years at irregular intervals, the *n*-butyl alcohol concentration in a factory in which photographic paper was coated with barite, and recorded the symptoms of the workers during this period. The irritative effects on the eyes were confirmed, though only at concentrations that were higher than those found in the study of Tabershaw *et al.* (1944), but not the formation of vacuoles in the cornea: at first the average concentrations were around 200 ml/m^3, later improved conditions of occupational hygiene led to a reduction to 100 ml/m^3. At the beginning of the study the physical examination of the between 16 and 100 workers exposed to butyl alcohol over the years yielded evidence of eye irritation, which in some cases was associated with inflammatory processes in the cornea, blurred vision, lacrimation, and increased sensitivity to light. The symptoms usually developed in the middle of the working week and worsened towards the end of the week. Slight to moderate oedema was observed in the cornea, in some cases also the conjunctiva were slightly oedematous. Over the work-free weekend the symptoms regressed. When the *n*-butyl alcohol concentration was reduced to an average of 100 ml/m^3, complaints of eye irritation were rare and attributed to exposure peaks. Only slight oedema was observed in the cornea and conjunctiva in these cases. A slight decrease in the erythrocyte count in the blood of the exposed workers at the first examination could not be confirmed later and was therefore not attributed to the exposure. The authors concluded from their observations that workplace concentrations of 100 ml/m^3 do not lead to adverse effects on health (Sterner *et al.* 1949).

Impaired hearing was reported in 9 of 11 workers of a cellulose acetate factory who had been exposed to *n*-butyl alcohol and other substances for 3 to 11 years. The *n*-butyl alcohol concentration at the time of the investigation was 80 ml/m^3; the noise level in the factory was between 70 and 80 dB A and below the then valid threshold limit value. The authors therefore excluded the noise level at the workplace as the reason for the findings and suspected *n*-butyl alcohol as being the active agent. The loss in hearing was limited to a frequency range between 3000 and 4000 hertz. This pattern was observed also in workers with damaged hearing who were exposed in another factory to 90 to 110 dB (A), but not to *n*-butyl alcohol (Velazquez *et al.* 1969). This makes noise seem the reason for the impaired hearing also in the workers exposed to *n*-butyl alcohol, and the suspected relationship with the exposure to *n*-butyl alcohol questionable. As there are no data available for the earlier levels of noise and *n*-butyl alcohol, and other possible confounders are not characterized for the collective investigated, this study cannot be included in the evaluation of a threshold limit value for *n*-butyl alcohol.

In other communications, the typical subjective symptoms of exposure to *n*-butyl alcohol, and *n*-butyl alcohol and isobutyl alcohol, such as eye irritation, headaches and dizziness, were confirmed. Clinically diagnosed were swelling of the conjunctiva and the corneal epithelium, and in some cases also epithelial defects. The levels of exposure, however, were not determined (ECB 1995, Peters 1958, Seitz 1972).

In a controlled study, 10 volunteers were exposed to *n*-butyl alcohol for 3 to 5 minutes and the subjective symptoms recorded. At concentrations of 25 ml/m^3, slight irritation of the throat, nose and eyes were reported. 50 ml/m^3 was found unpleasant. Irritation of the throat was more severe at the high concentration, and a few volunteers complained also of slight headaches (Nelson *et al.* 1943). As a result of the short exposure duration, these results cannot be taken into account in the evaluation of a threshold limit value for the workplace.

In a toxicokinetic study, 12 volunteers were exposed via breathing masks to *n*-butyl alcohol concentrations of 100 or 200 ml/m^3 for 2 hours, some of them during light physical activity. The specific determination of possible symptoms was not the aim of the study, but it is stated in the publication that none of the volunteers suffered adverse effects from the exposure (Astrand *et al.* 1976). By means of a special exposure system connected to an eye mask that allowed one eye to be exposed to *n*-butyl alcohol and the other to air or CO_2, in 8 volunteers subjective symptoms of eye irritation and objective clinical parameters (cytology of the conjunctival fluid, hyperaemia of the conjunctiva) were investigated. The *n*-butyl alcohol concentrations were 99, 314 and 990 ml/m^3. The volunteers were exposed to *n*-butyl alcohol first of all for 15 minutes, during which time the control eye was exposed to increasing concentrations of CO_2, and after a 5-minute pause for another 60 minutes, during which time the control eye was exposed to air. Sampling of the conjunctival fluid and photographic documentation of the conjunctiva was carried out before and after each whole exposure period. During the 60-minute exposure phase the volunteers were asked to note their subjective feelings of irritation every 5 minutes; notable eye irritation was not reported by any of the volunteers. Conjunctival hyperaemia was slightly increased at the two low concentrations, but this increase was not statistically significant. At 990 ml/m^3 the increase relative to the control value was statistically highly significant. Typying and counting the cells in the conjunctival fluid did not reveal any findings that could be attributed to the exposure (Hempel-Jørgensen *et al.* 1998). For exposure to *n*-butyl alcohol for about 2 seconds the threshold value for nasal irritation in anosmic persons was given as 1581 ml/m^3 (Abraham *et al.* 1996). An irritation threshold for *n*-butyl alcohol cannot be deduced from this result.

4.1 Allergenic effects

There are only a few reports of positive results obtained in patch tests with *n*-butyl alcohol. It seems probable that they were the result of cross-reactions because in these patients ethanol was very likely the sensitizing agent. In patch tests with such patients, positive results are frequently obtained with various primary and secondary alcohols and with acetone. In some of the persons, skin reactions also developed after the consumption of alcoholic drinks, also erythema, aphthous lesions and a burning sensation in the oral mucosa (Fisher 1983, Fregert *et al.* 1969, Ophaswongse and Maibach 1994, Stotts and Ely 1977, Van Ketel and Tan-Lim 1975). From these results, it cannot be concluded that *n*-butyl alcohol has a significant sensitizing potential, if any.

5 Animal Experiments and *in vitro* Studies

There are no data available for allergenic effects or carcinogenicity of *n*-butyl alcohol.

5.1 Acute toxicity

Like all butyl alcohol isomers, *n*-butyl alcohol leads via all routes of absorption to the typical symptoms of acute alcohol intoxication, from central nervous depression to narcosis. The severity of the effects of the individual isomers is about the same and overall greater than seen with ethanol (DECOS 1994, WHO 1987).

5.1.1 Inhalation

Exposure of rats to *n*-butyl alcohol concentrations of 8000 ml/m^3 for 4 hours was survived by all the animals, but led in half of them to narcosis. A concentration of 6000 ml/m^3 did not have narcotic effects. Exposure to 9000 ml/m^3 led to the death of 1/6 rats after 7 hours; repeat exposure the following day was lethal for another animal.

In mice, exposure to concentrations of 1650 or 3300 ml/m^3 for 7 hours had no externally visible adverse effects. Concentrations of 6600 ml/m^3 led after 2 hours to deep narcosis and after 3 hours to the death of some animals. After groups of 10 male Swiss mice were exposed to concentrations of 470, 548, 844 and 965 ml/m^3 for 4 hours, behavioural changes were observed in a swimming test even at the lowest concentration (EC_{50} 617 ml/m^3); these indicate central nervous effects (DECOS 1994, WHO 1987). In various strains of mouse, an RD_{50} value of 1268 ml/m^3 was determined for *n*-butyl alcohol with exposure for 5 minutes, 4784 ml/m^3 with exposure for 10 minutes, and 11696 ml/m^3 with exposure for 30 minutes (Bos *et al.* 1992). The effects on neuromuscular function were investigated in the rotarod test with male Wistar rats exposed to *n*-butyl alcohol for 4 hours; an EC_{50} of 7559 ml/m^3 was determined. The sensitivity to pain was evaluated by putting the animals on a hot-plate and measuring the time interval until they licked their paws. Exposure to *n*-butyl alcohol for 4 hours led to a concentration-dependent decrease in this reaction with an EC_{50} of 5901 ml/m^3. For Balb/C mice the RD_{50} value was found to be 4300 ml/m^3 on exposure for one minute and 1755 ml/m^3 on exposure for 6 minutes (Korsak and Rydzynski 1994).

5.1.2 Ingestion and dermal absorption

For the rat, oral LD_{50} values in the range from 700 to 2100 mg/kg body weight were given, for the mouse a value of 2700 mg/kg body weight and for the rabbit 3500 mg/kg body weight. For the rabbit, dermal LD_{50} values in the range from 3400 to 5300 mg/kg body weight were found. Detailed reviews have been published (DECOS 1994, ECB 1995, WHO 1987).

5.2 Subacute, subchronic and chronic toxicity

5.2.1 Inhalation

Groups of 12 male Wistar rats were exposed to *n*-butyl alcohol concentrations of 50 or 100 ml/m^3, 6 hours a day, 5 days a week for 90 days. The body weights of the treated animals after exposure for one and two months were significantly higher than those of the control animals; at the end of the study the differences were no longer statistically significant. In the rotarod test the animals of the 100 ml/m^3 group performed significantly worse than the control animals at the end of the study, while after exposure for one or two months the differences were not significant. The sensitivity to pain, measured as the time interval until the animals licked their paws after being placed on a hot-plate, was not impaired by exposure to *n*-butyl alcohol. The absolute and relative organ weights did not differ from the control values, and the most important biochemical parameters in plasma (enzyme, protein, albumin, glucose and electrolyte levels) were not abnormal. Histological examination was not carried out. In the haematological examination, in both concentration groups significantly decreased haemoglobin levels were found, and in addition in the high concentration group a significant decrease in the erythrocyte count and a significant increase in the number of white blood cells. The differential blood count revealed almost twice the number of eosinophils but only in the 100 ml/m^3 group. In the liver homogenate the level of malondialdehyde, an indicator of lipid peroxidation, was increased in the 50 ml/m^3 group by about 16 %, and in the 100 ml/m^3 group by about 30 %. The differences from the controls were significant. The levels of microsomal protein, cytochrome P450 enzymes and triglycerides were unchanged (Korsak *et al.* 1994). In an earlier study with guinea pigs, changes in the blood count were also described. Three animals were exposed to *n*-butyl alcohol concentrations of 100 ml/m^3, 4 hours a day, 6 days a week for 65 days. A decrease in the number of erythrocytes and lymphocytes was reported, and in 2 animals haemorrhagic areas in the lungs and transient albuminuria were observed. The experiment was repeated with 3 other animals which were exposed 28 times. The changes in the blood picture described in the first experiment were confirmed. In addition, the first signs of liver degeneration and more marked degeneration in the kidneys were found (Smyth and Smyth 1928).

It is questionable whether the haematological and biochemical changes in the studies described above represent adverse effects, particularly when no effects were found in the histopathological examination. In addition, the low concentrations used (100 ml/m^3 for 6 hours a day corresponds to a dose of about 50 mg/kg body weight and day) make it doubtful whether there is a causal relationship between the exposure and the findings, as in a valid 90-day study with oral administration, there were no histological or haematological findings even at the highest dose of 500 mg/kg body weight and day (IRIS 1998; see also Section 5.2.2). Also the poorer performance in the rotarod test observed with concentrations of 100 ml/m^3 seems questionable, as an EC_{50} of more than 7500 ml/m^3 was determined by the same research group for impaired performance in this test after short-term exposure for 4 hours (Korsak and Rydzynski 1994; see also Section 5.1.1) and it can be deduced from this publication that exposure to concentrations of 2500 ml/m^3 did not produce effects. These studies were therefore not included in the evaluation of the systemic-toxic potential of *n*-butyl alcohol.

In a series of Russian studies from the 1970s, effects were described after repeated exposure to *n*-butyl alcohol in much lower concentrations. As in these studies only selected effects were investigated and the adverse nature of the effects described is usually questionable, particularly when histopathological findings are lacking, they are also not suitable for deducing a threshold limit value for the workplace; the results are described elsewhere (WHO 1987) and therefore only summarised briefly here. Continuous exposure of rats and mice to 2.2 or 13.3 ml/m^3 for 4 months led to various externally visible central nervous effects and to increased activity of the thyroid gland. In rats exposed continuously to 7.2 ml/m^3 for 92 days, from week 4 of exposure the levels of RNA, DNA and catalase in blood were decreased and the luminescence of the leukocytes and the diastase activity increased. The lower concentration of 0.03 ml/m^3 was without effects. The sleeping time of mice exposed continuously to 4.5 or 13 ml/m^3 for 30 days was reduced. Transient effects on temperature regulation were observed in rats exposed to 72 ml/m^3 for 5 hours a day, on 6 days a week for 6 months. In addition, oxygen consumption was reduced.

5.2.2 Ingestion

Groups of 30 rats per dose and sex were given oral doses of *n*-butyl alcohol of 30, 125 or 500 mg/kg body weight and day for 90 days. During the last 6 weeks of the experiment ataxia and hypoactivity were observed in the high dose group. The gross pathological and histological examination at the end of the study did not reveal adverse effects in any dose group. The changes in some haematological parameters observed in the female animals of the 125 and 500 mg/kg groups at an interim killing after week 6 of the experiment were slight and could no longer be detected at the end of the study. They were therefore not regarded as adverse effects. A NOAEL (no observed adverse effect level) of 125 mg/kg body weight and day was thus deduced for central nervous effects in the rat. With regard to adverse effects on the organs, according to this study an even higher NOAEL can be assumed (IRIS 1998).

A group of 30 male Wistar rats was given the substance in the drinking water in a concentration of 69000 mg/l (about 6000 mg/kg body weight and day). According to the source mentioned below, 250 g sucrose was added per litre drinking water. After 5, 9 and 13 weeks, the liver was examined with an electron microscope. At the first interim examination, irregularly formed giant mitochondria were already observed in the liver cells, which the authors interpreted as an adaptation reaction. The same effects were induced with ethanol and *n*-propyl alcohol in drinking water in concentrations that were about 4 times higher (WHO 1987), and they have also been described in man e.g. after alcohol abuse (Svoboda and Manning 1964).

5.3 Local effects on skin and mucous membranes

Patch application of 405 or 500 mg *n*-butyl alcohol to rabbit skin for 24 hours produced moderate irritation. In the Draize test, undiluted *n*-butyl alcohol was found to be strongly irritative in the rabbit eye and was given an irritation index of 61 on a scale with a maximum value of 110. A 10 % solution was given an irritation index of 3. A single reading was taken 24 hours after application. Also in an earlier study *n*-butyl alcohol was found to have strongly irritative effects in the rabbit eye (irritation index 7 on a scale with a maximum value of 10). In another test for eye irritation carried out according to the OECD guidelines, in which the evaluation was based on the results of the readings 24, 48 and 72 hours after exposure, *n*-butyl alcohol was classed as not irritative to the eye, as the effects at the later readings were only slight (DECOS 1994).

5.4 Reproductive and developmental toxicity

5.4.1 Fertility

n-Butyl alcohol administered for 8 weeks to female rats with the drinking water in doses up to 5000 mg/kg body weight and day did not have any effects on the success of mating when the females were mated with untreated males. The duration of the oestrus cycle was also unchanged (Sitarek *et al.* 1994, Sitarek and Berlinska 1996; see also Section 5.4.2). In male Sprague-Dawley rats, exposure for 6 hours daily to an *n*-butyl alcohol concentration of 50 ml/m^3 led after day 1 of the experiment to a decrease in the level of testosterone in blood to 37 % of the control value; 18 hours later the level had increased to 52 % of the control value. The concentration of luteinizing hormone was not affected by the treatment, the corticosteroid level was, however, significantly increased. After exposure for one week, no significant differences from the controls could be detected (Cameron *et al.* 1985); thus impaired fertility is not to be expected as a result of this treatment. Other parameters were not investigated.

5.4.2 Developmental toxicity

Female Sprague-Dawley rats were exposed to air containing *n*-butyl alcohol in concentrations of 0, 3500, 6000 or 8000 ml/m^3 for 7 hours daily on days 1 to 19 of gestation. In the highest concentration group the following signs of maternal toxicity were reported: reduced body weight gains and reduced food consumption, narcotic effects in about 50 % of the dams, 2/18 dams died. Reduced food consumption was also detected in the dams of the 6000 ml/m^3 group; statistics for body weight development are not given in the publication. Whereas in the progeny of the 8000 ml/m^3 group, increased foetotoxicity with delayed ossification, reduced number of foetuses, slightly reduced foetal weights and rudimentary cervical ribs were found, in the 6000 ml/m^3 group not only slight reductions in foetal weights but also an increase in the incidence of skeletal variations was observed. No treatment-related effects were seen in the group exposed to 3500 ml/m^3.

The authors described the effects in the highest concentration group as teratogenic (Nelson *et al.* 1989a, 1990, Brightwell *et al.* 1987) although increases in the incidence of supernumerary ribs or cervical ribs is often considered to be merely a result of unspecific maternal stress (Kimmel and Wilson 1973, Wickramaratne 1988). Therefore the effects described in the clearly maternally toxic concentration range (8000 ml/m^3) are not clearly teratogenic effects but may be considered to be foetotoxic effects. In the middle concentration group (6000 ml/m^3) there was still evidence of foetotoxicity but without maternal toxicity.

Studies of the effects of exposure of male and female Sprague-Dawley rats to *n*-butyl alcohol on the behaviour of the progeny were carried out in the postnatal period. The animals were exposed for 7 hours daily to *n*-butyl alcohol concentrations of 0, 3000 or 6000 ml/m^3, the females from day 1 to day 20 of gestation and the males for 6 weeks before mating. The exposed males were mated with unexposed females. The 10-day old to 90-day-old progeny was subjected to the following tests: ascent on a wire mesh screen, rotorod, motor activity in a standard open field, running wheel, avoidance conditioning and operant conditioning. In addition, the brains of 21-day-old progeny were dissected and subjected to neurochemical examination. Occasional parameters in the exposed animals differed significantly from the control values but the differences were not dose-dependent and mostly occurred in only one sex or only on single testing days. All the results were considered to lie within the range of variation of the control data obtained in other studies in the same laboratory.

Effects on the percentage of pregnancies were not detected in the progeny of either exposed female or male parents. There was no evidence of infertility (Nelson *et al.* 1989b, Brightwell *et al.* 1988).

Thus it may be concluded that the substance has prenatal toxic effects only at high concentrations in and slightly below the maternally toxic range. A concentration of 3500 ml/m^3 did not have effects on the offspring or dams.

Data for a second species with *n*-butyl alcohol itself are not available, but an inhalation study with *n*-butyl acetate has been carried out in rabbits. At concentrations of 1500 ml/m^3 notable toxic effects on reproduction were not observed (see documentation for *n*-Butyl acetate in this volume). As a result of the rapid hydrolysis of *n*-butyl acetate in the organism to form *n*-butyl alcohol and acetate, it can be assumed that also *n*-butyl alcohol in concentrations of 1500 ml/m^3 would have no effects in rabbits.

The results of another study of the toxic effects of *n*-butyl alcohol on reproduction in rats after oral administration of the substance seem questionable, particularly in view of the above data, and are therefore not included in the evaluation. Groups of 11 to 17 female rats were given drinking water containing 0.24 %, 0.8 % or 4 % *n*-butyl alcohol 8 weeks before mating with untreated male animals, during the mating period and up to day 20 of gestation. The corresponding doses were given as 300, 1000 and 5000 mg/kg body weight and day. Externally visible symptoms of intoxication were not observed. Body weight gains and food and water consumption in the treated groups did not differ from the values in the controls. The cycle duration was also not changed by the treatment. The absolute and relative organ weights and the haemoglobin levels and haematocrit of the dams on day 20 of gestation corresponded with the control values. The success of mating, intrauterine mortality and body weights of the offspring were not changed by the treatment. Crown-rump lengths were, however, slightly, but significantly reduced in the high concentration group, and delayed ossification was significantly more frequent than in the control group. Wavy ribs or an additional 14th rib were observed in the low and high concentration groups in several foetuses from one litter, but not in any other litters. As these findings were not concentration-dependent and limited in each case to only one litter, it is uncertain whether the effects were substance-related. The investigation of visceral anomalies revealed from the lowest dose a dose-dependent increase in the number of foetuses and litters with extension of various ventricles of the central nervous system. Extension of the renal pelvis on one or both sides was observed in some foetuses from different litters, but only in the middle concentration group. External and internal hydrocephali were observed only in the two high concentration groups, but the findings were not concentration-dependent (Sitarek *et al.* 1994, Sitarek and Berlinska 1996). Thus, according to this study, also the lowest tested dose of 300 mg/kg body weight and day, which did not cause maternal toxicity, led to pathological changes in the central nervous system of the offspring. Higher doses also caused malformations in this target organ. Doses of 300 mg/kg body weight and day were reached, however, also after exposure to 600 ml/m^3 for about 7 hours (alveolar ventilation of the rat about 120 ml/min, absorption 60 %), and in the valid inhalation studies described above, toxic effects on reproduction did not occur at concentrations below 3500 ml/m^3. Effects on the brain were not observed up to the highest concentration of 8000 ml/m^3. In addition, in the offspring of parent animals exposed to 6000 ml/m^3, no neurochemical changes were detected in the brain and no unusual behaviour was observed in various behavioural tests.

5.5 Genotoxicity

5.5.1 *In vitro*

n-Butyl alcohol was not found to be mutagenic in the *Salmonella typhimurium* strains TA98, TA100 and TA102 (Jung *et al.* 1992) either with or without the addition of a metabolic activation system and yielded negative results in various indicator tests for DNA damage with *S. typhimurium.* In *Escherichia coli*, incubation with *n*-butyl alcohol led to the inhibition of DNA synthesis. A genotoxic potential cannot be deduced from this one finding, especially as it was obtained with a non-validated test system. In human lymphocytes the substance did not cause chromosomal aberrations. In a fibroblast cell line of the Chinese hamster it caused neither sister chromatid exchange nor micronuclei, and in CHO cells (a cell line derived from Chinese hamster ovary) was likewise not found to cause sister chromatid exchange (DECOS 1994, ECB 1995). In V79 cells, incubation with 0.1 M *n*-butyl alcohol led to a significantly increased number of aneuploid cells with a simultaneous reduction in the colony-forming ability to 36 %. The effects were interpreted as unspecific consequences of the high incubation concentration and the resulting changes in the physical-chemical properties of the medium (Önfelt 1987). They are therefore not to be considered as evidence of a genotoxic potential.

5.5.2 *In vivo*

In a Russian study all butyl alcohol isomers were tested with groups of 8 rats in a cytogenetic *in vivo* test. The dose in each case was one fifth of the LD_{50}. All isomers led to increased chromosomal aberrations. It is unusual that no aberrations were found in the controls. The incidence of polyploid bone marrow cells was also increased in the animals treated with *n*-butyl alcohol, *sec*-butyl alcohol and isobutyl alcohol. No chromosome breaks were found. As the study did not investigate whether the results were dose-dependent, they cannot be used as evidence of genotoxic effects of the butyl alcohol isomers (Barilyak and Kozachnuk 1988).

6 Manifesto (MAK value/classification)

According to earlier occupational-medical studies, the critical toxic effect of exposure to *n*-butyl alcohol is the irritative effect on the eyes. In a longitudinal study such symptoms occurred in persons exposed to average concentrations of 200 ml/m^3. When the *n*-butyl alcohol concentration was reduced to 100 ml/m^3, the workers were free of symptoms (Sterner *et al.* 1949). In the early observation phase, during which the workers were exposed to concentrations of 200 ml/m^3 or more, some workers developed corneal inflammation with slight to moderate corneal oedema, lacrimation and sensitivity to light. Volunteers in a toxicokinetic study exposed via breathing masks to concentrations up to 200 ml/m^3 for 2 hours did not feel adverse effects (Astrand *et al.* 1976), and exposure of one eye for one hour did not produce effects up to concentrations of 300 ml/m^3 (Hempel-Jørgensen *et al.* 1998). In an earlier study with 10 volunteers, however, moderate irritation of the nose, throat and eyes was reported after exposure to 25 ml/m^3 for 3 to 5 minutes. Concentrations of 50 ml/m^3 caused severe irritation of the throat in all the volunteers and in some cases slight headaches (Nelson *et al.* 1943). As a result of the contradictory findings for irritation of the throat and nose, and as marked irritation of the eye is documented at concentrations of twice the MAK value, an excursion factor of 1 has been provisionally set for peak limitation according to Category I.

Thus, with a MAK value of 100 ml/m^3, no eye irritation is to be expected. Likewise systemic-toxic effects are not to be expected under these conditions, as in the only relevant study with rats given *n*-butyl alcohol by oral administration, up to the highest tested dose of 500 mg/kg body weight and day no adverse effects were observed apart from central nervous depression from 250 mg/kg body weight and day (IRIS 1998). Assuming a body weight of 70 kg and a respiratory volume of 10 m^3 per shift with pulmonary absorption of 60 %, with a MAK value of 100 ml/m^3 a worker would absorb at most about 25 mg/kg body weight daily.

The MAK value for *n*-butyl alcohol of 100 ml/m^3 can therefore be confirmed.

With the exposure conditions expected at the workplace, the amount of *n*-butyl alcohol taken up by the skin cannot lead to adverse effects (see Section 3). Designation with an "H" is therefore not necessary.

The carcinogenic and germ cell mutagenic potential of the substance has not been investigated, a genotoxic potential is not recognisable from the studies available.

In rats, *n*-butyl alcohol in concentrations up to 3500 ml/m^3 did not have effects on the offspring. Studies with a second species are not available, but on the basis of an inhalation study with *n*-butyl acetate in rabbits, adverse effects on reproduction are also not to be expected for *n*-butyl alcohol in concentrations of 1500 ml/m^3. There is thus no reason to fear a risk of damage to the embryo or foetus when the MAK value is observed, and *n*-butyl alcohol has therefore been classified in Pregnancy risk group C.

The available data indicate that designation with an "S" (for substances which cause sensitization) is not necessary.

7 References

Abraham MH, Andonian-Haftvan J, Cometto-Muñiz JE, Cain WS (1996) An analysis of nasal irritation thresholds using a new solvation equation. *Fundam Appl Toxicol 31*: 71–76

Alarie Y, Schaper M, Nielsen GD, Abraham MH (1998) Structure-activity relationships of volatile organic chemicals as sensory irritants. *Arch Toxicol 72*: 125–140

Astrand I, Övrum P, Linquist T, Hultengren M (1976) Exposure to butyl alcohol. Uptake and distribution in man. *Scand J Work Environ Health 3*: 165–175

Barilyak IR, Kozachnuk YS (1988) Studies in cytogenetic action of some monohydric alcohols on the rat bone marrow cells (Russian). *Tsitol Genet 22*: 49–52

Boman A, Hagelthorn G, Magnusson K (1995) Percutaneous absorption of organic solvents during intermittent exposure in guinea pigs. *Acta Derm Venereol 75*: 114–119

Bos PMJ, Zwart A, Reuzel PGJ, Bragt PC (1992) Evaluation of the sensory irritation test for the assessment of occupational health risk. *Crit Rev Toxicol 21*: 423–450

Brightwell WS, Nelson BK, MacKenzie-Taylor DR, Burg JR, Khan A, Goad PT (1987) Lack of teratogenicity of three butanol isomers administered by inhalation to rats. *Teratology 35*: 56A

Brightwell WS, Robertson SK, Nelson BK, Khan A, Krieg EF jr, Massari VJ (1988) Behavioral teratology investigation of 1-butanol administered by inhalation to rats. *Teratology 37*: 523

Cameron AM, Zahlsen K, Haug E, Nilsen OG, Eik-Nes KB (1985) Circulating steroids in male rats following inhalation of *n*-alcohols. *Arch Toxicol, Suppl 8*: 422–424

Carlson GP (1994) Formation of esterified fatty acids in rats administered 1-butanol and 1-pentanol. *Res Comm Mol Pathol Pharmacol 86*: 111–117

Cogan DG, Grant WM (1945) An unusual type of keratitis associated with exposure to *n*-butyl alcohol (butanol). *Arch Ophthalmol 33*: 106–109

DECOS (Dutch Expert Committee on Occupational Standards) (1994) *1-, 2- and t-butanol, health-based recommended occupational exposure limit.* Publication No. 1994/10, Health Council of the Netherlands, Den Haag

ECB (European Chemicals Bureau) (1995) *IUCLID data sheet, butan-1-ol,* 23.10.95, Ispra, Italy

Fang Z, Ionescu P, Chortkoff BS, Kandel L, Sonner J, Laster MJ, Eger II EI (1997) Anesthetic potencies of *n*-alkanols: results of additivity and solubility studies suggest a mechanism of action similar to that for conventional inhaled anesthetics. *Anesth Analg (Cleveland) 84*: 1042–1048

Fiserova-Bergerova V, Pierce JT, Droz PO (1990) Dermal absorption potential of industrial chemicals: criteria for skin notation. *Am J Ind Med 17*: 617–635

Fisher AA (1983) Topically applied alcohol as a cause of contact dermatitis. *Cutis 31*: 588–600

Frantik E, Hornychova M, Horvath M (1994) Relative acute neurotoxicity of solvents: isoeffective air concentrations of 48 compounds evaluated in rats and mice. *Environ Res 66*: 173–185

Fregert S, Groth O, Hjorth N, Magnusson B, Rorsman H, Övrum P (1969) Alcohol dermatitis. *Acta Derm Venereol 49*: 493–497

Guy RH, Potts RO (1993) Penetration of industrial chemicals across the skin: a predictive model. *Am J Ind Med 23*: 711–719

Hansen LF, Nielsen GD (1994) Sensory irritation, pulmonary irritation and structure-activity relationships of alcohols. *Toxicology 80*: 81–99

Hempel-Jørgensen A, Kjærgaard SK, Mølhave L (1998) Cytological changes and conjunctival hyperemia in relation to sensory eye irritation. *Int Arch Occup Environ Health 71*: 225–235

Herzog H, Seitz RJ, Tellmann L, Schlaug G, Kleinschmidt A, Nebeling B, Stoecklin G, Mueller-Gaertner HW (1994) Pharmacokinetics and radiation dose of oxygen-15 labelled butanol in rCBF studies in humans. *Eur J Nucl Med 21*: 138–143

IRIS (Integrated Risk Information System) (1998) *Results of databank search on 23.10.98 for n-butyl alcohol*, National Library of Medicine, Bethesda, MD, USA

Jacobs RR, Phanprasit W (1993) An *in vitro* comparison of the permeation of chemicals in vapor and liquid phase through pig skin. *Am Ind Hyg Assoc J 54*: 569–575

Jaeger W (1955) Hornhautveränderungen durch Butylalkoholdämpfe (Corneal changes caused by butyl alcohol vapour) (German). *Graefes Arch Ophthalmol 156*: 480–483

Jung R, Engelhart G, Herbolt B, Jäckh R, Müller W (1992) Collaborative study of mutagenicity with *Salmonella typhimurium* TA102. *Mutat Res 278*: 265–270

Kaneko T, Wang PY, Sato A (1994) Partition coefficients of some acetate esters and alcohols in water, blood, olive oil, and rat tissues. *Occup Environ Med 51*: 68–72

Kawai T, Okada Y, Odachi T, Horiguchi S, Zhang ZW, Moon CS, Furuki K, Ukai H, Inui S, Ikeda M (1997) Monitoring of occupational exposure to 1-butanol by diffusive sampling and urinalysis. *Int Arch Occup Environ Health 69*: 266–272

Kimmel CA,Wilson JG (1973) Skeletal deviations in rats: malformations or variations? *Teratology 8*: 309–315

Korsak Z, Rydzynski K (1994) Effects of acute combined inhalation exposure to *n*-butyl alcohol and *n*-butyl acetate in experimental animals. *Int J Occup Med Environ Health 7*: 273–280

Korsak Z, Wisniewska-Knypl J, Swiercz R (1994) Toxic effects of subchronic combined exposure to *n*-butyl alcohol and *m*-xylene in rats. *Int J Occup Med Environ Health 7*: 155–166

Nelson KW, Ege Jr JF, Ross M, Woodman LE, Silverman L (1943) Sensory response to certain industrial solvent vapors. *J Ind Hyg Toxicol 25*: 282–285

Nelson BK, Brightwell WS, Khan A, Burg JR, Goad PT (1989a) Lack of selective developmental toxicity of three butanol isomers administered by inhalation to rats. *Fundam Appl Toxicol 12*: 469–479

Nelson BK, Brightwell WS, Robertson SK, Khan A, Krieg EF jr, Massari VJ (1989b) Behavioral teratology investigation of 1-butanol in rats. *Neurotoxicol Teratol 11*: 313–315

Nelson BK, Brightwell WS, Krieg EF jr (1990) Developmental toxicology of industrial alcohols: a summary of 13 alcohols administered by inhalation to rats. *Toxicol Ind Hlth 6*: 373–387

Önfelt A (1987) Spindle disturbances in mammalian cells. III. Toxicity, c-mitosis and aneuploidy with 22 different compounds. Specific and unspecific mechanisms. *Mutat Res 182*: 135–154

Ophaswongse S, Maibach HI (1994) Alcohol dermatitis: allergic contact dermatitis and contact urticaria syndrome. A review. *Contact Dermatitis 30*: 1–6

Peters T (1958) Augenschäden durch eine wasserlösliche, stark butanolhaltige Druckfarbe (Eye damage caused by a printing ink containing high levels of butanol) (German). *Zentralbl Arbeitsmed Arbeitsschutz 8*: 208

Poulin P, Krishnan K (1996) A tissue composition-based algorithm for predicting tissue:air partition coefficients of organic chemicals. *Toxicol Appl Pharmacol 136*: 126–130

Rowe VK, McCollister SB (1982) Alcohols. in: Clayton GD, Clayton FE (Eds) *Patty's Industrial Hygiene and Toxiology*, Vol. 2C, John Wiley & Sons, New York, 4571–4578

Seitz B (1972) Vertiges graves apparus après manipulation de butanol et d'isobutanol. A propos de trois cas (Severe vertigo following handling of butanol and isobutanol. Apropos of 3 cases) (French). *Arch Mal Prof Med Trav Secur Soc 33*: 393–395

Sitarek K, Berlinska B (1996) Effect of exposure to *n*-butanol in drinking water on fertility and prenatal development of rats. *Teratology 53*: 36A

Sitarek K, Berlinska B, Baranski B (1994) Assessment of the effect of *n*-butanol given to female rats in drinking water on fertility and prenatal development of their offspring. *Int J Occup Med Environ Health 7*: 365–370

Smyth HF, Smyth Jr HF (1928) Inhalation experiments with certain lacquer solvents. *J Ind Hyg 10*: 261–271

Sterner JH, Crouch HC, Brockmyre HF, Cusack M (1949) A ten-year study of butyl alcohol exposure. *Am Ind Hyg Assoc J 10*: 53–59

Stotts J, Ely WJ (1977) Induction of human skin sensitization to ethanol. *J Invest Dermatol 69*: 219–222

Svoboda DJ, Manning RT (1964) Chronic alcoholism with fatty metamorphosis of the liver. Mitochondrial alterations in hepatic cells. *Am J Pathol 44*: 645–662

Swiercz R, Korsak Z, Rydzynski K (1995) Kinetics of *n*-butyl alcohol and *m*-xylene in blood during single and combined exposure in rats. *Int J Occup Med Environ Health 8*: 361–365

Tabershaw IR, Fahy JP, Skinner JB (1944) Industrial exposure to butanol. *J Ind Hyg Toxicol 26*: 328–330

Van Ketel WG, Tan-Lim KN (1975) Contact dermatitis from ethanol. *Contact Dermatitis 1*: 7–10

Velasquez J, Escobar R, Almaraz A (1969) Audiologic impairment due to *n*-butyl alcohol exposition. in: *Proceedings of the XVI international congress on occupational health* (September 22–27, 1969, Tokyo), Excerpta Medica Foundation, 231–234

WHO (World Health Organization) (1987) *Butanols – four isomers: 1-butanol, 2-butanol, tert-butanol, isobutanol. IPCS – Environmental Health Criteria 65*, WHO, Genf

Wickramaratne GA (1988) The post-natal fate of supernumerary ribs in rat teratogenicity studies. *J Appl Toxicol 8*: 91–94

completed 25.04.2000

sec-Butyl alcohol

MAK value	**not yet established, see Section IIb of the *List of MAK and BAT Values***
Peak limitation	**–**
Absorption through the skin	**–**
Sensitization	**–**
Carcinogenicity	**–**
Prenatal toxicity	**–**
Germ cell mutagenicity	**–**
BAT value	**–**
Synonyms	2-hydroxybutane methyl ethyl carbinol 1-methyl-1-propanol 1-methyl propyl alcohol
Chemical name (CAS)	2-butanol
CAS number	78-92-2
Structural formula	CH_3–CHOH–CH_2–CH_3
Molecular formula	$C_4H_{10}O$
Molecular weight	74.12
Melting point	–114.7°C
Boiling point	99.5°C
Density at 20°C	0.81 g/cm^3
Vapour pressure at 20°C	16.6 hPa
log P_{ow}*	0.61
1 ml/m^3 (ppm) ≙ 3.03 mg/m^3	**1 mg/m^3 ≙ 0.330 ml/m^3 (ppm)**

* *n*-octanol/water distribution coefficient

The MAK value for *sec*-butyl alcohol of 100 ml/m^3 was established in 1969 in analogy to the TLV value at the time. The present documentation was based on reviews of toxicological data for *sec*-butyl alcohol (DECOS 1994, ECB 1995, WHO 1987) and supplemented with more recent data.

1 Toxic Effects and Mode of Action

According to data from *in vitro* experiments, absorbed *sec*-butyl alcohol is distributed mainly in the aqueous phases of the organism. The substance is probably oxidized via alcohol dehydrogenase to form 2-butanone, which can be eliminated via the lungs and kidneys or metabolized oxidatively or reductively. There is no information available about the effects in man. Allergic reactions of the skin to *sec*-butyl alcohol are very rare and usually interpreted as cross-reactions. The acute toxicity of the substance in experimental animals is low for all administration routes. The toxicity of the substance after repeated administration has not been investigated. On rabbit skin *sec*-butyl alcohol was not found to be irritative. In the rabbit eye, however, it was irritative or even caustic. *sec*-Butyl alcohol is not mutagenic in prokaryotes nor clastogenic in CHO cells (a cell line derived from Chinese hamster ovary). In rats, *sec*-butyl alcohol doses of about 3000 mg/kg body weight taken up with the drinking water over 2 generations have no effects on fertility. Exposure of rats to the maternally toxic concentration of 3500 ml/m^3 for the whole period of gestation has no effects on the offspring. Higher concentrations are foetotoxic. For *sec*-butyl alcohol administered with the drinking water over 2 generations, the NOAEL (no observed adverse effect level) in rats is about 1500 mg/kg body weight and day for both maternally toxic effects and toxic effects on reproduction.

2 Mechanism of Action

The irritative effects of the butyl alcohol isomers are probably the result of non-covalent interaction with the receptors of the sensory nerve endings in the mucous membranes of the respiratory tract and are a function of the physico-chemical properties (e.g. vapour pressure, lipophilia) of the substances (Abraham *et al.* 1996, Alarie *et al.* 1998, Hansen and Nielsen 1994). The central nervous effects which occur at higher concentrations are probably the result of the interaction of the relatively hydrophobic substances with neuronal membranes (Fang *et al.* 1997, Frantik *et al.* 1994).

3 Toxicokinetics and Metabolism

There are no studies available of the rate of absorption of *sec*-butyl alcohol via various routes. On the basis of the physico-chemical properties of *sec*-butyl alcohol, a penetration rate through human skin of 1.68 mg/cm^2 and hour (Fiserova-Bergerova *et al.* 1990) or 0.21 mg/cm^2 and hour (Guy and Potts 1993) has been estimated for a saturated aqueous *sec*-butyl alcohol solution. After oral administration of the substance to rats and rabbits, the *sec*-butyl alcohol concentrations in blood reached a maximum 1 to 2 hours after administration, and decreased rapidly within a few hours. There are no data available for the distribution of the substance, but *in vitro* experiments with various tissues showed that *sec*-butyl alcohol is distributed mainly in the aqueous phase.

In rats, about 97 % of an oral dose of *sec*-butyl alcohol was oxidized to 2-butanone, probably by hepatic alcohol dehydrogenase. The half-time for the elimination of *sec*-butyl alcohol from the plasma was 2.5 hours. With preparations of microsomes from rat liver and lung tissue it was shown that also various cytochrome P450 isoforms are involved in the oxidation of *sec*-butyl alcohol (Gadberry and Carlson 1994). 2-Butanone is oxidized to 3-hydroxy-2-butanone, which is then reduced to 2,3-butanediol (see also "2-Butanone" in Volume 12 of the present series). The increase in the concentrations of these metabolites in the blood of rats given oral doses of *sec*-butyl alcohol corresponded with the decrease in the concentration of the substance itself. The elimination half-time for 3-hydroxy-2-butanone from the blood was given as 47 minutes, that for 2,3-butanediol was 3.5 hours.

The elimination of *sec*-butyl alcohol and its metabolites has not been sufficiently investigated. Male rabbits eliminated 3.3 % of a single oral dose of *sec*-butyl alcohol unchanged via the lungs and 2.6 % unchanged with the urine. 22.3 % of the dose was exhaled as 2-butanone, 4 % was eliminated renally in this form. In another study with rabbits given oral doses of 618 mg/kg body weight, around 14 % of the dose was detected as the glucuronic acid conjugate in the 24-hour urine. The form in which the rest of the dose was eliminated and the routes of elimination involved were not determined in either study (DECOS 1994, WHO 1987). For the elimination of the main metabolite 2-butanone, the reader is referred to the chapter "2-Butanone" in Volume 12 of the present series.

4 Effects in Man

With the exception of the threshold value for nasal irritation of 9440 ml/m^3 determined during exposure of anosmic persons for about 2 seconds (Abraham *et al.* 1996), there are no data available for the effects of exposure to *sec*-butyl alcohol in man.

4.1 Allergenic effects

There are only a few reports of positive results obtained in patch tests with *sec*-butyl alcohol. It seems probable that they were the result of cross-reactions because in these patients ethanol was very likely the sensitizing agent. In patch tests with such patients, positive results are frequently obtained with various primary and secondary alcohols and with acetone. In some of the persons, skin reactions also developed after the consumption of alcoholic drinks, also erythema, aphthous lesions and a burning sensation in the oral mucosa (Fisher 1983, Fregert *et al.* 1969, Ophaswongse and Maibach 1994, Stotts and Ely 1977, Van Ketel and Tan-Lim 1975). From these results, it cannot be concluded that *sec*-butyl alcohol has a significant sensitizing potential, if any.

5 Animal Experiments and *in vitro* Studies

There are no data available for the subacute, subchronic and chronic toxicity of *sec*-butyl alcohol, or the carcinogenic effects of the substance.

5.1 Acute toxicity

Like all butyl alcohol isomers, *sec*-butyl alcohol leads via all routes of absorption to the typical symptoms of acute alcohol intoxication, from central nervous depression to narcosis. The severity of the effects of the individual isomers is about the same and overall greater than that of ethanol (DECOS 1994, WHO 1987).

5.1.1 Inhalation

In rats exposed to *sec*-butyl alcohol concentrations of 3500 ml/m^3 for 7 hours, no signs of narcotic effects were observed. Concentrations of 5000 ml/m^3 caused transient narcotic effects and restricted motor activity. 7000 ml/m^3 led in all animals to deep narcosis and 10000 ml/m^3 was lethal for all 5 animals. In another study in which 6 rats were exposed to 16000 ml/m^3 for 4 hours, 5 animals died within the observation period of 14 days.

Exposure to 10670 ml/m^3 for 3.7 hours or to 16000 ml/m^3 for 2.7 hours was lethal for mice. Exposure to 1650 ml/m^3 for 7 hours had no effects; at a concentration of 3300 ml/m^3, the first signs of central nervous depression were observed after 50 to 100 minutes (DECOS 1994, WHO 1987).

5.1.2 Ingestion and dermal absorption

For the rat, oral LD_{50} values of 2200 mg/kg body weight (Shell 1994) and 6500 mg/kg body weight were given, for the rabbit a value of 4900 mg/kg body weight. More details may be found in other reviews (DECOS 1994, ECB 1995, WHO 1987). In a study of acute dermal toxicity in rabbits, even the highest dose tested of 2000 mg/kg body weight was not lethal (Shell 1994).

5.2 Local effects on skin and mucous membranes

In a study from 1954, *sec*-butyl alcohol applied to rabbit skin was given an irritation index of 1 on a scale with a maximum value of 10 and was thus not regarded as irritative (DECOS 1994, WHO 1987). Semi-occlusive application of the undiluted substance to the shaved dorsal skin of rabbits for 4 hours did not lead to effects on the skin at any time during the experiment (4, 24, 48 and 72 hours, and 7 days after application), and was regarded as not irritative (Shell 1994).

In a study from 1954, *sec*-butyl alcohol instilled into the rabbit eye was given an irritation index of 4 on a scale with a maximum value of 10 and was thus regarded as irritative (DECOS 1994, WHO 1987). In a test for irritative effects on the eye carried out according to OECD guideline 405, moderate pain reactions were observed even while the undiluted substance was being dropped into the rabbit eye. In all animals the conjunctiva were moderately inflamed, swollen and red. In some animals the reactivity of the iris was reduced and the cornea clouded. Because the symptoms in one animal were not reversible even after 7 days, *sec*-butyl alcohol was regarded as caustic (Shell 1994).

5.3 Allergenic effects

In a Magnusson and Kligman maximization test with guinea pigs, *sec*-butyl alcohol was not found to be sensitizing (Shell 1994).

5.4 Reproductive and developmental toxicity

5.4.1 Fertility

In the 2-generation study with Wistar rats (ECB 1995, Gallo *et al.* 1977; see also Section 5.4.2), even at the highest tested dose of about 3000 mg/kg body weight—administered with the drinking water—the mating success was not affected.

5.4.2 Developmental toxicity

Female Sprague-Dawley rats were exposed to *sec*-butyl alcohol concentrations of 3500, 5000 or 7000 ml/m^3 from days 1 to 19 of gestation, for 7 hours a day. Body weight gains were significantly decreased in the two lower concentration groups in weeks 1 and 3, and in the high concentration group during the whole exposure period. At the same time also food consumption was decreased in the treated groups. The body weights of the foetuses in the middle and high concentration groups were significantly reduced. In the high concentration group the number of resorptions per litter and the frequency of skeletal variations were significantly increased, the number of living foetuses per litter significantly decreased. Visceral and skeletal malformations were not found (Nelson *et al.* 1989, 1990).

In 1975, a modified 2-generation study was carried out with Wistar rats. The F_0 animals of both sexes were given *sec*-butyl alcohol with the drinking water in concentrations of 0.3 %, 1 % and 2 %; the treatment was begun 8 weeks before mating. The doses were estimated from the concentrations in the drinking water to be about 500, 1500 and 3000 mg/kg body weight and day. In the two lower concentration groups the mating success and growth of the parent generation and offspring were not impaired. In the 2 % group the growth of the offspring was significantly delayed. The F_1 generation was treated further and mated again, and after the pups were delivered on day 20 of gestation underwent clinico-chemical, haematological, gross pathological and histopathological examination. There were no pathological findings in the two low concentration groups. In the high concentration group slight changes in the kidneys were observed (no other details), which the authors, however, did not interpret as an indication of substance-specific toxicity, but as an unspecific stress reaction. The mating success of the F_1 generation was not impaired by the treatment. Survival of the offspring in all groups corresponded with that in the control group. Only in the 2 % group were the weights of the foetuses reduced relative to those of the controls. Skeletal changes were seen, but the type and frequency corresponded with the spontaneously occurring pattern of such changes in this strain of rat. Visceral malformations were not detected. From this study, the NOAEL (no observed adverse effect level) for *sec*-butyl alcohol in rats is about 1500 mg/kg body weight and day for both maternally toxic effects and toxic effects on reproduction (ECB 1995, Gallo *et al.* 1977, Union Carbide 1992).

5.5 Genotoxicity

5.5.1 *In vitro*

In the *Salmonella* mutagenicity test with the strains TA98, TA100, TA1535, TA1537 and TA1538, *sec*-butyl alcohol was not found to be mutagenic up to the highest concentration tested of 10 mg/plate. The test was carried out according to OECD guideline 471. Also in *Escherichia coli* no mutagenic effects were observed up to the maximum concentration tested of 4 mg/plate. A mitotic gene conversion test with *Saccharomyces cerevisiae* yielded negative results up to the highest concentration of 5 mg/ml. No chromosomal aberrations were observed in CHO cells after incubation with up to 5 mg/ml. All studies were carried out with and without the addition of a metabolic activation system (ECB 1995). *sec*-Butyl alcohol was tested also with *Schizosaccharomyces pombe*; no mutagenic effects were detected either with or without added mouse liver microsomes (WHO 1987).

5.5.2 *In vivo*

In a Russian study all butyl alcohol isomers were tested with groups of 8 rats in a cytogenetic *in vivo* test. The dose in each case was one fifth of the LD_{50}. All isomers led to increased chromosomal aberrations. It is unusual that no aberrations were found in the controls. The incidence of polyploid bone marrow cells was also increased in the animals treated with *n*-butyl alcohol, *sec*-butyl alcohol and isobutyl alcohol. No chromosome breaks were found. As the study did not investigate whether the effects were dose-dependent, they cannot be used as evidence of genotoxic effects of the butyl alcohol isomers (Barilyak and Kozachnuk 1988).

6 Manifesto (MAK value/classification)

There are no relevant studies in man or experiments with animals with repeated inhalation or oral exposure. Therefore a scientifically grounded threshold limit value for *sec*-butyl alcohol cannot be evaluated and the substance is listed in Section IIb of the *List of MAK and BAT Values*.

Because there are no data for systemic toxicity of *sec*-butyl alcohol, it cannot be decided whether or not designation with an "H" is necessary.

The carcinogenic and germ cell mutagenic potential of the substance has not been investigated, a genotoxic potential is not recognizable from the studies available.

The available data indicate that designation with an "S" (for substances which cause sensitization) is not necessary.

7 References

Abraham MH, Andonian-Haftvan J, Cometto-Muñiz JE, Cain WS (1996) An analysis of nasal irritation thresholds using a new solvation equation. *Fundam Appl Toxicol 31*: 71–76

Alarie Y, Schaper M, Nielsen GD, Abraham MH (1998) Structure-activity relationships of volatile organic chemicals as sensory irritants. *Arch Toxicol* 72: 125–140

Barilyak IR, Kozachnuk YS (1988) Studies of the cytogenetic action of some monohydric alcohols on rat bone marrow cells (Russian). *Tsitol Genet 22*: 49–52

DECOS (Dutch Expert Committee on Occupational Standards) (1994) *1-, 2- and t-butanol, health-based recommended occupational exposure limit.* Publication No. 1994/10, Health Council of the Netherlands, Den Haag

ECB (European Chemicals Bureau) (1995) IUCLID data sheet, *butan-2-ol*, 23.10.95, Ispra, Italy

Fang Z, Ionescu P, Chortkoff BS, Kandel L, Sonner J, Laster MJ, Eger II EI (1997) Anesthetic potencies of *n*-alkanols: results of additivity and solubility studies suggest a mechanism of action similar to that for conventional inhaled anesthetics. *Anesthesia Analgesia 84*: 1042–1048

Fiserova-Bergerova V, Pierce JT, Droz PO (1990) Dermal absorption potential of industrial chemicals: criteria for skin notation. *Am J Ind Med 17*: 617–635

Fisher AA (1983) Topically applied alcohol as a cause of contact dermatitis. *Cutis 31*: 588–600

Frantik E, Hornychova M, Horvath M (1994) Relative acute neurotoxicity of solvents: isoeffective air concentrations of 48 compounds evaluated in rats and mice. *Environ Res 66*: 173–185

Fregert S, Groth O, Hjorth N, Magnusson B, Rorsman H, Övrum P (1969) Alcohol dermatitis. *Acta Derm Venereol 49*: 493–497

Gadberry MG, Carlson GP (1994) 2-Butanol metabolism by rat hepatic and pulmonary cytochromes P450. *Toxicol Lett 74*: 203–209

Gallo MA, Oser BL, Cox GE, Bailey DE (1977) Studies on the long-term toxicity of 2-butanol (Abstract). *Toxicol Appl Pharmacol 41*: 135

Guy RH, Potts RO (1993) Penetration of industrial chemicals across the skin: a predictive model. *Am J Ind Med 23*: 711–719

Hansen LF, Nielsen GD (1994) Sensory irritation, pulmonary irritation and structure-activity relationships of alcohols. *Toxicology 80*: 81–99

Nelson BK, Brightwell WS, Khan A, Burg JR, Goad PT (1989) Lack of selective developmental toxicity of three butanol isomers administered by inhalation to rats. *Fundam Appl Toxicol 12*: 469–479

Nelson BK, Brightwell WS, Krieg Jr EF (1990) Developmental toxicology of industrial alcohols: a summary of 13 alcohols administered by inhalation to rats. *Toxicol Ind Health 6*: 373–387

Ophaswongse S, Maibach HI (1994) Alcohol dermatitis: allergic contact dermatitis and contact urticaria syndrome. A review. *Contact Dermatitis 30*: 1–6

Shell (1994) *Toxicity studies of secondary butyl alcohol with cover letter dated 12/23/94.* NTIS/OTS 0557575, EPA/OTS 86950000062, NTIS, Springfield, VA, USA

Stotts J, Ely WJ (1977) Induction of human skin sensitization to ethanol. *J Invest Dermatol 69*: 219–222

Union Carbide (1992) *Initial submission: toxicity studies in rats with 2-butanol including growth, reproduction, and teratologic observations with cover letter dated 092492.* NTIS/OTS 0571256, EPA/OTS 88-920009599, NTIS, Springfield, VA, USA

Van Ketel WG, Tan-Lim KN (1975) Contact dermatitis from ethanol. *Contact Dermatitis 1*: 7–10

WHO (World Health Organization) (1987) *Butanols – four isomers: 1-butanol, 2-butanol, tert-butanol, isobutanol. IPCS – Environmental health criteria 65*, WHO, Genf

completed 15.01.1999

tert-Butyl alcohol

MAK value (1999)	**20 ml/m^3 (ppm) ≙ 62 mg/m^3**
Peak limitation (2002)	**Category II, excursion factor 4**
Absorption through the skin	**–**
Sensitization	**–**
Carcinogenicity	**–**
Prenatal toxicity (1999)	**Pregnancy risk group D**
Germ cell mutagenicity	**–**
BAT value	**–**
Synonyms	trimethyl carbinol 2-methyl-2-propanol
Chemical name (CAS)	2-methyl-2-propanol
CAS number	75-65-0
Structural formula	$(CH_3)_3COH$
Molecular formula	$C_4H_{10}O$
Molecular weight	74.12
Melting point	25.6°C
Boiling point	82.4°C
Density at 20°C	0.79 g/cm^3
Vapour pressure at 20°C	40.8 hPa
log P_{ow}*	0.37
1 ml/m^3 (ppm) ≙ 3.03 mg/m^3	**1 mg/m^3 ≙ 0.330 ml/m^3 (ppm)**

A MAK value of 100 ml/m^3 was established in 1958 for *tert*-butyl alcohol in analogy to the TLV value at the time. The present documentation is based on reviews of toxicological data for *tert*-butyl alcohol (DECOS 1994, ECB 1995, WHO 1987) and supplemented with more recent data.

---

* *n*-octanol/water distribution coefficient

1 Toxic Effects and Mode of Action

tert-Butyl alcohol is absorbed completely after ingestion. There are no data available for absorption after inhalation or dermal exposure. The absorbed substance is distributed rapidly in the organism, mainly in aqueous compartments. The half-time for the elimination from the plasma of rodents is at least 4 hours. *tert*-Butyl alcohol is not transformed by alcohol dehydrogenase. Conjugation with sulfuric acid or glucuronic acid takes place, or oxidative demethylation by cytochrome P450 isoenzymes to form acetone and formaldehyde, for example. The metabolites and the substance itself are eliminated via the kidneys or the lungs, but quantitative data are not available. Oral administration of doses of 500 mg/kg body weight and above led in rats to saturation of metabolism or excretion.

Data for effects in man are sparse. Allergic reactions of the skin to *tert*-butyl alcohol are rare and usually interpreted as cross-reactions. The acute toxicity of the substance in experimental animals is low for all administration routes. In rodents, the target organ after repeated exposure is the urinary tract. The most sensitive experimental animal was found to be the male rat; nephrotoxic effects were observed from concentrations of 135 ml/m^3 and reduced body weights from doses of 85 mg/kg body weight and day. On rabbit skin *tert*-butyl alcohol was not found to be irritative. In the rabbit eye the substance was also not irritative, or only moderately so. In maximization tests with guinea pigs, the substance was not found to be sensitizing in one case and to cause weak sensitization in another. *tert*-Butyl alcohol was found to be foetotoxic in rats from concentrations of 2000 ml/m^3, a concentration below the toxic range for the dams. In the mouse, embryotoxic and foetotoxic effects were observed after oral administration of doses from 1600 mg/kg body weight and day. The substance is not mutagenic in *Salmonella typhimurium* or in mouse lymphoma cells, and not clastogenic in CHO cells (a cell line derived from Chinese hamster ovary). In the rat and mouse, treatment with *tert*-butyl alcohol did not lead to the formation of micronuclei in the polychromatic erythrocytes. A long-term drinking water study with the rat and mouse did not yield evidence of carcinogenic effects with doses below the maximum tolerated dose.

2 Mechanism of Action

The irritative effects of the butyl alcohol isomers are probably the result of non-covalent interaction with the receptors of the sensory nerve endings in the mucous membranes of the respiratory tract and are a function of the physico-chemical properties (e.g. vapour pressure, lipophilia) of the substances (Abraham *et al.* 1996, Alarie *et al.* 1998, Hansen and Nielsen 1994). The central nervous effects which occur at higher concentrations are probably the result of the interaction of the relatively hydrophobic substances with neuronal membranes (Fang *et al.* 1997, Frantik *et al.* 1994).

The mechanism of the nephropathy observed in male rats after medium-term inhalation of *tert*-butyl alcohol (NTP 1997) is unknown. A species-specific and sex-specific accumulation of hyaline droplets (α2u-globulin nephropathy) was detected in male rats only after medium-term oral administration of *tert*-butyl alcohol (Lindamood *et al.* 1992, Takahashi *et al.* 1993). As, however, after long-term oral administration there was evidence in both sexes of an increase in spontaneous age-related nephropathy, it is assumed that in addition to α2u-globulin nephropathy, which is not relevant for man, *tert*-butyl alcohol can have toxic effects on the kidneys also as a result of the acceleration or amplification of chronic age-related nephropathy (NTP 1997).

3 Toxicokinetics and Metabolism

There are no quantitative data for the absorption of *tert*-butyl alcohol after inhalation or dermal exposure. Ingested *tert*-butyl alcohol was almost completely absorbed by rats. Only about 1 % of the administered radioactivity was thus detected in the faeces within 24 hours after administration of single doses of 350 mg/kg body weight of the ^{14}C-labelled substance. In rats exposed by inhalation to concentrations of 2000 ml/m^3 for 6 hours, the levels in the blood increased linearly up to the fifth hour and reached their maximum level at the end of exposure. The same maximum levels in the blood were also produced after oral doses of 350 mg/kg body weight.

tert-Butyl alcohol in the bloodstream is quickly distributed in the organism. Two minutes after intravenous injection of 350 mg/kg body weight of the ^{14}C-labelled substance only 15 % of the administered radioactivity was detected in the blood of the rats (Arco 1994a). In F344 rats given intravenous injections of 37, 75, 150 or 300 mg/kg body weight, there was also a rapid initial distribution phase with a half-time of 3 minutes for the concentrations of *tert*-butyl alcohol in blood (Poet *et al.* 1997). At 37°C, distribution coefficients were found for *tert*-butyl alcohol of 462 for human blood/air, 603 for water/air and 167 for olive oil/air. From these and from the water and fat content of various tissues, distribution coefficients between individual tissues (brain, kidneys, liver, lungs, muscle) and blood of 1.02 to 1.06 were calculated. The distribution coefficient calculated for fat/blood was 0.65 (Nihlen *et al.* 1995). With blood and tissue from rats, other authors determined concurring distribution coefficients of 481 for blood/air, 400 for liver/air, 543 for kidneys/air, 490 for muscle/air and 191 for adipose tissue/air. The distribution coefficients for physiological saline/air and oil/air were 556 and 150 (Borghoff *et al.* 1996). The absorbed substance is thus distributed mainly in aqueous compartments. In accordance with this, a distribution volume was determined after oral administration of *tert*-butyl alcohol to rats which corresponded more or less to the volume of water in the body (Arco 1994a). Other authors, however, reported a distribution volume for *tert*-butyl alcohol which was about 4.5 times higher than the volume of water in the body (Poet *et al.* 1997); this suggests accumulation of the substance.

The elimination half-time of *tert*-butyl alcohol and its metabolites from the blood of rats was at least 9 hours after inhalation, ingestion, and intraperitoneal and intravenous injection of relatively high doses and concentrations. The half-times were longer in animals given oral doses of 1500 mg/kg body weight (Arco 1994a, DECOS 1994, WHO 1987). In F344 rats given intravenous injections of *tert*-butyl alcohol doses between 37 and 300 mg/kg body weight, the elimination half-time from the plasma was 3.8 hours for the low dose; for 300 mg/kg body weight it increased in the male animals to 5 and in the females to 4.3 hours. This was explained as saturation of metabolism or excretion (Poet *et al.* 1997). For mice a value of about 5 hours was found (DECOS 1994, WHO 1987).

The metabolism of *tert*-butyl alcohol is not catalysed by alcohol dehydrogenase, as has been demonstrated in various studies with the rat and mouse and in *in vitro* studies. Sulfate and glucuronide conjugates, derivatives of *tert*-butyl alcohol hydroxylated on the methyl side-chains, acetone and formaldehyde are described as metabolites. *In vitro* studies with rat liver microsomes show that the small amounts of acetone and formaldehyde are formed during the transformation of *tert*-butyl alcohol by the mixed-function monooxygenase system by demethylation and with the involvement of reactive oxygen species. In rats, pretreatment with phenobarbital for the induction of cytochrome P450 enzymes did not increase the formation of acetone after treatment with *tert*-butyl alcohol. Pretreatment with *tert*-butyl alcohol increased the rate of elimination of the substance after further exposure in mice, but not in rats (Arco 1994a, DECOS 1994, WHO 1987). In a more recent study with 3 rats given single oral doses of ^{13}C-labelled *tert*-butyl alcohol of 250 mg/kg body weight, the metabolite profile described above was confirmed. In the 24-hour pooled urine samples of the animals the sulfate conjugate of *tert*-butyl alcohol was identified as the main metabolite; in addition, the substance itself, the glucuronide conjugate, 2-hydroxyisobutyric acid, 2-methyl-1,2-propanediol and small amounts of acetone were found. In the urine of a volunteer given doses of ^{13}C-labelled *tert*-butyl alcohol of 5 mg/kg body weight, the main metabolites were 2-methyl-1,2-propanediol and 2-hydroxyisobutyric acid. In addition, free *tert*-butyl alcohol, the glucuronide conjugate and also traces of the sulfate conjugate were detected. Another metabolite could not be characterized; more exact quantification of the metabolites was not possible (Bernauer *et al.* 1998). In the blood of rats given intravenous injections of *tert*-butyl alcohol doses between 37 and 300 mg/kg body weight, the concentration of acetone was not significantly increased (Poet *et al.* 1997).

The elimination of *tert*-butyl alcohol has not been sufficiently investigated, at least from a quantitative point of view. In the urine of rats and a volunteer, both the substance itself and also various metabolites (see above) were eliminated (Bernauer *et al.* 1998). Rats given oral doses of the ^{14}C-labelled substance of 1 to 500 mg/kg body weight excreted about 25 % of the radioactivity with the urine within 24 hours. Only 9 % of a dose of 1500 mg/kg body weight was excreted. Over 97 % of renal excretion involved polar and relatively volatile metabolites. With increasing dose, the proportion of the renally eliminated radioactivity made up by unchanged *tert*-butyl alcohol increased from 0.3 % to 5.5 %. This was interpreted as saturation of metabolism (Arco 1994a). In the exhaled air of rats given isotope-labelled *tert*-butyl alcohol, correspondingly labelled acetone and CO_2 were detected (DECOS 1994, WHO 1987). Metabolites of *tert*-butyl alcohol are also excreted with the faeces, and the substance itself is also exhaled after oral administration (no other details) (ECB 1995).

It was concluded from the toxicokinetic data for *tert*-butyl alcohol that the results of studies with rats and oral doses of over 500 mg/kg body weight and day may not be suitable for deducing the effect profile of the substance in the low dose range as a result of the saturation of excretion or metabolism (Arco 1994a).

inflammatory processes and hyperplasia of the transitional epithelium were observed significantly more often in the bladder in male animals of the high concentration group (and to a lesser extent also in the females) than in the control groups. In the thyroid gland also neoplastic changes were found, which are listed in Table 1. The tumour incidence was significantly increased only in female animals and at doses which—according to the data for body weight gains and mortality—were above the MTD (maximum tolerated dose) (Cirvello *et al.* 1995, NTP 1995).

Table 1. Two-year drinking water study with *tert*-butyl alcohol (Cirvello *et al.* 1995, NTP 1995)

Species:	rat (F344), groups of 60 ♂/♀, interim killing of 10 animals/group after 15 months
Administration route:	with the drinking water (♂: 1.25, 2.5 and 5 g/l; ♀: 2.5, 5 and 10 g/l)
Dose:	♂: 85, 195 and 420 mg/kg body weight and day ♀: 175, 330 und 650 mg/kg body weight and day
Duration:	104 weeks

Hyperplasia and tumours in the kidneys of male animals [1]:

	tert-butyl alcohol (mg/kg body weight and day) 0	85	195	420
survivors (after 101 weeks)	12/50	10/50	4/50	2/50 [2]
tubule cell hyperplasia	14/50	20/50	17/50	25/50 [3]
tubule cell adenomas	7/50	7/50	9/50	9/50
tubule cell adenomas, multiple	1/50	4/50	10/50 [3]	4/50
tubule cell carcinomas	0/50	2/50	1/50	1/50
tubule cell adenomas or carcinomas	8/50	13/50	19/50 [3]	13/50

Species:	mouse ($B6C3F_1$), groups of 60 ♂/♀
Administration route:	with the drinking water (5, 10 and 20 g/l)
Dose:	♂: 535, 1035 and 2065 mg/kg body weight and day ♀: 510, 1015 and 2105 mg/kg body weight and day
Duration:	104 weeks

Hyperplasia and tumours of the thyroid gland:

		tert-butyl alcohol (mg/kg body weight and day) 0	about 520	about 1020	about 2070
survivors (after 101 weeks)	♂	34/60	44/60	41/60	23/60
	♀	29/60	28/60	26/60	17/60
follicle cell hyperplasia	♂	5/60	18/59 [3]	15/59 [2]	18/57 [3]
	♀	19/58	28/60	33/59 [2]	47/59 [3]
follicle cell adenomas	♂	1/60	0/60	4/59	1/57
	♀	2/60	3/60	2/59	9/59 [2]
follicle cell carcinomas	♂	0/60	0/59	0/59	1/57

[1] incidences based on evaluation of an increased number of sections
[2] significantly different from the controls ($p < 0.05$, Fisher exact test)
[3] significantly different from the controls ($p < 0.01$, Fisher exact test)

As in this study in rats and male mice non-neoplastic effects were found even at the lowest dose tested in each case, the NOEL for systemic-toxic effects for male rats is below 85 mg/kg body weight and day, for female rats below 175 mg/kg body weight and day and for male mice below 535 mg/kg body weight and day. For female mice the NOEL was 515 mg/kg body weight and day.

6 Manifesto (MAK value/classification)

In a long-term drinking water study with *tert*-butyl alcohol the incidence of tumours in the kidney was increased in male rats, and in the thyroid gland in female rats. In view of the high incidence of kidney tumours also in the control group it is questionable whether these were substance-related. Thyroid tumours occurred only in one sex and above the maximum tolerated dose. There is therefore no reason to suspect carcinogenic potential in man. According to the studies available, *tert*-butyl alcohol is not genotoxic.

Results of studies of exposed persons which would be suitable for the derivation of a MAK value are not available. Drinking water studies with the rat and mouse did not yield a NOEL, at least for the male rat. Medium-term inhalation studies (NTP 1997) yielded for female rats and for mice of both sexes a NOEC of 540 ml/m^3. For male rats, as a result of nephrotoxic effects, the NOEC was below the lowest concentration tested of 135 ml/m^3. As the effects on the kidneys were slight and they were limited to one species and one sex, this concentration can be seen as the LOEC, close to the NAEC. The MAK value for *tert*-butyl alcohol has therefore been reduced to 20 ml/m^3. Assuming a respiratory volume of 10 m^3 during the shift, complete pulmonary retention and a body weight of 70 kg, if the MAK value is observed the daily dose is at most 10 mg/kg body weight, and is therefore lower by a factor of 8 than the LOEL for male rats from the 2-year study.

The critical effect of *tert*-butyl alcohol is the nephrotoxicity observed in male rats exposed to concentrations equivalent to 8 times the MAK value. A NOEL has not been reported; the mechanism of the toxicity is not known. Because it seems plausible that the acids produced in metabolism of *tert*-butyl alcohol are involved in the nephrotoxicity and tracer studies suggest that their half-lives are long, *tert*-butyl alcohol is classified in Peak limitation category II with an excursion factor of 4.

There are no studies available of the dermal penetration of *tert*-butyl alcohol. The acute toxicity after dermal absorption is, however, much lower than after oral absorption. The substance has therefore not been designated with an "H".

tert-Butyl alcohol led in 2 studies to embryotoxic or foetotoxic effects. A conclusive evaluation is, however, not possible, in particular because in these studies concentrations or doses were used which may have led to the saturation of metabolism or of elimination, which is not to be expected if the MAK value is observed. Inhalation studies with two species in lower concentration ranges may clarify the relevance of the above findings. Until such data are available, *tert*-butyl alcohol will be listed in Pregnancy risk group D.

The potential of the substance for germ cell mutagenicity has not been investigated, but in view of the lack of genotoxic effects is not to be expected.

To date there is only one case report available on the sensitizing effects on the skin of *tert*-butyl alcohol. In an experiment with animals the substance was not found to be sensitizing, while another study revealed only a weak potential for skin sensitization. Designation with an "S" (for sensitizing substances) is therefore not necessary.

7 References

Abel EL, Bilitzke PJ (1992) Effects of prenatal exposure to methanol and *t*-butanol in Long Evans rats. *Am J Obstet Gynecol 166*: 433

Abraham MH, Andonian-Haftvan J, Cometto-Muñiz JE, Cain WS (1996) An analysis of nasal irritation thresholds using a new solvation equation. *Fundam Appl Toxicol 31*: 71–76

Acharya S, Mehta K, Rodrigues S, Pereira J, Krishnan S, Rao CV (1995) Administration of subtoxic doses of *t*-butyl alcohol and trichloroacetic acid to male Wistar rats to study the interactive toxicity. *Toxicol Lett 80*: 97–104

Acharya S, Mehta K, Rodrigues S, Pereira J, Krishnan S, Rao CV (1997) A histopathological study of liver and kidney in male Wistar rats treated with subtoxic doses of *t*-butyl alcohol and trichloroacetic acid. *Exp Toxicol Pathol 49*: 369–373

Alarie Y, Schaper M, Nielsen GD, Abraham MH (1998) Structure-activity relationships of volatile organic chemicals as sensory irritants. *Arch Toxicol 72*: 125–140

Arco (Arco Chemical Company) (1994a) *Toxicologist's report on metabolism and pharmacokinetics of radiolabeled TBA 534 tertiary butyl alcohol with cover letter dated 03/24/94.* NTIS/OTS 0572366, EPA/OTS 86940000263, NTIS, Springfield, VA, USA

Arco (Arco Chemical Company) (1994b) *Acute dermal toxicity (LD_{50}) study in rabbits with arconol with cover letter dated 03/24/94.* NTIS/OTS 0572359, EPA/OTS 86940000256, NTIS, Springfield, VA, USA

Barilyak IR, Kozachnuk YS (1988) Studies in cytogenetic action of some monohydric alcohols on the rat bone marrow cells (Russian). *Tsitol Genet 22*: 49–52

BASF (1994) *Letter from BASF Corp to USEPA regarding submission of health and safety data subject to 8D reporting with attachements, dated 05/05/94 (sanitized).* NTIS/OTS 0572606, EPA/OTS 86940001380S, NTIS, Springfield, VA, USA

Bernauer U, Amberg A, Scheutzow D, Dekant W (1998) Biotransformation of ^{12}C- and 2-^{13}C-labeled methyl *tert*-butyl ether, ethyl *tert*-butyl ether, and *tert*-butyl alcohol in rats: identification of metabolites in urine by ^{13}C nuclear magnetic resonance and gas chromatography/mass spectrometry. *Chem Res Toxicol 11*: 651–658

Borghoff SJ, Murphy JE, Medinsky MA (1996) Development of physiologically based pharmacokinetic model for methyl tertiary-butyl ether and tertiary-butanol in male Fisher-344 rats. *Fundam Appl Toxicol 30*: 264–275

Cirvello JD, Radovsky A, Heath JE, Farnell DR, Lindamood 3rd C (1995) Toxicity and carcinogenicity of *t*-butyl alcohol in rats and mice following chronic exposure in drinking water. *Toxicol Ind Health 11*: 151–165

DECOS (Dutch Expert Committee on Occupational Standards) (1994) *1-, 2- and t-butanol, health-based recommended occupational exposure limit.* Publication No. 1994/10, Health Council of the Netherlands, Den Haag

ECB (European Chemicals Bureau) (1995) *IUCLID data sheet, 2-methylpropan-2-ol, 23.10.95*

Fang Z, Ionescu P, Chortkoff BS, Kandel L, Sonner J, Laster MJ, Eger II EI (1997) Anesthetic potencies of *n*-alkanols: results of additivity and solubility studies suggest a mechanism of action similar to that for conventional inhaled anesthetics. *Anesthesia Analgesia 84*: 1042–1048

Faulkner TP, Wiechart JD, Hartman DM, Hussain AS (1989) The effects of prenatal tertiary butanol administration in CBA/J and C57BL/6J mice. *Life Sci 45*: 1989–1995

Fisher AA (1983) Topically applied alcohol as a cause of contact dermatitis. *Cutis 31*: 588–600

Frantik E, Hornychova M, Horvath M (1994) Relative acute neurotoxicity of solvents: isoeffective air concentrations of 48 compounds evaluated in rats and mice. *Environ Res 66*: 173–185

Fregert S, Groth O, Hjorth N, Magnusson B, Rorsman H, Övrum P (1969) Alcohol dermatitis. *Acta Derm Venereol 49*: 493–497

Hansen LF, Nielsen GD (1994) Sensory irritation, pulmonary irritation and structure-activity relationships of alcohols. *Toxicology 80*: 81–99

Lindamood 3rd C, Farnell DR, Giles HD, Prejean JD, Collins JJ, Takahashi K, Maronpot RR (1992) Subchronic toxicity studies of *t*-butyl alcohol in rats and mice. *Fundam Appl Toxicol 19*: 91–100

Nelson BK, Brightwell WS, Khan A, Burg JR, Goad PT (1989) Lack of selective developmental toxicity of three butanol isomers administered by inhalation to rats. *Fundam Appl Toxicol 12*: 469–479

Nelson BK, Brightwell WS, Krieg Jr EF (1990) Developmental toxicology of industrial alcohols: a summary of 13 alcohols administered by inhalation to rats. *Toxicol Ind Health 6*: 373–387

Nelson BK, Brightwell WS, Khan A, Shaw PB, Krieg Jr EF, Massari VJ (1991) Behavioral teratology investigation of tertiary-butanol administered by inhalation to rats. *Pharmaco-psycho-ecologia 4*: 1–7

Nihlen A, Löf A, Johanson G (1995) Liquid/air partition coefficients of methyl and ethyl *t*-butyl ethers, *t*-amyl methyl ether, and *t*-butyl alcohol. *J Exposure Anal Environ Epidemiol 5*: 573–582

NTP (National Toxicology Program) (1995) *Toxicology and carcinogenesis studies of t-butyl alcohol (CAS No. 75-65-0) in F344/N rats and B6C3F$_1$ mice (drinking water studies)*, Technical report No. 436, NIH Publication No. 95-3167, National Institutes of Health, Bethesda, MD, USA

NTP (National Toxicology Program) (1997) *t-Butyl alcohol (CAS No. 75-65-0). Administration by inhalation to F344/N rats and B6C3F$_1$ mice.* Toxicity report No. 53, NIH Publication No. 97-3942, National Institutes of Health, Bethesda, MD, USA

Ophaswongse S, Maibach HI (1994) Alcohol dermatitis: allergic contact dermatitis and contact urticaria syndrome. A review. *Contact Dermatitis 30*: 1–6

Poet TS, Valentine JL, Borghoff SJ (1997) Pharmacokinetics of tertiary butyl alcohol in male and female Fisher 344 rats. *Toxicol Lett 92*: 179–186

Stotts J, Ely WJ (1977) Induction of human skin sensitization to ethanol. *J Invest Dermatol 69*: 219–222

Takahashi K, Lindamood 3rd C, Maronpot RR (1993) Retrospective study of possible α-2μ-globulin nephropathy and associated cell proliferation in male Fisher 344 rats dosed with *t*-butyl alcohol. *Environ Health Perspect 101, Suppl 5*: 281–285

Van Ketel WG, Tan-Lim KN (1975) Contact dermatitis from ethanol. *Contact Dermatitis 1*: 7–10

WHO (World Health Organization) (1987) *Butanols – four isomers: 1-butanol, 2-butanol, tert-butanol, isobutanol. IPCS – Environmental health criteria 65*, WHO, Genf

completed 02.03.1999, peak limitation category 28.02.2002

Demeton

MAK value	**not yet established, see Section IIb of the *List of MAK and BAT Values***
Peak limitation	–
Absorption through the skin (1969)	**H**
Sensitization	–
Carcinogenicity	–
Prenatal toxicity	–
Germ cell mutagenicity	–
BAT value (for acetylcholinesterase inhibitors) (1986)	**inhibition of acetylcholinesterase activity in erythrocytes by 30 %**
Synonyms	mixture of demeton-S: *O,O*-diethyl-*S*-(2-(ethylthio)ethyl)phosphorthioate and demeton-O: *O,O*-diethyl-*O*-(2-(ethylthio)ethyl)phosphorthioate isosystox mercaptophos systox
Chemical name (CAS)	phosphorothioic acid *O,O*-diethyl *O*-[2-(ethylthio)ethyl] ester mixture with tetramethylthioperoxydicarbonic diamide
CAS number	126-75-0 (demeton-S) 298-03-3 (demeton-O) 8065-48-3 (mixture)
Structural formula	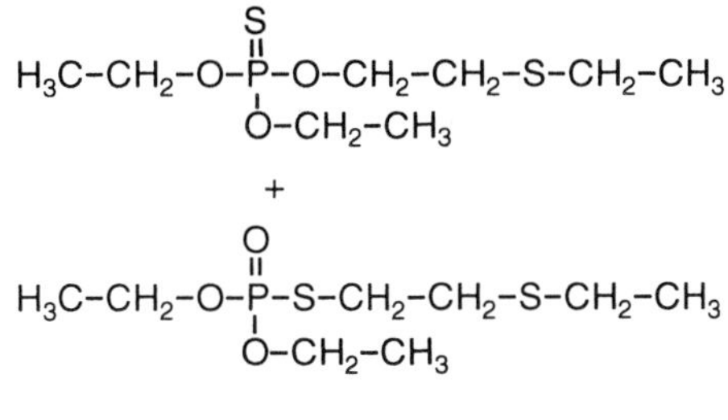
Molecular formula	$C_8H_{19}O_3PS_2$
Molecular weight	179.2

Melting point	–25°C
Boiling point at 2.7 hPa	134°C
Density	1.18 g/cm^3
Vapour pressure at 20°C	0.00034 hPa
Solubility in water	60 mg/l (demeton-S) 2000 mg/l (demeton-O)

The MAK value for demeton which was valid until 1999 (0.01 ml/m^3, 0.11 mg/m^3) was established in 1969 in analogy to the TLV value at the time.

1 Toxic Effects and Mode of Action

Demeton belongs to the group of organophosphates and is acutely toxic after inhalation, ingestion and dermal absorption. Intoxication causes miosis, accommodation disorders, salivation, lacrimation and raving fits, and also muscle cramps, cramps in the gastrointestinal tract, bronchoconstriction, bronchial secretion, nausea, weakness and paralysis. In addition, it can cause respiratory depression and circulatory insufficiency. Death occurs as a result of peripheral respiratory arrest, suffocation and heart failure.

The main routes of metabolism are the formation of sulfoxides and sulfones at the ethylthioethyl residue, the oxidation of demeton-O to form phosphate and the hydrolysis of the P-O and P-S bonds to form alcohols and acids.

Repeated oral administration to volunteers of about 5 mg demeton (about 0.07 mg/kg body weight), caused inhibition of cholinesterase in the erythrocytes by a maximum of about 20 %. Demeton concentrations of 1 mg/kg diet or more (doses of about 0.05 mg/kg body weight and day) led in rats to the dose-dependent inhibition of acetylcholinesterase activity in the brain and blood. From doses of 0.9 mg/kg body weight and day clinical symptoms were observed.

Foetotoxic effects were seen in mice given intraperitoneal doses of demeton in the range of the LD_{50}.

Demeton increases the incidence of sister chromatid exchange (SCE) in V79 cells and induces chromosomal aberrations in the bone marrow of hamsters while prolonging the cell cycle. In the mouse the substance does not produce dominant lethal mutations.

2 Mechanism of Action

All the clinico-toxic effects of the compound are attributed to alkyl phosphorylation and the resulting acetylcholinesterase inhibition (Ballantyne and Marrs 1992, Henschler and

Lehnert 1986). This inactivation leads to the accumulation of acetylcholine in the synapses of the nervous system, the smooth and striated muscles and the secretory glands with the typical symptoms of organophosphate intoxication. For demeton there is no evidence of the delayed peripheral neuropathy (OPIDN: organophosphorus-ester-induced delayed neurotoxicity) described for other organophosphates (Richardson 1992).

3 Toxicokinetics and Metabolism

During metabolism, the sulfur bound to the phosphorus atom with a double bond can generally be replaced by oxygen; in this way the more active phosphates are formed. Organophosphates are absorbed rapidly via the mucous membranes of the respiratory and digestive tracts. Lipophilic organophosphates can also be absorbed via the skin. Many organophosphates are metabolized by monooxygenases, hydrolases and transferases (Ballantyne and Marrs 1992).

Mice given oral doses (12.5–50 mg/kg body weight) of the two ^{32}P-labelled demeton components excreted 50 % to 70 % of the radioactivity in the urine within 24 hours. The main routes of metabolism were the formation of sulfoxides and sulfones at the ethylthioethyl residue, the oxidation of demeton-O to form phosphate and the hydrolysis of the P-O and P-S bonds to form alcohols and acids (March *et al.* 1955).

The sulfoxides and sulfones of demeton-S are of the same toxicity as the initial compound (DuBois *et al.* 1956, Wirth 1958), while the sulfoxides and sulfones of demeton-O are less toxic (Wirth 1958).

4 Effects in Man

Four deaths and several cases of severe to slight accidental intoxication have been reported. The symptoms are typical of organophosphate intoxications and generally appear immediately, but can be delayed by up to 12 hours. After inhalation, the first effects appear in the respiratory tract and eyes, often within minutes. The effects on the respiratory tract can be a feeling of oppression in the chest and whistling noises while breathing as a result of bronchoconstriction and excessive bronchial secretion. The effects on the eyes include miosis, unclear recognition of distant objects and lacrimation. After oral uptake, the symptoms which developed within 15 minutes to 2 hours included loss of appetite, nausea, vomiting, cramps in the stomach area and diarrhoea (Proctor *et al.* 1988).

In 5 volunteers given demeton in increasing oral doses of up to 5.25 mg for 30 days, the activity of the cholinesterases in plasma and in the erythrocytes was reduced by up to 20 % (Moeller and Rider 1963). After doses of 7.125 mg the reduction in the cholinesterase activity in plasma was 39.8 % and in the erythrocytes 15.9 % (Moeller and Rider 1965).

5 Animal Experiments and *in vitro* Studies

There are no data available for the effects of demeton on the skin and mucous membranes, or the allergenic effects or carcinogenicity of the substance.

5.1 Acute toxicity

Six rats were exposed once to demeton concentrations of 18 mg/m^3. The animals died within 50 to 90 minutes (Deichmann and Rakoczy 1955).

Table 1 shows the LD_{50} values for demeton found in various animal species after different modes of administration.

The greater toxicity for the dog of the mixture compared to that of demeton-S is attributed to toxic effects of unknown impurities. The use of corn oil as the vehicle for oral administration to rats leads to delayed absorption of demeton-O and thus to much lower toxicity (Deichmann and Rakoczy 1955) than after administration in an aqueous solution (Wirth 1958).

Comparison of the intraperitoneal and oral LD_{50} values reveals that the substance is absorbed well after oral administration. Symptoms of intoxication were excitability, salivation, lacrimation, defaecation, muscle twitches, convulsions, prostration, paralysis, coma and peripheral respiratory arrest. Depending on the dose, the animals died between one hour and four days after administration (Deichmann and Rakoczy 1955, DuBois *et al.* 1956). Local cardiotoxic effects are thought to be partially responsible for the lethal effects (Deichmann and Rakoczy 1955).

5.2 Subacute, subchronic and chronic toxicity

5.2.1 Inhalation

Seventeen rats were exposed in whole animal exposure chambers to demeton concentrations of 3 mg/m^3 for 2 hours a day. Effects were not observed during the first exposure. During the second exposure tremor was observed, which was more pronounced at the third exposure. In addition, lacrimation was seen. During the fourth exposure 10 of the 17 animals died. Twenty rats were exposed to demeton concentrations of 3 mg/m^3 for 1 hour a day. The cholinesterase activity in the brain, erythrocytes and plasma was determined in 2 rats after 2, 4, 6 and 12 exposures. As the number of exposures increased, the cholinesterase activity decreased in the brain from 72 % to 42 %, in the erythrocytes from 73 % to 53 % and in plasma from 84 % to 63 % of the normal values. After 2 exposures, no clinical signs of intoxication were detected, after 4 exposures slight tremor was observed. Pronounced tremor and lacrimation were observed after the sixth exposure, and one animal died. Of the remaining 19 animals 7 died after the seventh to twelfth exposures (Deichmann and Rakoczy 1955).

Table 1. LD_{50} values for various species after administration of the demeton components either individually or as a mixture

Species	Administration route	LD_{50} (mg/kg body weight)			References
		demeton-O	demeton-S	mixture	
rat	peroral	♀ 117	♀ 1.5	♀ 3–5 * ♂ 5–6 *	Deichmann and Rakoczy 1955
rat	peroral	♂ 7.5	♂ 1.5		Wirth 1953
rat	peroral	♂ 30	♂ 1.5		Wirth 1958
rat	peroral			♂ 10 ♀ 4	Barnes and Denz 1954
rat	peroral			♂ 6.2 ♀ 2.5	Gaines 1960
rat	peroral		♂/♀ 1.7 *		DuBois *et al.* 1956
rat	epicutaneous, without occlusion			♂ 14 ♀ 8.2	Gaines 1960
rat	intraperitoneal			♂/♀ 9 *	Deichmann and Rakoczy 1955
rat	intraperitoneal		♂/♀ 1.5 *		DuBois *et al.* 1956
mouse	intraperitoneal			♀ 14	Budreau and Singh 1973
mouse	intraperitoneal		♂ 5.6 *		DuBois *et al.* 1956
mouse	subcutaneous	♂/♀ 120 *	♂/♀ 7 *	♂/♀ 10 *	Deichmann and Rakoczy 1955
guinea pig	subcutaneous			♂/♀ 5 *	Deichmann and Rakoczy 1955
guinea pig	intraperitoneal		♂ 5.5 *		DuBois *et al.* 1956
rabbit	peroral	♂/♀ 50 *			Deichmann and Rakoczy 1955
dog	subcutaneous	♂/♀ 37 *	♂/♀ 16 *	♂/♀ 6–8 *	Deichmann and Rakoczy 1955
sheep	peroral			> 5	McCarty *et al.* 1967

* approximate lethal dose

5.2.2 Ingestion

Groups of 5 rats were given gavage doses of demeton of 0.4, 0.66, 0.9 or 1.89 mg/kg body weight 65 times in 90 days. The animals of the control group received corn oil. Symptoms of intoxication (hyperexcitability and tremor from day 21) were observed only in the animals of the two highest dose groups. One animal from the highest dose group died after 17 doses. Gross pathological and histological examination of the heart, liver, lungs, spleen and a kidney from each of two animals per dose group yielded no evidence of damage. According to this study, demeton doses of 0.66 mg/kg body weight and day did not cause adverse effects. The acetylcholinesterase activity was not investigated (Deichmann and Rakoczy 1955).

Groups of 18 female rats were given demeton in concentrations of 1, 3 or 10 mg/kg in the diet over a period of 11 weeks. With an assumed average daily uptake of 10 g food and average body weights of 200 g, this corresponds to doses of 0.05, 0.15 or 0.5 mg/kg body weight and day. Even in the lowest dose group a reduction in the cholinesterase activity of 7 % in the brain and of 10 % in the blood was recognizable. After 10 mg/kg diet the activity was reduced by 80 % and 79 %, respectively. Clinical symptoms of intoxication were not detected. Another 12 female rats per group were given demeton in concentrations of 10, 20 or 50 mg/kg diet over a period of 16 weeks. The animals of the 50 mg/kg group ingested demeton doses of 2.8 mg/kg body weight and day in week 4, that is 70 % of the acute lethal dose, and doses of 4.1 mg/kg body weight in week 9, which indicates habituation to the effects of demeton after repeated doses. In the first weeks the animals consumed less food than the controls, towards the end of the study, however, more than the controls, but were nevertheless of lower weight. In this group demeton caused pronounced symptoms, so that 3 animals had to be killed. The remaining animals recovered completely. At the end of the study the cholinesterase activity in the brains of the animals from the highest dose group was only 7 % of the control value and that in the blood only 8 %. These animals, however, did not have any symptoms of intoxication. Histopathological examination did not reveal any effects (no other details). In the animals of the other dose groups the inhibition of cholinesterase was correspondingly lower and also did not cause clinical symptoms (Barnes and Denz 1954).

Demeton concentrations of 50 mg/kg diet administered to rats for 112 days caused the typical cholinergic effects but no histopathological effects (no other details) (FAO 1964).

Groups of 4 guinea pigs were given daily doses of 0.0025 to 10 mg on 6 days a week for up to 8 weeks. Cholinesterase activity was inhibited at doses of 2.5 mg (about 5 mg/kg body weight) and more; at the next lower dose of 0.25 mg (about 0.5 mg/kg body weight) no inhibition was detected (Bär 1954).

One female and one male dog per dose group were given demeton in concentrations of 1, 2 or 5 mg/kg in the diet (corresponding to 0.025, 0.05 and 0,125 mg/kg body weight) for 24 weeks. The cholinesterase was inhibited in the erythrocytes from week 12 after 5 mg/kg and in plasma from week 16 after 2 mg/kg or more. In no case, however, was the value lower than 70 % of the control value (Frawley and Fuyat 1957).

5.2.3 Intraperitoneal injection

In mice given intraperitoneal doses of demeton-S of 0.75 mg/kg body weight on three consecutive days, the inhibition of acetylcholinesterase increased with time (Du Bois *et al.* 1956).

5.3 Reproductive and developmental toxicity

First of all, the LD_{50} was determined as 14 mg/kg body weight in a study with 12 female mice given demeton by intraperitoneal injection. In the subsequent experiments, female mice (4–9 animals per group) were given single intraperitoneal doses of demeton of 7, 10 or 14 mg/kg body weight on one day between days 7 and 12 of gestation. The foetal weights were significantly reduced after treatment on day 9 or later and after doses of 10 mg/kg body weight or more. In another experiment, single intraperitoneal doses (10 mg/kg body weight) were administered on three consecutive days (days 7–9, days 8–10 or days 9–11 of gestation). No macroscopic structural anomalies were detected. After staining with alizarin red, three slight skeletal anomalies were found. After treatment of the dams with demeton on days 8 to 10 or 9 to 11 of gestation, the weights of the 16-day-old foetuses were significantly lower than those of the controls. No effects of demeton on the body weight gains of the dams were observed in these experiments (Budreau and Singh 1973).

5.4 Genotoxicity

With demeton concentrations of 10 to 160 µg/ml medium, a significant concentration-dependent increase in the incidence of SCE was observed in V79 cells and a delay in the cell cycle from 80 µg/ml (Chen *et al.* 1981). Groups of 6 female hamsters were given intraperitoneal doses of demeton of 1, 2, 5 or 10 mg/kg body weight. After 24 hours a dose-dependent increase in chromosomal aberrations (without gaps) was found in bone marrow from animals given doses of 2 mg/kg body weight or more (Dzwonkowska and Hübner 1986). 20 adult male mice per group were used in a test for the induction of dominant lethal mutations. The animals were given single intraperitoneal doses of demeton-O or treated on five consecutive days. The dose of 2 mg/kg body weight corresponded to about 1/5 of the LD_{50}. Each treated animal was mated 12 times over 4 days with one female each time. The numbers of total implantations, live implantations and dead implantations were not affected by the treatment (Dzwonkowska and Hübner 1991).

6 Manifesto (MAK value/classification)

The critical effects of occupational exposure to demeton are thought to be the inhibition of acetylcholinesterase and its resulting symptoms. Adverse clinical effects are to be expected on inhibition of more than 30 % of the acetylcholinesterase activity in erythrocytes (Henschler and Lehnert 1986). The relationship between the external exposure concentration of demeton and the extent of acetylcholinesterase inhibition in man is unknown. Also the results from experiments with animals cannot be used to evaluate a MAK value. Therefore the MAK value for demeton has been withdrawn and the substance is listed in Section IIb of the *List of MAK and BAT Values*.

There is no information available for the carcinogenic potential of demeton. In V79 cells demeton increased the incidence of SCE. In hamsters chromosomal aberrations were found in bone marrow. A test for the induction of dominant lethal mutations did not yield evidence of germ cell mutagenicity.

Designation with an "H" has been retained as a result of the high dermal toxicity of the substance in the rat and the ready penetration of the skin known for other organophosphates.

As there are insufficient data for the sensitizing potential of demeton, it cannot be decided whether the substance should be designated with an "S".

7 References

Ballantyne B, Marrs TC (1992) Overview of the biological and clinical aspects of organophosphates and carbamates. in: Ballantyne B, Marrs TC (Eds) *Clinical and experimental toxicology of organophosphates and carbamates*, Butterworth-Heinemann, Oxford, 3–14

Bär F (1954) Fütterungsversuche mit Systox-behandelten Zuckerrüben (Feeding studies with sugar beet treated with systox) (German). *Arzneim-Forsch 4*: 668–672

Barnes JM, Denz FA (1954) The reaction of rats to diets containing octamethyl pyrophosphoramide (Schradan) and *OO*-diethyl-*S*-ethylmercaptoethanol thiophosphate (Systox). *Br J Ind Med 11*: 11–19

Budreau CH, Singh RP (1973) Teratogenicity and embryotoxicity of Demeton and Fenthion in CF1 mouse embryos. *Toxicol Appl Pharmacol 24*: 324–332

Chen HH, Hsueh JL, Siriani SR, Huang CC (1981) Induction of sister-chromatid exchanges and cell cycle delay in cultured mammalian cells treated with eight organophosphorus pesticides. *Mutat Res 88*: 307–316

Deichmann WB, Rakoczy R (1955) Toxicity and mechanism of action of Systox. *Arch Ind Health 11*: 324–331

DuBois KP, Murphy SD, Thursh DR (1956) Toxicity and mechanism of action of some metabolites of Systox. *Arch Ind Health 13*: 606–612

Dzwonkowska A, Hübner H (1986) Induction of chromosomal aberrations in the Syrian hamster by insecticides tested *in vivo*. *Arch Toxicol 58*: 152–156

Dzwonkowska A, Hübner H (1991) Studies on commercial insecticides with the dominant lethal mutations test. *Pol J Occup Med 4*: 43–53

FAO (Food and Agricultural Organization of the united Nations) (1964) *Evaluation of the toxicity of pesticide residues in food*. FAO meeting report No. PL/1963/13, WHO, Rome 1964, 75–80

Frawley JP, Fuyat HN (1957) Effect of low dietary levels of parathion and Systox on blood cholinesterase of dogs. *J Agric Food Chem 5*: 346–348

Gaines TB (1960) The acute toxicity of pesticides to rats. *Toxicol Appl Pharmacol 2*: 88–99

Henschler D, Lehnert G (Eds) (1986) Acetylcholinesterase-Hemmer. *Biologische Arbeitsstoff-Toleranz-Werte (BAT-Werte) und Expositionsäquivalente für krebserzeugende Arbeitsstoffe (EKA)*, 3rd issue, and in English translation (Acetylcholinesterase inhibitors) in *Biological Exposure Values for Occupational Toxicants and Carcinogens* Volume 2, VCH-Verlagsgesellschaft, Weinheim

March RB, Metcalf RL, Fukuto TR, Maxon MG (1955) Metabolism of systox in the white mouse and American cockroach. *J Econ Entomol 48*: 355–363

McCarty RT, Haufler M, McBeth CA (1967) Acute toxicity of Carbophenothion and Demeton in sheep. *Am J Vet Res 28*: 507–510

Moeller HC, Rider JA (1963) Further studies on the toxicity of systox and methyl parathion. *Fed Proc 22*: 189

Moeller HC, Rider JA (1965) Further studies on the anticholinesterase effect of systox and methylparathion in humans. *Fed Proc 24 (2, Part 1)*: 641

Proctor, NH, Hughes JP, Fischman ML (1988) *Chemical hazards of the workplace, JB Lippincott Company*, PD, USA, 176–179

Richardson RJ (1992) Interactions of organophosphorus compounds with neurotoxic esterase. in: Chambers JE, Levi PE (Eds) *Organophosphates. Chemistry, fate, and effects*, Academic Press, San Diego, CA, USA, 299–323

Wirth W (1953) Zur Pharmakologie der Phosphorsäureester (Pharmacology of phosphoric acid esters) (German). *Naunyn-Schmiedebergs Arch Exp Pathol Pharmakol 217*: 144–152

Wirth W (1958) Zur Wirkung System-insecticider Phosphorsäure-Ester im Warmblüter-Stoffwechsel (The action of systemic insecticidal phosphoric acid esters in the metabolism of warm-blooded animals) (German). *Naunyn-Schmiedebergs Arch Exp Pathol Pharmakol 234*: 352–363

completed 02.03.1999

4,6-Dinitro-*o*-cresol

MAK value (1997)	**not yet established, see Section IIb of the *List of MAK and BAT Values***
Peak limitation	**–**
Absorption through the skin (1958)	**H**
Sensitization	**–**
Carcinogenicity	**–**
Prenatal toxicity	**–**
Germ cell mutagenicity	**–**
BAT value	**–**
Synonyms	4,6-dinitro-*ortho*-cresol DNOC
Chemical name (CAS)	2-methyl-4,6-dinitrophenol
CAS number	534-52-1
Structural formula	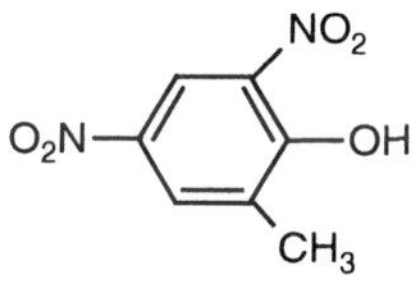
Molecular formula	$C_7H_6N_2O_5$
Molecular weight	198.13
Melting point	88.2–89.9°C
Boiling point	312°C

The MAK value of 0.2 ml/m^3 for 4,6-dinitro-*o*-cresol (as inhalable dust) which was valid until 1997 was established in 1958 in analogy to the TLV value at the time. The present documentation is based on reviews of 4,6-dinitro-*o*-cresol (ACGIH 1991, DFG 1976, Hayes and Laws 1991, SEG 1992) and on a preliminary draft "Environmental Health Criteria" document for the substance (WHO 1997).

1 Toxic Effects and Mode of Action

4,6-Dinitro-*o*-cresol is readily absorbed after inhalation or ingestion and especially readily on skin contact; it is eliminated mainly via the urine. The elimination half-time varies from species to species; it can be as much as 7 days for man, considerably longer than the 1.5 days for the rat.

4,6-Dinitro-*o*-cresol is used as a herbicide; it is an uncoupler of the mitochondrial respiratory system. It causes an increase in basal metabolic rate with raised temperature and weight loss in man and animals. In man, the first symptoms can develop after ingestion of a 4,6-dinitro-*o*-cresol dose of about 1 mg/kg body weight or after attainment of blood concentrations of about 20 ppm. Doses between 5 and 60 mg/kg body weight or blood concentrations of 70 ppm or more are lethal. In a long-term study in which 4,6-dinitro-*o*-cresol was administered to rats in the diet, the no observed effect level (NOEL) was given as 0.59 mg/kg body weight and day for males and 0.75 mg/kg body weight and day for females because higher doses resulted in increased food consumption.

Only when 4,6-dinitro-*o*-cresol is administered in doses which are highly toxic for the dams does it cause embryotoxic effects such as increased resorption rates and malformations in the rat and rabbit.

After metabolic activation, 4,6-dinitro-*o*-cresol has mutagenic potential *in vitro. In vivo*, evidence of clastogenic effects was obtained with a herbicide containing 4,6-dinitro-*o*-cresol but not with the pure substance. A long-term study with rats yielded no evidence of carcinogenic effects.

2 Mechanism of Action

As with other dinitrophenol compounds, the effects of 4,6-dinitro-*o*-cresol are a result of the decoupling of cell respiration and phosphorylation (respiratory chain phosphorylation) (see DFG 1976).

3 Toxicokinetics and Metabolism

4,6-Dinitro-*o*-cresol is very rapidly absorbed after inhalation, ingestion and dermal contact; more than 90 % of the substance is found in the plasma.

In the rat and rabbit, 4,6-dinitro-*o*-cresol is metabolized mainly by reduction of the nitro groups to amino groups (6-amino-4-nitro-*o*-cresol and further to 4,6-diamino-*o*-cresol), by acetylation and *O*-conjugation (DFG 1976, Ingebrigtsen and Froslie 1980, Leegwater *et al.* 1982).

4,6-Dinitro-*o*-cresol is eliminated mainly in the urine. The elimination half-time depends on the species. In the rabbit it is about 6.5 hours (WHO 1997), in the rat after oral administration of ^{14}C-4,6-dinitro-*o*-cresol doses of 0.4 mg/kg body weight between 1 and 1.5 days (Leegwater *et al.* 1982), and in man after high level exposure 4 days (Thiele 1983) or 6 to 7 days (Jastroch *et al.* 1978).

4 Effects in Man

The symptoms of acute poisoning result from the effects on cell metabolism and include severe thirst, painful colic, diarrhoea and vomiting. Dyspnoea, cyanosis, tachycardia, adynamia and cardiac insufficiency with anginal pain have also been described. The first effect on the central nervous system is typically euphoria, later the persons suffer from dizziness, a tendency to collapse, anxiety and restlessness, confusion, unconsciousness and terminal spasms. Chronic intoxication is characterized by headaches, weakness, anorexia and marked weight loss. Later degenerative changes develop in the heart muscle, liver and kidneys, neuritis, leukopenia, Heinz bodies, and methaemoglobin; agranulocytosis and severe toxic dermatitis are sometimes seen as well. Liver damage is the main effect after ingestion of the substance. Subcapsular cataract formation can be a complication (DFG 1976).

In one case, exposure to 4,6-dinitro-*o*-cresol concentrations of 4.7 mg/m^3 in the air resulted in symptoms of intoxication with fever, raised basal metabolic rate, increased respiration rate and pulse, perspiration, shortness of breath and coughing. When the concentration in the air was reduced to 2.5 mg/m^3, the symptoms did not develop (ACGIH 1991).

There are a number of early reports of lethal poisoning of persons using 4,6-dinitro-*o*-cresol as herbicide and insecticide (DFG 1976).

Studies with volunteers have demonstrated that ingestion of 75 mg 4,6-dinitro-*o*-cresol (0.92–1.27 mg/kg body weight) was sufficient to produce symptoms such as raised temperature, perspiration, and increased respiration rate and pulse in some persons. A dose of 3.5 mg/kg body weight increased the basal metabolic rate by more than 30 % after 6 to 8 hours. The lethal dose for an adult is given as 0.35 to 3.0 g 4,6-dinitro-*o*-cresol (about 5–66 mg/kg body weight). The use of 4,6-dinitro-*o*-cresol to treat adiposity has resulted in deaths on several occasions (DFG 1976).

Because 4,6-dinitro-*o*-cresol is very readily absorbed through the skin, the concentration of the substance in blood is the best parameter for the establishment of dose-response relationships.

Toxic effects were not seen either in healthy employees of a chemical factory whose blood 4,6-dinitro-*o*-cresol levels averaged 3 ppm (1.0–8.7 ppm) (Hayes and Laws 1991) or in volunteers whose blood levels were in the range up to 10 ppm after single doses of 75 mg of the substance. However, repeated administration of doses of 75 mg resulted after 3 to 5 days in an increase in the blood level to 20 ppm and in discoloration of the conjunctiva. After 5 to 7 days of dosing, the blood levels in two persons were transiently over 40 ppm and caused tiredness, headaches and nausea. Even 40 days later the levels in blood were still in the range between 1 and 1.5 ppm (Harvey *et al.* 1951). In persons using the substance in agriculture, even higher blood levels have occasionally been described; the persons did not always develop symptoms, presumably because of cold outdoor temperatures. Blood levels of 70 ppm and higher have been described as potentially lethal (Hayes and Lawes 1991).

In a patient who developed generalized psoriasiform and lichenoid dermatitis after percutaneous 4,6-dinitro-*o*-cresol poisoning, no evidence of an immunological genesis of the disorder was found (Mali and Grosfeld 1965).

Also in the case of another patient who developed extensive exfoliative contact dermatitis after prolonged intensive contact with a pesticide concentrate (98 % 4,6-dinitro-*o*-cresol), the report provided no evidence of allergenic effects of the substance (Jung 1979). The same applies for the case of a man with onychopathy after several hours handling a pesticide containing 5 % 4,6-dinitro-*o*-cresol (Baran 1974) and for an undocumented case report of dermatitis in an agricultural worker after contact with 2,4-dinitro-*o*-cresol (Szegö 1965).

In patch tests carried out on 200 and 492 patients with suspected allergic contact dermatitis or other not allergic skin disorders, no reactions to 0.5 % or 1 % 4,6-dinitro-*o*-cresol were seen (Lisi *et al.* 1986, 1987). Only one of 303 patients produced a reaction in the patch test with 2,4-dinitrophenol, a substance which is structurally very similar to 4,6-dinitro-*o*-cresol; the latter substance was not tested in this group of patients (IVDK 1997).

A patient sensitized to dinitrohalobenzene derivatives did not produce a reaction to 0.1 % 4,6-dinitro-*o*-cresol (Björnberg and Mobacken 1973). Likewise, no reactions to 2,4-dinitro-*o*-cresol (1 % to 2 %) were seen in patch tests on 27 persons sensitized experimentally to dinitrochlorobenzene or on 15 patients with chloramphenicol sensitization (Eriksen 1978).

There are no data available for allergenic or irritative effects on the respiratory passages.

5 Animal Experiments and *in vitro* Studies

5.1 Acute toxicity

The 4-hour LC_{50} determined for 4,6-dinitro-*o*-cresol in the rat was 230 mg/m^3 (Dey-Hazra and Heisler 1981) and in the cat < 40 mg/m^3 (WHO 1982). The oral LD_{50} for the rat is given as 25 to 85 mg/kg body weight (Ben-Dyke *et al.* 1970, DFG 1976). The dermal LD_{50} for the rat was found to be 200 to 600 mg/kg body weight (Ben-Dyke *et al.* 1970) and for the rabbit 1000 mg/kg body weight (WHO 1982).

5.2 Subacute, subchronic and chronic toxicity

The results of studies of subacute, subchronic and chronic toxicity of 4,6-dinitro-*o*-cresol are shown in Table 1.

Repeated administration of 4,6-dinitro-*o*-cresol to rats, mice and dogs in the diet did not produce any abnormal findings in the clinical or histopathological examinations. Administered to rats, doses of about 5 mg/kg body weight and day resulted in reduced body weights and occasional deaths. Doses of at least 100 mg/kg body weight and day were required to reduce body weight gains in mice. Biochemical studies revealed reductions in T_3 and T_4 levels, increased blood urea levels and changes in other biochemical parameters in the rat and mouse.

The NOEL for the rat after dosing for 104 weeks was given as 0.59 mg/kg body weight and day. Doses of 4.12 mg/kg body weight and day caused increased food consumption by male animals. The no observed adverse effect level (NOAEL) for dogs given the substance for 90 days was 0.89 mg/kg body weight and day because food consumption was increased at 4.83 mg/kg body weight and day.

5.3 Local effects on skin and mucous membranes

Application of 105 mg of 4,6-dinitro-*o*-cresol during 9 days caused slight irritation of rabbit skin. 20 mg applied for 24 hours was moderately irritating (Lewis 1992).

5.4 Allergenic effects

There are no data available for allergenic effects of 4,6-dinitro-*o*-cresol.

Three to four weeks after intracutaneous injection of 0.1 ml 0.01 % 2,4-dinitrophenol, daily on 10 consecutive days, no reactions were produced in "at least two" guinea pigs (no other details) by application without occlusion of 1 % 2,4-dinitrophenol in arachis oil (Griepentrog 1959).

In an earlier, incompletely documented study, animals sensitized to 2,4-dinitrochlorobenzene by intracutaneous induction produced only weak or questionable cross-reactions to 2,4-dinitrophenol. Attempts to produce sensitization with 2,4-dinitrophenol also yielded questionable results (no other details) (Landsteiner and Jacobs 1935).

Groups of 5 male Hartley guinea pigs were sensitized to 2,4-dinitrochlorobenzene by one of three different induction procedures (maximization test, single topical induction, or five topical induction treatments); provocation with 0.4 % 4,6-dinitro-*o*-cresol or 0.4 % 2,4-dinitrophenol did not produce any reactions (Nishioka *et al.* 1983).

Table 1. Effects of repeated exposure of experimental animals to 4,6-dinitro-*o*-cresol

Species strain sex	Administration route and dose	Effects	References
cat 3	inhalation, **60, 90 days,** 0.2 mg/m^3 (4 h/day); **30 days,** 2 mg/m^3 (4 h/day)	**0.2 mg/m^3:** body temperature slightly raised, slight increases in leukocyte count and erythrocyte sedimentation rate, slight decreases in haemoglobin concentration, erythrocyte count, catalase and peroxidase activities **2 mg/m^3:** mortality increased (2/3)	Burkatskaya 1965
rat (from Charles River) 5 ♂, 5 ♀	oral (diet), **6 weeks,** 0, 5, 13, 32, 80, 200 mg/kg diet (0, 0.44, 1.17, 2.89, 7.24, 18.6 mg/kg body weight and day)	**1.17 mg/kg:** NOEL **2.89 mg/kg:** no data **from 7.24 mg/kg:** decreased body weight gain (♀), decreased alanine amino transferase (ALT) activity, increased blood urea (♀)	Broadmeadow 1988
rat Wistar 10 ♂ 10 ♀	oral (diet), **90 days,** 0, 50, 100, 200, 400 mg/kg diet (0, 2.5, 5, 10, 20 mg/kg body weight and day)	**from 2.5 mg/kg:** pyruvate, T_3 and T_4 decreased **from 5 mg/kg:** decreased body weights, increased mortality, increased blood urea (♂) **from 10 mg/kg:** decreased food consumption, increased blood glucose, increased blood urea **20 mg/kg:** increased ALT, inactive thyroid, Langerhans' cells atrophic, no *corpora lutea* in the ovaries, reduced spermatogenesis, thymus atrophic	den Tonkelaar *et al.* 1983
rat F344 50 ♂ 50 ♀	oral (diet), **104 weeks,** 0, 2.5, 15, 100 mg/kg diet (0, 0.12, 0.75, 5.3 (♀) and 0, 0.10, 0.59, 4.12 mg/kg body weight and day (♂))	**0.59 mg/kg:** NOEL **4.12 mg/kg:** increased food consumption (♂)	Broadmeadow 1991
mouse n.s. 12 ♂ 12 ♀	oral (diet), **4 weeks,** 0, 10, 50, 100 mg/kg body weight and day	**from 10 mg/kg:** alkaline phosphatase and aspartate aminotransferase (AST) slightly increased (♂), T_3 and T_4 changed (not dose-dependent) **50 mg/kg:** decreased blood glucose **100 mg/kg:** decreased body weights	Broadmeadow 1996
mouse n.s. 18 ♂ 18 ♀	oral (diet), **13 weeks,** 0, 1, 5, 10 mg/kg body weight and day	**1 mg/kg:** no effects **from 5 mg/kg:** T_4 slightly decreased (♂), T_3 decreased (♀)	Kelly 1995
dog beagle 16 ♂ 16 ♀	oral (diet), **90 days,** 0, 4, 20, 100 ppm in the diet (0, 0.17, 0.89, 4.82 mg/kg body weight and day)	**0.17 mg/kg:** increased prothrombin time (on day 49; relevance questionable) **0.89 mg/kg:** NOAEL **4.83 mg/kg:** increased food consumption	Til 1980

Marzin (1991d) *Etude de l'activité génotoxique du produit dinitro-ortho-cresol par la recherche d'aberrations chromosomiques par analyse de métaphases sur lymphocytes humains en culture* (Study of the genotoxic activity of the product dinitro-*ortho*-cresol by testing for chromosomal aberrations in cultured human lymphocytes in metaphase) (French). Institut Pasteur de Lille study No. IPL-R 911019, Lille, France, unpublished report prepared for Elf Atochem Agri SA; cited in WHO (1997)

Nehéz M, Selypes A, Páldy A, Berencsi G (1978a) The mutagenic effect of a dinitro-*o*-cresol containing pesticide on mice germ cells. *Ecotoxicol Environ Saf 2*: 401–405

Nehéz M, Selypes A, Páldy A, Berencsi G (1978b) Recent data on the examination of the mutagenic effect of a dinitro-*o*-cresol-containing pesticide by different test methods. *Ecotoxicol Environ Saf 2*: 243–248

Nehéz M, Páldy A, Selypes A, Scheufler H, Berencsi G, Freye H-A (1981) The teratogenic and mutagenic effects of dinitro-*o*-cresol-containing herbicide on the laboratory mouse. *Ecotoxicol Environ Saf 5*: 38–44

Nehéz M, Selypes A, Páldy A, Mazzag E, Berencsi G, Jarmay K (1982) The effects of five weeks treatment with dinitro-*o*-cresol or trifluralin containing pesticides on the germ cells of male mice. *J Appl Toxicol 2*: 179–180

Nehéz M, Selypes A, Mazzag E, Berencsi G (1984) Additional data on the mutagenic effect of dinitro-*o*-cresol-containing herbicides. *Ecotoxicol Environ Saf 8*: 75–79

Nishioka K, Asagami C, Kurata M, Fujita H (1983) Sensitivity to the weed killer DNA-nitralin and cross-sensitivity to dinitrochlorobenzene. *Arch Dermatol 119*: 304–306

SEG (Scientific Expert Group on Occupational Exposure Limit) (1992) *4,6-Dinitro-o-cresole.* SEG/CDO/29, Commission of the European Communities, Directorate General Employment, Industrial Relations and Social Affairs, Health and Safety Directorate, Luxembourg

Szegö L (1965) Landwirtschaftlicher Bildbericht (Agricultural report) (German). *Z Haut Geschlechtskr 38*: I–IV

Thiele E (1983) Dose-effect relations in the occupational handling of DNOC. *Wiss Z Ernst-Moritz-Arndt-Univ Greifsw Med Reihe 32*: 7–10

Til HP (1980) Sub-chronic (90-day) oral toxicity study with DNOC in dogs. CIVO TNO Instituut study No. B 80/0359, Zeist, The Netherlands, unpublished report prepared for Pennwalt Holland bv; cited in WHO (1997)

den Tonkelaar EM, Van Leeuwen FXR, Kuiper C (1983) Semichronic toxicity testing of DNOC in the rat. *Med Fac Landbouww, Rijksuniv. Gent, 48*: 1015–1022; cited in SEG (1992)

WHO (World Health Organization) (1982) *Recommended health limits in occupational exposure to pesticides. Technical report series 677*, WHO, Genf

WHO (World Health Organization) (1997) *Draft Environmental Health Criteria (EHC) for DNOC (dinitro-ortho-cresol).* First draft 21st June 1996, unpublished

completed 07.04.1997, allergenic effects 29.01.1998

Ferbam

MAK value (1999)	**not yet established, see Section IIb of the *List of MAK and BAT Values***
Peak limitation	–
Absorption through the skin	–
Sensitization	–
Carcinogenicity	–
Prenatal toxicity	–
Germ cell mutagenicity	–
BAT value	–
Synonyms	tris(dimethyldithiocarbamato)iron dimethyldithiocarbamic acid iron salt ferric dimethyldithiocarbamate
Chemical name (CAS)	tris(dimethylcarbamodithioato-*S,S'*)iron
CAS number	14484-64-1
Structural formula	$[(CH_3)_2N–CS–S]_3Fe$
Molecular formula	$C_9H_{18}FeN_3S_6$
Molecular weight	416.5
Melting point	180°C

The MAK value for ferbam which was valid until 1999 (15 mg/m^3 (I)) was established in 1957 in analogy to the TLV value at the time. In the meantime, studies have appeared which made it necessary to re-evaluate the substance. The present documentation is based on reviews of the toxicological data for ferbam (ACGIH 1992, Savolainen 1991).

1 Toxic Effects and Mode of Action

Ferbam is irritative to the eyes, skin and respiratory passages in man. In rats, ferbam causes ataxia, reduced muscle tone, reduced motor activity, clonic cramps, paralysis of the hind extremities and an increase in the incidence of thyroid metaplasia. The studies available have not yielded evidence of carcinogenic effects. Under slightly acidic conditions, as found in the stomach, ferbam can react with nitrite to form *N*-nitrosodimethylamine, which has been shown to be carcinogenic in various species. In feeding studies with maternally toxic doses, ferbam was found to be embryotoxic.

There are no data available for the mode of action.

2 Toxicokinetics and Metabolism

2.1 Absorption, distribution, elimination

In the gastrointestinal tract of rats, 40 % to 70 % of an oral dose of ^{35}S-ferbam of 500 mg/kg body weight was absorbed within 24 hours; 22.7 %, 18.1 % and 16.9 % of the radioactivity, respectively, was detected in the urine, in the exhaled air and in the faeces. After rats were given oral ^{14}C-ferbam doses of 500 mg/kg body weight, 42.9 % of the radioactivity was recovered in the urine, 20 % in the faeces, 9.8 % in the gastrointestinal tract and 0.6 % in the exhaled air. Radioactivity was detected also in the mother's milk, foetuses and amniotic fluid (Hodgson *et al.* 1975). The dose-dependent accumulation of iron in bone marrow was demonstrated in a 2-year feeding study with rats (Hodge *et al.* 1956).

2.2 Metabolism

After rats were given oral ^{14}C-ferbam doses of 500 mg/kg body weight, the metabolites found in the exhaled air were CS_2 and traces of COS (Hodgson *et al.* 1975).

3 Effects in Man

Exposure to ferbam at the workplace caused irritation of the eyes and respiratory tract and in some cases induced dermatitis. High doses caused damage in the gastrointestinal tract. Experiments with animals suggest that high doses can have effects on the central nervous system (ACGIH 1992).

4 Animal Experiments and *in vitro* Studies

There are no data available for the toxicity of inhaled ferbam.

4.1 Acute toxicity

One of ten rats died after gavage doses of ferbam of 17 g/kg body weight. For guinea pigs lethal concentrations are given in the range from 450 to 2000 mg/kg body weight, for rabbits from 2000 to 3000 mg/kg body weight (ACGIH 1992), for rats from 2700 to 3500 mg/kg body weight and for mice from 3100 to 3700 mg/kg body weight (Savolainen 1991).

4.2 Subacute, subchronic and chronic toxicity

Groups of 20 male rats were given daily ferbam doses of 23, 66, 109, or 331 mg/kg body weight with the diet for 13 weeks. In the highest dose group all the animals died within the first week. From the lowest concentration, body weights and food consumption were reduced. In the two medium dose groups, pigmentation of the reticulo-endothelial cells in the spleen and the mesenteric lymph nodes developed.

Groups of 20 male and 20 female rats were exposed to ferbam doses of 9, 37 or 96 mg/kg body weight in the diet for 80 weeks. Body weights and food consumption were reduced in all dose groups, in the males more so than in the females. From doses of 37 mg/kg body weight, the spleen, testis and thyroid weights of the male animals were reduced. Dose-dependent metaplasia was observed in the thyroid gland and fat deposition in the pancreas. Ferbam doses of 96 mg/kg body weight caused ataxia, paralysis of the hind extremities, alopecia and severe weight loss in three female rats. Clinicochemical parameters were not affected (Lee *et al.* 1978)

4.3 Reproductive and developmental toxicity

Groups of 24 male rats were given ferbam in concentrations of 0.5 %, 0.12 % or 0.25 % (23, 66, 109 mg/kg body weight) in the diet for 13 weeks and were then mated with untreated females. Female animals were given ferbam doses of 15 or 51 mg/kg body weight with the diet two weeks before mating with untreated male animals. In the highest dose groups in the two experiments body weights and food consumption were reduced and 6 male animals died. In neither experiment were effects observed on the fertility and gestation index, or on the survival of the offspring (Short *et al.* 1976).

Male F_1 mice were given either oral or intraperitoneal ferbam doses of 25, 50 or 100 mg/kg body weight once a day for 5 days. After 35 days no effects were observed on the testis weights or sperm count. A significant increase in sperm anomalies was determined only in the highest dose group after oral administration (Quinto *et al.* 1989).

Rats were given gavage doses of ferbam of 11 or 114 mg/kg body weight from days 6 to 15 of gestation. In the 114 mg/kg group the body weights of the dams and the litter sizes were reduced, and the number of resorptions was increased. The foetal weights decreased, and the following anomalies were observed: breastbone not ossified (20/31), slightly collapsed and deformed skull (9/31), hydrocephalus (2/35) and haematomas (2/73). In mice given ferbam doses of 23 or 228 mg/kg body weight on days 6 to 12 of gestation, body weights and survival were unchanged. Also, no effects were seen on the litter size, the number of resorptions or the foetal weights. In the high exposure group the following anomalies were seen: thin ventricle walls (8/88), hydroureter (13/88), hydronephrosis, slightly collapsed skull, extra ossification of the supra-occipital and deformation of the breastbone (Short *et al.* 1976).

Pregnant rats were given ferbam doses of 7.6 or 41 mg/kg body weight with the diet for 16 days before and 23 days after giving birth or 152 mg/kg body weight for 21 days after giving birth. In the high dose group, food consumption and body weights were reduced. Survival and body weights of the litters were reduced (Short *et al.* 1976).

Pregnant rats were given ferbam doses of 150 mg/kg body weight from days 6 to 15 of gestation. Some animals died, the number of resorptions increased, foetal weights decreased and the number of animals with skeletal alterations increased. No teratogenic effects were observed in pregnant mice given 30 or 300 mg/kg body weight from days 6 to 14 of gestation (Minor *et al.* 1974).

4.4 Genotoxicity

In *Salmonella* mutagenicity tests with the strains TA98, TA100, TA1535, TA1537 and TA1538, 10 to 1500 µg ferbam per plate was not found to be mutagenic either with or without an added metabolic activation system (De Lorenzo *et al.* 1978). In *Salmonella* mutagenicity tests with the strains TA100 and TA1535, 1000 to 5000 µg ferbam per plate was found to be mutagenic without metabolic activation, but not with strains TA1537, TA1538, TA98 or *Escherichia coli* WP2 (Moriya *et al.* 1983).

4.5 Carcinogenicity

Groups of 18 male and 18 female mice were given daily gavage doses of ferbam of 10 mg/kg body weight in gelatine from days 7 to 28 of life. From day 29 they received ferbam doses of 32 mg/kg body weight with the diet for 78 weeks. There was no significant increase in the tumour incidence relative to that in the control animals (Innes *et al.* 1969).

After administration of ferbam to rats for 80 weeks, no significant increases in tumour incidences were observed (see Section 4.2; Lee *et al.* 1978). The study is, however, insufficiently documented and therefore unsuitable for evaluating the carcinogenic effects of ferbam.

Groups of 25 male and 25 female rats received ferbam in concentrations of 25, 250 or 2500 mg/kg diet (doses of about 2.5, 25, 250 mg/kg body weight) for 2 years. In the highest dose group, body weights were reduced and mortality was increased. After 3 to 4 months, neurological effects were seen in all dose groups including the lowest. The blood and urine parameters, and organ weights were in the range of those of the control animals. In the highest dose group, lesions were observed in the brain and were characterized as cystic regions in the cerebellum. These changes in the brain were, however, not correlated with the neurological effects. Chronic lung and kidney inflammation were frequent in the older rats. Histopathological examination yielded no evidence of increased tumour incidences (Hodge *et al.* 1956).

In a one-year feeding study with beagle dogs (0.5, 5, 25 mg/kg body weight), neither histopathological changes nor increased tumour incidences were observed (Hodge *et al.* 1956).

In the feeding studies, ferbam was, therefore, not found to be carcinogenic.

5 Manifesto (MAK value/classification)

A NOEL (no observed effect level) for ferbam cannot be deduced from the data available from experiments with animals and information about the effects of the substance in man. Earlier studies did not yield evidence of carcinogenic effects. In addition to negative results in *Salmonella* mutagenicity tests, also positive results have been obtained. There are no data for the genotoxicity of ferbam *in vivo*. A conclusive evaluation of the genotoxicity of the substance is, therefore, not possible. The former MAK value of 15 mg/m^3 (I) has, therefore, been withdrawn. Ferbam has been listed in Section IIb of the *List of MAK and BAT Values*.

6 References

ACGIH (American Conference of Governmental Industrial Hygienists) (1992) *Ferbam, Documentation of BEIs and TLVs*, ACGIH, Cincinnati, OH, USA

De Lorenzo F, Staiano N, Silengo L, Cortese R (1978) Mutagenicity of diallate, sulfallate, and triallate and relationship between structure and mutagenic effects of carbamates used widely in agriculture. *Cancer Res 38*: 13–18

Hodge HC, Maynard EA, Downs WL, Coye RD, Steadman LT (1956) Chronic toxicity of ferric dimethyldithio-carbamate (ferbam) and zinc dimethyldithio-carbamate (ziram). *J Pharmacol Exp Ther 118*: 174–181

Hodgson JR, Hoch JC, Castles TR, Helton DO, Lee CC (1975) Metabolism and disposition of ferbam in the rat. *Toxicol Appl Pharmacol 33*: 505–513

Innes JRM, Ulland BM, Valerio MG, Petrucelli L, Fishbein L, Hart ER, Pallotta AJ (1969) Bioassay of pesticides and industrial chemicals for tumorigenicity in mice: a preliminary note. *J Nat Cancer Inst 42*: 1101–1114

Lee CC, Russell JQ, Minor JL (1978) Oral toxicity of ferric dimethyl-dithiocarbamate (ferbam) and tetramethylthiuram disulfide (thiram) in rodents. *J Toxicol Environ Health 4*: 93–106

Minor JL, Russell JQ, Lee CC (1974) Reproduction and teratology studies with the fungicide ferbam. *Toxicol Appl Pharmacol 29*: 120

Moriya M, Ohta T, Watanabe K, Miyazawa T, Kato K, Shirasu Y (1983) Further mutagenicity studies on pesticides in bacterial reversion assay systems. *Mutat Res 116*: 185–216

Quinto I, De Marinis E, Mallardo M, Arcucci A, Della Morte R, Staiano N (1989) Effect of DNOC, ferbam and imidan exposure on mouse sperm morphology. *Mutat Res 224*: 405–408

Savolainen K (1991) *Thiurams and dimethyldithiocarbamates*. in: Beije B, Lundberg P (Eds) Arbeite och Hälsa, 1991:2, Criteria documents from the Nordic Expert Group 1990, Arbeitsmiljöinstitutet, Solna, Schweden, 7–66

Short RD, Russel JQ, Minor JL, Lee CC (1976) Developmental toxicity of ferric dimethyldithiocarbamate and bis(dimethylthiocarbamoyl) disulfide in rats and mice. *Toxicol Appl Pharmacol 35*: 83–94

completed 01.03.1999

Formic Acid

MAK value (1966)	**5.0 ml/m^3 (ppm) ≙ 9.5 mg/m^3**
Peak limitation (2000)	**Category I, excursion factor 2**
Absorption through the skin	–
Sensitization	–
Carcinogenicity	–
Prenatal toxicity (1997)	**Pregnancy risk group C**
Germ cell mutagenicity	–
BAT value	–
Synonyms	aminic acid formylic acid hydrogen carboxylic acid methanoic acid
Chemical name (CAS)	formic acid
CAS number	64-18-6
Structural formula	HCOOH
Molecular formula	CH_2O_2
Molecular weight	46.03
Melting point	8°C
Boiling point	101°C
Density at 20°C	1.22 g/cm^3
Vapour pressure at 20°C	42 hPa
log P_{ow}*	–0.54
1 ml/m^3 (ppm) ≙ 1.91 mg/m^3	**1 mg/m^3 ≙ 0.524 ml/m^3 (ppm)**

The MAK value for formic acid of 5 ml/m^3 was established in 1966 in line with the TLV value at the time. In the meantime studies have appeared which make re-evaluation of the substance necessary. This documentation is based on a short report on formic acid from the Beratergremium für umweltrelevante Altstoffe (BUA 1991).

---

* *n*-octanol/water distribution coefficient

1 Toxic Effects and Mode of Action

Formic acid is a strong acid which has irritative to caustic effects on the mucous membranes, eyes and skin. Inhalation exposure of rats and mice for 13 weeks led to degeneration of the olfactory nasal epithelium after concentrations of 32 ml/m^3 and more, and also to squamous metaplasia of the respiratory nasal epithelium from 128 ml/m^3. The amount of formic acid taken up during inhalation exposure is too small to cause metabolic acidosis, as described after ingestion of large amounts of formic acid.

After absorption, formic acid is metabolized to carbon dioxide or excreted unchanged with the urine. Depending on the species of animal, the half-life of the substance is 15 minutes to 1 hour. For man it was given as 2.5 hours.

In vitro investigations with rat and mouse embryo cultures revealed increased mortality after concentrations of 11.8 mmol/l and more, due to reduction in pH ($pH < 7.05$). Parameters of growth and development were also changed from this concentration even without a reduction in the pH value. In a 2-generation study with administration of formic acid doses of 150 to 200 mg/kg body weight with the drinking water, no toxic effects on reproduction were observed.

Administration of formic acid in the drinking water for 27 weeks in a study which does not meet present-day standards did not result in tumour development. Genotoxic effects were observed only because of the reduction in the pH value.

2 Mechanism of Action

Formic acid is a strong acid with a pKa value of 3.75 at 25°C; accordingly formic acid causes irritation of the upper respiratory tract.

Ingestion of large amounts of formic acid results in metabolic acidosis. In rats metabolic acidosis was described at formic acid concentrations of 8 to 15 mmol/l blood (Eells *et al.* 1996).

The haemolysis occurring in man after ingestion of formic acid is attributed to the acidity of the solution. Methaemoglobin was not detected during the course of intoxication after ingestion of this substance (BUA 1991).

The damage to the eyes observed after methanol intoxication is attributed to the metabolite formic acid; permanent damage occurred after persistence for longer periods (> 24 hours) of blood formic acid concentrations over 7 mmol/l. Evidence of disorders of the retina without histopathological changes was seen in rats with blood formic acid concentrations of 4 to 6 mmol/l (Eells *et al.* 1996).

3 Toxicokinetics and Metabolism

Formic acid can be taken up via all administration routes. The substance also occurs as an endogenous metabolic intermediate. Some of the formic acid is oxidized via the tetrahydrofolate system to form carbon dioxide which is exhaled; the rest is eliminated unchanged with the urine. In human urine, formic acid concentrations of 11.7 to 60 mg/l are found under physiological conditions (BUA 1991). As a result of the lower tetrahydrofolate level in primates than in rodents (50–75 %), oxidation is saturated more rapidly in primates (Johlin *et al.* 1987; Kavet and Nauss 1990) and formic acid can accumulate in blood.

The biological half-life, depending on the main route of elimination, the species of animal and amount taken up, is between 15 minutes and 1 hour; for man a value of 2.5 hours was reported in a case of intoxication (BUA 1991).

In 10 employees from a formic acid bottling plant and the production of urea-formaldehyde resin who were exposed to methanol concentrations of 40 to 160 ml/m^3 and formic acid concentrations of 2 to 5.5 ml/m^3, formic acid concentrations of 21.2 to 118 mg/g creatinine were found in urine 16 hours after exposure (Liesivuori and Savolainen 1987). It was not stated whether the measured concentrations caused irritation.

4 Effects in Man

Depending on the concentration, formic acid can cause hyperaemia, oedema and the formation of blisters, or even necrosis of the skin and the mucous membranes of the eyes, nose, throat and gastrointestinal tract, and corneal clouding. In persons who had ingested the substance accidentally, severe local acid burns were observed in the gastrointestinal tract with oesophageal stricture, perforation of the stomach, haemorrhage and systemic effects. In addition to nausea, vomiting, difficulties in swallowing and unconsciousness, acidosis, haemolysis, coagulation disorders, liver function disorders, reduced blood pressure, acute kidney failure and inflammation of the lungs were observed. Of 16 persons who swallowed 45 to 200 g formic acid, 9 died as a result of perforation in the gastrointestinal tract and 5 of acute kidney failure (BUA 1991).

A case-control study was carried out with 3730 patients with tumours at 19 different localities. Four persons were exposed to formic acid and had non-Hodgkin's lymphomas. In two patients the exposure to formic acid was described as "non-substantial"; the odds ratio (OR) was 2.2 [95 % confidence interval (CI) 0.4–11.3]. Two patients were exposed to "substantial" concentrations of formic acid. The OR was given as 1.5 [95 % CI 0.3–8.0] (Fritschi and Siemiatycki 1996). Because the number of persons exposed to formic acid was small and the description of the exposure conditions inadequate, the findings cannot be included in the evaluation of the carcinogenic effects of formic acid in man.

Biological monitoring

To investigate the suitability of formic acid excretion in the urine as a parameter for biological monitoring and factors influencing the physiological excretion of formic acid, the urine of non-exposed persons was analysed. In the urine of 70 non-exposed persons, the median level of formic acid was found to be 13.64 mg/l (range 1.00–95.12 mg/l) and 14.06 mg/g creatinine (range 1.22–278.71 mg/g creatinine). Statistically significant differences in urinary formic acid levels were found between pregnant (19.00 mg/g creatinine) and not pregnant women (8.98 mg/g creatinine) and between children (22.87 mg/g creatinine) and adults (8.54 mg/g creatinine). The upper limit of the normal range (95th percentile) of the urinary formic acid level in adults is given as 23 mg/g creatinine. Analysis of urine samples collected at 3 hour intervals during the day revealed enormous variation in formic acid concentration in the course of the day in some persons; the variation was put down to the intake of large amounts of formic acid with food. In one woman the urinary formic acid concentration increased by a factor of almost 30 (from 2.0 to 57.1 mg/g creatinine) within 6 hours. Formic acid concentrations in morning urine on three consecutive days differed at most by a factor of 2.7 (9.3–24.8 mg/g creatinine) (Schmid *et al.* 1994). In summary, excretion of formic acid in the urine is not a specific parameter for assessing the exposure to formic acid by inhalation.

One research group has carried out several studies on biological monitoring of formic acid in the urine of workers exposed to formic acid.

13 farmers were assigned to one of three groups on the basis of their urinary formic acid concentrations (group I: 0.1–0.3 mmol/l; group II: 0.5–0.7 mmol/l; group III: 0.9–1.0 mmol/l). The urinary formic acid levels in the farmers of groups II and III, who worked with silage, increased with increasing time after the end of exposure. In group II farmers about 14 hours after the end of exposure, the level was about 1 mmol/l; in group III 16 hours after the end of exposure about 1.6 mmol/l (Liesivuori 1986).

In urine collected 16 hours after the end of exposure (Friday morning) from ten persons exposed at work during the week to formic acid concentrations of 2.0 to 5.5 ml/m^3, the formic acid levels were 21.2 to 118 mg/g creatinine (about 60–340 mmol/mol creatinine). The urinary formic acid levels in the 18 control persons were 15.1 ± 6.1 mg/g creatinine (about 43 mmol/mol creatinine). There was a linear correlation between the formic acid concentrations in the air (personal samplers) and the concentration in the urine. At the threshold concentration of 5 ml/m^3, the urinary formic acid level was about 90 mg/g creatinine (about 260 mmol/mol creatinine) (Liesivuori and Savolainen 1987a).

Analyses of the concentrations of formic acid, ammonium and calcium ions in the urine collected 16 hours after the end of exposure (Friday morning) were carried out for 19 workers exposed to methanol and 4 exposed during the week to formic acid (air concentrations were not determined). With increasing urinary formic acid levels (from 27 ± 5 to 101 ± 22 mmol/mol creatinine), the concentration of ammonium ions decreased (from 2.4 ± 0.2 to 1.9 ± 0.4 mmol/mol creatinine) and that of calcium ions increased (from about 200 to about 750 mmol/mol creatinine). No correlation was found between the concentrations of ammonium and calcium ions for individual persons (Liesivuori and Savolainen 1987b).

Twelve farmers were exposed to formic acid for several days. On the day of the study, they were exposed for 8 hours to formic acid in a concentration of 7.2 ± 2.2 mg/m^3. Urine samples were obtained from the workers at the end of exposure and 15 and 30 hours later. The level of formic acid in the urine increased with increasing time after the end of exposure and was highest 30 hours after the end of exposure (control: 26 ± 14 mmol/mol creatinine; 0 hours: 31 ± 17, 15 hours: 65 ± 45, 30 hours: 104 ± 60); later times after the end of exposure were not studied. There was a linear correlation between the formic acid concentration in the urine and that in the air: a formic acid concentration in the air of about 200 nmol/l (5 ml/m^3; MAK value) corresponded to a urinary formic acid concentration of about 150 mmol/mol creatinine. The pH of the urine was not changed, but 30 hours after the end of exposure the ammonium ion concentrations were significantly increased (control: 1.4 ± 0.4 mmol/mol creatinine; 0 hours: 1.5 ± 0.6, 15 hours: 1.7 ± 0.6, 30 hours: 2.3 ± 0.8) and calcium ions (control: 227 ± 72 mmol/mol creatinine; 0 hours: 226 ± 129, 15 hours: 241 ± 122, 30 hours: 370 ± 175). The authors interpreted this as an effect on the kidneys (Liesivuori *et al.* 1992).

The studies of the research group of Liesivuori reveal a linear increase in the level of urinary formic acid for up to 30 hours after the end of exposure to formic acid. This indicates that the half-life of formic acid in man is considerably longer than that in animals (15 to 60 minutes, depending on the species and administration route; BUA 1991). Liesivuori and Savolainen (1987b) found an inverse correlation between urinary formic acid and ammonium ion levels whereas Liesivuori *et al.* (1992) showed that exposure to formic acid increased renal ammoniagenesis; these contradictory results do not suggest that exposure to formic acid affects the urinary excretion of ammonium ions. However, the two studies produced consistent results for an increased excretion of calcium ions 16 and 30 hours after the end of exposure. This finding requires further study, but it is not considered to be an adverse effect relevant for the establishment of a threshold value. It must be pointed out that the kind of food consumed by the persons studied by this research group was not documented. As was shown in the study of non-exposed persons by Schmid *et al.* (1994), food consumption has a marked effect on the urinary formic acid levels. Physiological variation is also observed.

5 Animal Experiments and *in vitro* Studies

5.1 Acute toxicity

The LC_{50} for rats after inhalation exposure for 4 hours is 7400 mg/m^3 (BUA 1991). The LD_{50} values after oral administration were given as 730 to 1830 mg/kg body weight for rats and 1100 mg/kg body weight for mice (BUA 1991, ECB 1995).

Exposure of guinea pigs (at least 7 animals per group) for one hour to formic acid concentrations of 0.34, 1.0, 2.8, 6.6, 13.5 or 42.5 ml/m^3 resulted in increased airway resistance and reduced dynamic compliance in all the exposed groups relative to the controls. At 42.5 ml/m^3 the respiration rate and minute volume were also reduced (Amdur 1960). Guinea pigs, because of their extensive bronchial smooth musculature, are very sensitive to irritant substances and react with bronchoconstriction (Fowler and Dodd 1986). It is not possible at present to make a quantitative comparison with man. It must be pointed out that the guinea pigs used in this study were not anaesthetized. It has been reported that when these animals are exposed to irritant substances, they change from nasal to oronasal breathing which can result in changes in lung function parameters (Dodd *et al.* 1986). In addition, data such as origin of the animals, age, weight, distribution of the animals in the exposure groups and state of health, were not given. The respiratory minute volumes given in the publication suggest a wide range of ages and body weights. Seen as a whole, in spite of its deficiencies, the study does show effects on airway parameters. The relevance of the result for man and its quantitative interpretation for man are unclear.

5.2 Subacute, subchronic and chronic toxicity

5.2.1 Inhalation

Groups of 5 male and 5 female F344 rats and B6C3F$_1$ mice were exposed for 2 weeks to formic acid concentrations of 0, 31, 62.5, 125, 250 or 500 ml/m^3 (6 hours daily, 5 days per week). Body weight gains were reduced by at least 10 % in mice exposed to 250 ml/m^3 or more and in rats after exposure to 500 ml/m^3. Concentrations of 250 ml/m^3 were lethal for one of 10 mice, 500 ml/m^3 for all mice and for 4 of 10 rats. Local effects in the form of squamous metaplasia of the respiratory epithelium were observed in both species from 62.5 ml/m^3. From concentrations of 125 ml/m^3 inflammation of the respiratory epithelium developed and from 250 ml/m^3 necrosis of the respiratory and olfactory epithelium. At the highest concentration, squamous metaplasia and inflammation of the larynx were also found and in mice necrosis of the pharynx. A systemic effect, a slight increase in kidney weights, was observed in male mice after exposure to 62.5 ml/m^3 or more and in female mice after 250 ml/m^3. After exposure to concentrations of 250 ml/m^3, the thymus weights were decreased in male and female mice and the lung weights increased. Concentrations of 500 ml/m^3 led in rats to decreased thymus weights. At 31 ml/m^3 neither local nor systemic effects were observed (Leach *et al.* 1989, NTP 1992).

After inhalation exposure for 13 weeks of groups of 10 male and 10 female F344 rats and B6C3F$_1$ mice to formic acid concentrations of 0, 8, 16, 32, 64 or 128 ml/m^3, in the rats neither body weight gains nor mortality were changed. In female mice from 64 ml/m^3 and male mice after 128 ml/m^3 body weight gains were reduced by more than 10 %. 128 ml/m^3 was lethal for one male and one female mouse. Local effects in the form of minimal degeneration of the olfactory epithelium were observed in one male rat after exposure to 32 ml/m^3 and in one male rat and 2 female mice after 64 ml/m^3, and in rats and mice of both sexes after 128 ml/m^3 (Table 1). In animals exposed to 128 ml/m^3

squamous metaplasia of the respiratory epithelium of rats also developed. Systemic effects were observed from 32 ml/m^3 in male rats and in mice of both sexes in the form of increased relative liver weights. In all exposed rats the relative and absolute lung weights were also decreased relative to the corresponding control values. This is due, however, to the fact that lung weights in the control animals were increased because of inflammation and is therefore not to be regarded as the result of exposure (Leach *et al.* 1989, NTP 1992). The NOAEC (no observed adverse effect concentration) for histopathological lesions is given by the NTP (1992) as 32 ml/m^3, including the 2-week inhalation study. The NOEC (no observed effect concentration) for histopathological and systemic changes is 16 ml/m^3.

Table 1. Number of animals with histopathological lesions of the nasal epithelium after inhalation of formic acid for 13 weeks (NTP 1992)

		Exposure concentration (ml/m^3)					
		0	8	16	32	64	128
Number of animals	♂	10	10	10	10	10	10
	♀	10	10	10	10	10	10
rats							
respiratory epithelium squamous metaplasia	♂	0	0	0	0	0	9
	♀	0	0	0	0	0	6
olfactory epithelium degeneration	♂	0	0	0	1	1	9
	♀	0	0	0	0	0	5
mice							
olfactory epithelium degeneration	♂	0	0	0	0	0	2
	♀	0	0	0	0	2	5

5.2.2 Ingestion

After repeated administration of formic acid doses of 8.2, 10.3, 90, 160 or 360 mg/kg body weight and day in the drinking water for 27 weeks to groups of 6 rats, no toxic effects were observed up to doses of 160 mg/kg body weight. A reduction in food consumption and growth occurred at the highest dose of 360 mg/kg body weight (Sollmann 1921).

Administration of 0.2 % calcium formate in the drinking water to 10 Wistar rats for 3 years (corresponding to 150–200 mg/kg body weight and day) or of 1 % sodium formate to 6 Wistar rats for 1.5 years (corresponding to 730 mg/kg body weight and day) did not lead to toxic effects (Malorny 1969).

5.3 Local effects on skin and mucous membranes

Depending on the concentration and method of application, formic acid was found to be irritative or even caustic for rabbit skin and in the rabbit eye (BUA 1991, ECB 1995).

There are no studies available of the allergenic effects of formic acid.

5.4 Reproductive toxicity

5.4.1 Fertility

In a 13-week inhalation study (see Section 5.2.1) investigations were carried out of sperm morphology and vaginal cytology with groups of 10 rats and mice exposed to formic acid concentrations of 0, 8, 32 and 128 ml/m^3. In male animals, no relevant effects on sperm motility, sperm concentration or on testis or epididymis weights were found. In female animals, the oestrus cycle was not affected (NTP 1992).

5.4.2 Developmental toxicity

In vitro

Rat embryos explanted on day 9 of gestation and mouse embryos explanted on day 8 of gestation, were incubated in culture ("whole embryo culture") for 24 or 48 hours in serum with formic acid concentrations of 5.9 to 44 mmol/l. At 5.9 mmol/l (pH 7.11–7.41) statistically significant effects were not yet observed. From 11.8 mmol/l (pH 7.07–7.23) mortality was slightly increased, from 17.6 mmol/l (pH 6.87–7.05) significantly increased. Comparable concentrations of sodium formate, however, did not lead to changes in the pH value, nor to increased mortality (Andrews *et al.* 1995). If the pH value was lowered, however, a significant increase in mortality was also observed with sodium formate (27 % at pH 6.5 and 100 % at pH 6.0) (Andrews *et al.* 1993). The increased mortality observed with formic acid can therefore be attributed to a lowering of the pH value. As embryo cultures have no buffer systems, the pH value is probably already lowered at low formic acid concentrations, unlike the situation *in vivo*.

In a comparable study with rat embryo cultures severe embryotoxicity was observed at formic acid concentrations of 18.66 mmol/l and more (pH value 6.94). The NOEC is given as 3.74 mmol/l (Brown-Woodman *et al.* 1995).

Parameters of growth and development such as the diameter of the yolk sac, crown-rump length, head length, number of somites and degree of development of the mouse and rat embryos were decreased in a concentration-dependent manner after exposure to formic acid concentrations of 11.8 and 17.6 mmol/l and more. The incidence of embryos with anomalies, mainly dysraphic disturbances, was significantly increased in mouse embryos from 11.8 mmol/l and in rat embryos at the highest concentration of 17.6 mmol/l after increasing the exposure time to 48 hours. No effects were observed after 5.9 mmol/l

(Andrews *et al.* 1995). As similar results were obtained for the parameters of growth and development and incidence of anomalies with sodium formate even without a change in pH (Andrews *et al.* 1995, Dorman *et al.* 1995), these results indicate that formic acid has embryotoxic potential *in vitro*.

In vivo

Administration of formic acid in the drinking water to Wistar rats over 5 generations at a concentration of 0.2 % and over 2 generations at a concentration of 0.4 % (corresponding to 150–200 mg/kg body weight and day) yielded no evidence of effects on fertility or the development of the embryos, the number of litters or the body weight gains of the offspring (Malorny 1969).

Single gavage doses of sodium formate of 750 mg/kg body weight were administered to CD-1 mice on day 8 of gestation (sensitive phase of methanol-induced exencephaly). After administration, the formate concentrations determined in plasma were 1.05 mmol/l and in the decidua 2 mmol/l. The foetuses were examined on days 10 and 18 of gestation. No effects were found on the neural tube, head or body lengths, or weight of the foetuses (Dorman *et al.* 1995). Comparison with the concentrations used *in vitro* indicates that 1 to 2 mmol/l would not be expected to cause such changes.

5.5 Genotoxicity

Valid *in vitro* genotoxicity studies such as several mutagenicity tests in *Salmonella*, an SOS chromotest with *Escherichia coli* and a test for SCE (sister chromatid exchange) with V79 cells yielded negative results. A chromosomal aberration test with CHO cells (a cell line derived from Chinese hamster ovary) revealed clastogenic effects only when the pH value of the exposure medium was reduced (pH < 6). A weak positive result was observed in a reverse mutation test with *Escherichia coli* carried out in 1951, but only with a high bacterial density with constant substance concentration. A test for X chromosomal recessive lethal mutations with *Drosophila melanogaster* carried out in 1964 yielded positive results after inhalation exposure and administration with the food (BUA 1991).

Taking all the available studies into consideration, there is, however, no convincing evidence that formic acid is mutagenic.

5.6 Carcinogenicity

5.6.1 Short-term studies

Up to the highest tested concentration of 300 µg/ml, formic acid was not found to have effects on metabolic cooperation in V79 cells (Malcolm *et al.* 1985). Thus this study yielded no evidence of a tumour-promoting effect of formic acid.

Painting the mouse ear with 8 % formic acid twice a week for 50 days revealed no evidence of histological or histomorphometric changes, unlike with croton oil and Tween 60 (Frei and Stephens 1968).

5.6.2 Long-term studies

In chronic toxicity studies (see Section 5.2.2) which, however, do not meet present-day standards, the incidence of tumours was not increased in rats.

6 Manifesto (classification/MAK value)

The studies of workers exposed to formic acid by the research group of Liesivuori and Savolainen are not relevant for evaluation of the MAK value. There are no further studies in man from which a MAK value could be derived. In a long-term study with rats which, however, does not meet present-day standards, no tumours were found. Genotoxic effects were only observed when the pH value was lowered. Because formic acid has strongly irritative effects, these are used as a relevant endpoint for the setting of a MAK value. A NOAEC of 32 ml/m^3 for histopathological lesions of the nasal epithelium was determined in a 13-week inhalation study with rats. The NOEC for histopathological lesions of the nasal epithelium and for systemic effects is 16 ml/m^3. This study provides no grounds for lowering the MAK value for formic acid (5 ml/m^3); this must, however, be verified in studies with man. The results of an acute inhalation study with guinea pigs are not suitable for the evaluation of the MAK value. Because of its irritant effects, formic acid remains classified in Peak limitation category I. Because the NOEC derived in the 13-week inhalation study with rats for histological reactions caused by local irritation is considerably higher than the MAK value, an excursion factor of 2 is acceptable.

The formic acid concentration in blood and the pH value is of importance for potential embryotoxic effects of formic acid. With a formic acid concentration of 5 ml/m^3 (9.6 mg/m^3) and a respiratory volume of about 1.25 m^3 per hour with 100 % absorption, 96 mg formic acid (2.09 mmol) is absorbed per 8 hours or 12 mg (0.26 mmol) per hour. For a body weight of 70 kg, the amount of formic acid absorbed would be 0.17 mg/kg body weight (0.003 mmol/kg body weight) and hour. As in primates the transformation rate of formic acid to carbon dioxide is 34 mg/kg body weight (0.75 mmol/kg body weight) and hour (Johlin *et al.* 1987, Kavet and Nauss 1990), that is, 2 orders of magnitude higher than this absorption rate, accumulation of formic acid is not to be expected.

From the amount of formic acid taken up per hour (12 mg (0.26 mmol)) and for a blood volume in man of about 4.5 l, a calculated formic acid concentration in blood of 0.0027 mg/ml (0.058 mmol/l) is obtained. Assuming 100 % dissociation, 0.058 mmol H^+ ions/l blood reduce the bicarbonate concentration from 24 to 23.942 mmol/l blood. According to the Henderson-Hasselbalch equation, the physiological pH value of 7.4

(Lentner 1984) is not markedly changed by this. Toxic effects on reproduction are observed *in vitro* only at 250 times higher formic acid concentrations (at 11.8 mmol/l and more). Toxic effects on reproduction are therefore not to be expected given observance of a MAK value of 5 ml/m^3. Formic acid is therefore classified in Pregnancy risk group C.

There are no studies of dermal absorption or of allergenic effects of formic acid, so that no decision can be made regarding designation with an "H" (for substances which can be absorbed by the skin) or "S" (for substances with sensitizing potential).

7 References

Amdur MO (1960) The response of guinea pigs to inhalation of formaldehyde and formic acid alone and with a sodium chloride aerosol. *Int J Air Pollut 3*: 201–220

Andrews JE, Ebron-McCoy M, Kavlock RJ, Rogers JM (1993) Lowering pH increases embryonic sensitivity to formate in whole embryo culture. *Toxicol In Vitro 7*: 757–762

Andrews JE, Ebron-McCoy M, Kavlock RJ, Rogers JM (1995) Developmental toxicity of formate and formic acid in whole embryo culture: a comparative study with mouse and rat embryos. *Teratology 51*: 243–251

Brown-Woodman PDC, Huq F, Hayes L, Herlihy C, Picker K, Webster WS (1995) *In vitro* assessment of the effect of methanol and the metabolite, formic acid, on embryonic development of the rat. *Teratology 52*: 233–243

BUA (Beratergremium für umweltrelevante Altstoffe der Gesellschaft Deutscher Chemiker) (1991) *Reports 81 (Formic acid, Formates) and 82 (Diethyleneglycol-bis-chloroformate).* VCH, Weinheim

Dodd DE, Fowler EH, Snellings WM, Pritts IM, Baron RL (1986) Acute inhalation studies with methyl isocyanate vapor. I. Methodology and LC_{50} determinations in guinea pigs, rats, and mice. *Fundam Appl Toxicol 6*: 747–755

Dorman DC, Bolon B, Struve MF, LaPerle KMD, Wong BA, Elswick B, Welsch F (1995) Role of formate in methanol-induced exencephaly in CD-1 mice. *Teratology 52*: 30–40

ECB (European Chemicals Bureau) (1995) *IUCLID data sheet, Formic acid*, 23.10.1995

Eells JT, Salzman MM, Lewandowski MF, Murray TG (1996) Formate-induced alterations in retinal function in methanol-intoxicated rats. *Toxicol Appl Pharmacol 140*: 58–69

Fowler EH, Dodd DE (1986) Acute inhalation studies with methyl isocyanate vapor. II. Respiratory tract changes in guinea pigs, rats, and mice. *Fundam Appl Toxicol 6*: 756–771

Frei JV, Stephens P (1968) The correlation of promotion of tumour growth and of induction of hyperplasia in epidermal two-stage carcinogenesis. *Br J Cancer 22*: 83–92

Fritschi L, Siemiatycki J (1996) Lymphoma, myeloma and occupation: Results of a case-control study. *Int J Cancer 67*: 498–503

Johlin FC, Fortman CS, Nghiem DD, Tephly TR (1987) Studies on the role of folic acid and folate-dependent enzymes in human methanol poisoning. *Mol Pharmacol 31*: 557–561

Kavet R, Nauss KM (1990) The toxicity of inhaled methanol vapors. *Crit Rev Toxicol 21*: 21–50

Leach CL, Abdo KM, Chou BJ, Roycroft JH, Mellnick PW (1989) 13-Week inhalation toxicity study of formic acid in rats and mice. *Toxicologist 9*: 144

Lentner C (Ed.) (1984) *Geigy scientific tables. Vol. 3, Physical chemistry composition of blood hematology somatometric data.* CIBA-GEIGY, Basel, 70–71

Liesivuori (1986) Slow urinary elimination of formic acid in occupationally exposed farmers. *Ann Occup Hyg 30*: 329–333

Liesivuori J, Savolainen H (1987a) Urinary formic acid as an indicator of occupational exposure to formic acid and methanol. *Am Ind Hyg Assoc J 48*: 32–34

Liesivuori J, Savolainen H (1987b) Effect of renal formic acid excretion on urinary calcium and ammonia concentrations. *Klin Wochenschr 65*: 860–863

Liesivuori J, Laitinen J, Savolainen H (1992) Kinetics and renal effects of formic acid in occupationally exposed farmers. *Arch Toxicol 66*: 522–524

Malcolm AR, Mills LJ, Trosko JE (1985) Effects of ethanol, phenol, formaldehyde, and selected metabolites on metabolic cooperation between Chinese hamster V79 lung fibroblasts. *Carcinogenesis 8*: 305–318

Malorny G (1969) Die akute und chronische Toxizität der Ameisensäure und ihrer Formiate (Acute and chronic toxicity of formic acid and formates) (German). *Z Ernährungswiss 9*: 332–339

NTP (National Toxicology Program) (1992) *Formic Acid. Administration by inhalation to F344/N rats and B6C3F1 mice.* Toxicity report series number 19, NIH Publication 92-3342

Schmid K, Schaller K-H, Angerer J, Lehnert G (1994) Untersuchungen zur Dignität der Ameisensäureausscheidung im Harn für umwelt- und arbeitsmedizinische Fragestellungen (The importance of formic acid excretion in the urine for environmental and occupational medicine questions) (German). *Zentralbl Hyg 196*: 139–152

Sollmann T (1921) Studies of chronic intoxications on albino rats. III. Acetic and formic acids. *J Pharmacol Exp Ther 16*: 463–474

completed 07.04.1997

Formic acid ethyl ester

MAK value (1961)	**100 ml/m³ (ppm) ≙ 3.10 mg/m³**
Peak limitation (2000)	**Category I, excursion factor 1**
Absorption through the skin (1997)	**H**
Sensitization	**–**
Carcinogenicity	**–**
Prenatal toxicity (1997)	**Pregnancy risk group C**
Germ cell mutagenicity	**–**
BAT value	**–**
Synonyms	ethyl formate
Chemical name (CAS)	formic acid ethyl ester
CAS number	109-94-4
Structural formula	$HCOO–CH_2–CH_3$
Molecular formula	$C_3H_6O_2$
Molecular weight	74.08
Melting point	–80°C
Boiling point	54.3°C
Density at 20°C	0.917 g/cm³
Vapour pressure at 20°C	256 hPa
$\log P_{ow}$*	not specified
1 ml/m³ (ppm) ≙ 3.07 mg/m³	**1 mg/m³ ≙ 0.325 ml/m³ (ppm)**

The MAK value for formic acid ethyl ester of 100 ml/m³ was established in 1961 in analogy to the TLV value at the time. In the meantime studies have appeared which make it necessary to re-evaluate the substance. The present document is based on documentation of the American Conference of Governmental Industrial Hygienists on formic acid ethyl ester (ACGIH 1991) and an evaluation by the Food and Drug Administration (FDA 1976).

---

* *n*-octanol/water distribution coefficient

1 Toxic Effects and Mode of Action

Formic acid ethyl ester is a liquid with a pleasant odour. The odour threshold is 31 ml/m^3. Formic acid ethyl ester in concentrations of 330 ml/m^3 caused slight irritation of the eyes of test persons, and rapidly increasing irritation of the nose, which was still severe after 4 hours (no other details, Flury and Zernik 1931). Compared with formic acid methyl ester, formic acid ethyl ester produces less pronounced irritative effects, but more pronounced narcotic effects.

Formic acid ethyl ester is metabolized to ethanol and formic acid. After repeated exposure, these two substances are thought to be mainly responsible for the systemic effects. In persons who have absorbed large amounts of formic acid ethyl ester, metabolic acidosis can result from formation of formic acid and acetic acid.

Formic acid ethyl ester was not found to be mutagenic in bacteria and yeasts.

2 Mechanism of Action

Formic acid ethyl ester causes irritation which is attributed to the hydrolysis product formic acid.

The metabolic acidosis is caused by the lowering of the pH value in blood by the formation of formic acid and acetic acid.

3 Toxicokinetics and Metabolism

Formic acid ethyl ester is enzymatically hydrolysed to formic acid and ethanol. Formic acid is oxidized via the tetrahydrofolate system to form carbon dioxide which is exhaled. As a result of the lower (50–75 %) tetrahydrofolate level, in primates this oxidation process is saturated more rapidly than in rodents (Johlin *et al.* 1987, Kavet and Nauss 1990) so that formic acid can accumulate. Ethanol is oxidized in the liver by alcohol dehydrogenase to form acetaldehyde and by aldehyde dehydrogenase to form acetic acid. The acetic acid is oxidized via the tricarboxylic acid cycle to form mainly carbon dioxide. To a lesser extent acetic acid can also enter intermediary metabolism.

4 Effects in Man

In man, formic acid ethyl ester in a concentration of 330 ml/m^3 caused slight irritation of the eyes and rapidly increasing irritation of the nose which was still present to a marked degree after 4 hours (Flury and Zernik 1931). Concentrations of 10500 ml/m^3 caused moderate, yet progressive irritation of the eyes and nose. The occlusive application of 4 % formic acid ethyl ester to the skin of volunteers did not produce visible irritation (Snyder 1992).

There is no information available about the exposure concentrations which can lead to metabolic acidosis via the metabolites formic acid and acetic acid in man. It is estimated below whether accumulation of metabolites and metabolic acidosis can occur when the MAK value is observed.

With formic acid ethyl ester concentrations of 100 ml/m^3 (300 mg/m^3) and a respiratory volume of about 1.25 m^3 per hour, assuming 100 % absorption, 3000 mg formic acid ethyl ester (40.5 mmol) per 8 hours or 375 mg (5.06 mmol) per hour is absorbed. For a person of 70 kg body weight, the amount absorbed is 5.4 mg/kg body weight and hour (0.072 mmol/kg body weight and hour). Assuming complete hydrolysis, this corresponds to the same molar amounts of formic acid and ethanol. For formic acid, the metabolic rate in primates to form carbon dioxide is given as 34 mg/kg body weight (0.75 mmol/kg body weight) and hour (Johlin *et al.* 1987, Kavet and Nauss 1990). The rate of elimination of ethanol for women is 85 mg/kg body weight (1.15 mmol/kg body weight) and hour and for men 100 mg/kg body weight (1.35 mmol/kg body weight) and hour (Forth *et al.* 1992). Accumulation of the substances or their metabolites is therefore not to be expected.

With a blood volume in man of 4.5 l and amounts of formic acid ethyl ester of 375 mg (5.06 mmol) absorbed per hour, the concentrations of formic acid and acetic acid formed are each 1.12 mmol/l blood. Assuming 100 % dissociation of the formic and acetic acids, 2.24 mmol H^+ ions per litre blood causes a decrease in the bicarbonate concentration from 24 to 21.76 mmol/l blood. A pH value of 7.38 is calculated according to the Henderson-Hasselbalch equation (pK value of bicarbonate 6.1; concentration of the undissociated acid in an open system 1.2 mmol/l), which is at the lower end of the physiological range of the blood pH value of 7.37 to 7.46 (Lentner 1984). Overloading of the body's buffer systems is, however, not to be expected as the extracellular space and other physiological buffer systems were not taken into consideration in the above calculation and an hourly bolus uptake was assumed. Therefore in persons exposed to formic acid ethyl ester concentrations of 100 ml/m^3 metabolic acidosis is not to be expected.

5 Animal experiments and *in vitro* Studies

There are no data available on the allergenic effects of formic acid ethyl ester.

5.1 Acute toxicity

Exposure of mice and cats for 20 minutes to formic acid ethyl ester concentrations of 5000 ml/m^3 and more led to eye irritation. Exposure of cats to 10000 ml/m^3 for 75 minutes caused deep narcosis and, after 90 minutes, death. Exposure to 8000 ml/m^3 for 4 hours was lethal for rats (Smyth *et al.* 1954), and 10000 ml/m^3 was lethal for dogs (Flury and Zernik 1931).

The oral LD_{50} for rats was 1850 mg/kg body weight, for guinea pigs 1110 mg/kg body weight and for rabbits 2075 mg/kg body weight (FDA 1976, Sandmeyer and Kirwin 1981). In rabbits the LD_{50} after dermal application was found to be over 20 ml/kg body weight (> 18000 mg/kg body weight) (Smyth *et al.* 1954).

5.2 Subacute, subchronic and chronic toxicity

After administration of formic acid ethyl ester for 17 weeks to groups of 10 male and 10 female Osborne-Mendel rats in concentrations of 1000, 2500 or 10000 ppm with the diet, corresponding to 100, 250 or 1000 mg/kg body weight and day, adverse effects were not reported (FDA 1976). Because formic acid ethyl ester is volatile, however, administration with the diet is not a suitable method.

5.3 Local effects on skin and mucous membranes

Formic acid ethyl ester was not found to cause irritation of rabbit skin (a score of 1 on a scale up to 10); in the rabbit eye, however, inflammation of the cornea was observed (4 points on a scale up to 10) (Smyth *et al.* 1954).

5.4 Reproductive toxicity

There are no studies available of the reproductive toxicity of formic acid ethyl ester. Formic acid ethyl ester was previously classified in Pregnancy risk group D, in analogy to methanol (Henschler 1989). Methanol was re-classified in 1995 in Pregnancy risk group C (see MAK documentation for Methanol in Volume 16 of the present series).

5.5 Genotoxicity

Formic acid ethyl ester was not found to be mutagenic in plate tests or suspension tests with *Saccharomyces cerevisiae* D4 or *Salmonella typhimurium* TA1535, TA1537 and TA1538 in concentrations up to 5 % with and without the addition of a metabolic activation system made from mouse, rat or monkey liver homogenates (FDA 1976).

5.6 Carcinogenicity

In a short-term carcinogenicity test, two groups of 15 male and 15 female strain A mice were given a total dose of formic acid ethyl ester of 2400 or 12000 mg/kg body weight 3 times per week for 8 weeks by intraperitoneal injection. This corresponded to formic acid ethyl ester doses per injection of 100 or 500 mg/kg body weight and formic acid doses of 66 or 330 mg/kg body weight (Stoner *et al.* 1973). No effects on survival or the incidence of lung adenomas were detected. Other organs were not investigated.

Valid long-term studies of the carcinogenicity of formic acid ethyl ester are not available.

6 Manifesto (MAK value/classification)

Formic acid ethyl ester in concentrations of 330 ml/m^3 caused slight irritation of the eyes of test persons, and rapidly increasing irritation of the nose, which was still severe after 4 hours. After brief exposures to 200 ml/m^3 systemic effects (acidosis) are not to be expected. The previous MAK value of 100 ml/m^3 can, therefore, be retained. No NOEL for the irritant effects has been determined. On the basis of these data, formic acid ethyl ester is classified in Peak limitation category I and an excursion factor of 1 is established provisionally.

In view of the fact that the MAK value for ethanol is 1000 ml/m^3 and the substance is classified in Pregnancy risk group C, provided the MAK value for formic acid ethyl ester of 100 ml/m^3 is observed, no danger to reproduction is to be expected from the ethanol formed from formic acid ethyl ester. Also the concentration of formic acid formed given observation of the MAK value (see Section 4; 1.12 mmol/l blood) is not expected to lead to toxic effects on reproduction as *in vitro* effects were first observed after formic acid concentrations of 11.8 mmol/l had been achieved (see MAK documentation for formic acid in this volume). Therefore, provided the MAK value for formic acid ethyl ester of 100 ml/m^3 is observed, toxic effects on reproduction are not to be expected. Formic acid ethyl ester is therefore classified in Pregnancy risk group C.

Data for the dermal absorption of formic acid ethyl ester are not available. In analogy to formic acid methyl ester, formic acid ethyl ester is also designated with an “H”.

Studies of the allergenic effects of formic acid ethyl ester are not available; for this reason no decision can be made on the designation of the substance with an “S”.

7 References

ACGIH (American Conference of Governmental Industrial Hygienists) (1991) *Documentation of the threshold limit values and biological exposure indices, ethyl formate*, ACGIH, Cincinnati, Ohio

FDA (Food and Drug Administration) (1976) *Evaluation of the health aspects of formic acid, sodium formate, and ethyl formate as food ingredients*, FDA, Washington, DC, NTIS PB-266 282

Flury F, Zernik F (1931) *Schädliche Gase* (Hazardous gases) (German), J Springer, Berlin, 375–376

Forth W, Henschler D, Rummel W, Starke K (Eds) (1992) *Allgemeine und spezielle Pharmakologie und Toxikologie* (German), 6th edition, Wissenschaftsverlag, Mannheim, Wien, Zürich, 795–800

Henschler D (1989) Methanol, Methylacetat, Methylformiat, Ethylformiat (MAK-Werte und Schwangerschaft). *Toxikologisch-arbeitsmedizinische Begründungen von MAK-Werten*, 15th issue, VCH-Verlagsgesellschaft mbH, Weinheim

Johlin FC, Fortman CS, Nghiem DD, Tephly TR (1987) Studies on the role of folic acid and folate-dependent enzymes in human methanol poisoning. *Mol Pharmacol 31*: 557–561

Kavet R, Nauss KM (1990) The toxicity of inhaled methanol vapors. *Crit Rev Toxicol 21*: 21–50

Lentner (Ed.) (1984) *Geigy Scientific Tables. Vol. 3, Physical chemistry, composition of blood, hematology, somatometric data.* CIBA-GEIGY, Basel, 70–71

Sandmeyer EE, Kirwin Jr CJ (1981) Esters. in: Clayton GD, Clayton FE (Eds) *Patty's Industrial Hygiene and Toxicology, Vol. 2A*, John Wiley & Sons, New York, NY, 2263–2267

Smyth HF, Carpenter CP, Weil CS, Pozzani UC (1954) Range-finding toxicity data. List V. *Arch Ind Hyg Occup Med 10*: 61–68

Snyder (Ed.) (1992) *Ethel Browning's Toxicity and Metabolism of Industrial Solvents. 2nd ed, Vol. 3: Alcohols and Esters*, Elsevier, New York, NY

Stoner GD, Shimkin MB, Kniazeff AJ, Weisburger JH, Weisburger EK, Gori GB (1973) Test for carcinogenicity of food additives and chemotherapeutic agents by the pulmonary tumor response in strain A mice. *Cancer Res 33*: 3069–3085

completed 07.04.1997; peak limitation category completed 25.04.00

Furfuryl alcohol

MAK value (1992)	**10 ml/m³ (ppm) ≙ 41 mg/m³**
Peak limitation (2002)	**Category I, excursion factor 1**
Absorption through the skin (1998)	**H**
Sensitization	**–**
Carcinogenicity	**–**
Prenatal toxicity	**–**
Germ cell mutagenicity	**–**
BAT value	**–**
Synonyms	2-furancarbinol furfuralcohol 2-furylcarbinol 2-hydroxymethylfuran
Chemical name (CAS)	2-furanmethanol
CAS number	98-00-0

The toxicology of furfuryl alcohol was reviewed in Volume 7 of the present series; this chapter is a supplement to the earlier review.

The toxic effects of furfuryl alcohol have not been studied in detail. Information about kinetics or metabolism is not available.

In rabbits the dermal LD_{50} was found to be 657 mg/kg body weight and the intravenous LD_{50} 650 mg/kg body weight. Dermal application of four times 300 to 1400 mg/kg body weight was lethal for 11/12 rats (ACGIH 1992). It is not clear whether testing was carried out occlusively.

Furfuryl alcohol is designated with an "H", as the dermal toxicity in rabbits is as high as the intravenous toxicity, and repeated dermal application was observed to be lethal for rats.

The critical effect of furfuryl alcohol is eye irritation. The substance is therefore classified in Peak limitation category I. Exposure at the workplace to a concentration of 15.8 ml/m³ caused severe lacrimation in two persons within 15 minutes. The persons were exposed simultaneously to formaldehyde (0.33 ml/m³) and possibly to other substances as well. Other evidence of irritative effects is not available. In another worker who was exposed to a concentration of 16 ml/m³ (exposure period not specified), no adverse effects were observed (ACGIH 1992). Because the possibility cannot be excluded that exposure to 20 ml/m^3 causes irritation, an excursion factor of 1 has been established for furfuryl alcohol.

References

ACGIH (American Conference of Governmental Industrial Hygienists) (1992) Furfuryl alcohol. in: *Documentation of TLVs and BEIs*, ACGIH, Cincinnati, OH, USA

dermal toxicity completed 31.03.1998, peak limitation category completed 28.02.2002

Hydrogen cyanide, potassium cyanide and sodium cyanide

MAK value (2001)	**2 mg/m³ I (inhalable fraction) as CN^-** **HCN: 1.9 ml/m³ (ppm) ≙ 2.1 mg/m³** **KCN: 5.0 mg/m³ I** **NaCN: 3.8 mg/m³ I**
Peak limitation (2001)	**Category II** **HCN: excursion factor 2** **KCN: excursion factor 1** **NaCN: excursion factor 1**
Absorption through the skin (2001)	**H**
Sensitization	**–**
Carcinogenicity	**–**
Prenatal toxicity (2001)	**Pregnancy risk group C**
Germ cell mutagenicity	**–**
BAT	**–**

Chemical name (CAS)	Synonyms	CAS number	Formula	Molecular weight	Melting point (°C)	Boiling point (°C)	log P_{OW}
hydrogen cyanide	hydrocyanic acid	74-90-8	HCN	27.03	–13.24	25.70	0.66
potassium cyanide		151-50-8	KCN	65.11	634.5	1625	–0.44 (calculated)
sodium cyanide		143-33-9	NaCN	49.02	563.7	1496	no data

The MAK values which applied until 2001, 5 mg/m³ I for cyanides, 10 ml/m³ (11 mg/m³) for HCN were established, respectively, in 1958 on the basis of the then valid TLV and in 1971 (Henschler 1971). The present document is based in part on other reviews of the toxicological data for cyanides (Ballantyne and Marrs 1987, EPA 1992, NL Health Council 1999).

1 Toxic Effects and Mode of Action

Cyanide salts and HCN are very poisonous. The toxic effects include headaches, dizziness, nausea, a feeling of suffocation, metabolic acidosis, hyperventilation, oxygen saturation of the venous blood (pink skin colour), convulsions, loss of consciousness, respiratory arrest and dilated pupils. The cause of death is usually central respiratory paralysis caused by inhibition of cytochrome oxidase. The lethal oral dose of KCN and NaCN for humans is between 0.5 and 3.5 mg/kg body weight. Detoxification takes place mainly by reaction of cyanide with endogenous sulfur donors to yield thiocyanate, a reaction which is catalysed by the hepatic enzyme rhodanese. A small part of the dose is exhaled unchanged. Survivors of acute cyanide poisoning may later develop neurological damage. Unspecific CNS effects and enlarged thyroid glands have been found in chronically exposed workers. Cyanide concentrations above 5 mg/m^3 cause nasal irritation. Central and peripheral neurotoxicity is also observed in experimental animals given repeated doses just below the acutely toxic range. Cyanide is less toxic when administered in the feed or drinking water than when administered by gavage.

2 Mechanism of Action

The systemically active principle of the three compounds is the cyanide ion. It reacts with the trivalent iron in cytochrome oxidase to give a relatively stable complex. This inhibits the enzyme and blocks the last step in oxidative phosphorylation. The result is a deficiency of mitochondrial ATP and cell death (Szinicz 1996). Particularly sensitive tissues are those of the CNS and heart (EPA 1992).

Acute cyanide exposure leads to acidosis (increase in lactate concentrations), reduced carbon dioxide concentrations, and increase in the oxygen concentration, increasing catabolism via the pentose phosphate pathway, reduction in catabolism via the Embden-Meyerhof pathway and the citrate cycle, and an increase in glucose and inorganic phosphates in the blood (Ballantyne 1987, Isom *et al.* 1975, Katsumata *et al.* 1980, 1983).

Cyanide forms reversible complexes with metal ions and thus inhibits many other metalloenzymes (NTP 1993).

The thiocyanate produced from cyanide can reduce iodide uptake into the thyroid and therefore prolonged exposure, especially of persons with iodine deficiency, can increase the incidence of goitre. It is very probable that tropical ataxic neuropathy is a result of chronic intake of cyanide with a staple food, cyanide-containing cassava roots (manioc) (see Section 4.2.2).

3 Toxicokinetics and metabolism

3.1 Intake

The retention determined after exposure of volunteers breathing HCN (4–20 mg/m^3) via the mouth for 1 to 3 minutes was 39 % to 77 %. Nasal retention was 13 % to 19 %. It is unclear whether steady state was reached during these short exposure times. The data therefore probably represent maximum retention levels (NL Health Council 1999). For the salts, it must be assumed that intake via the respiratory tract occurs by inhalation of particles or droplets of aqueous solutions, and absorption is 100 % because any substance cleared from the respiratory tract may be swallowed.

Ingested cyanide salts react with the hydrochloric acid in the stomach to yield HCN which, because it is largely undissociated, is rapidly absorbed in the gastrointestinal tract. 94.7 % of the radioactivity ingested as $K^{14}CN$ by rats was recovered in the urine (collected for up to 14 days after dosing), which suggests complete absorption (see Section 3.4 Excretion).

The presence of moisture is a prerequisite for dermal absorption of the salts (Section 5.1). It has been shown with human skin *in vitro* that penetration of sodium cyanide in aqueous solution decreases with increasing pH (increasing dissociation). This means that undissociated HCN is absorbed more rapidly than is the cyanide ion: the permeability constant measured for the cyanide ion in aqueous solution was 3.5×10^{-4} cm/h, and that calculated for HCN was 1×10^{-2} cm/h (Dugard 1987). The level of absorption of gaseous HCN through the skin of volunteers was low (Schütze 1927, O'Donnell *et al.* 1940). Epicutaneous exposure of dogs to gaseous HCN (6500–16900 mg/m^3) for up to 180 minutes caused respiratory distress and even death (EPA 1992).

3.2 Distribution

After absorption, cyanide is distributed in all organs. After fatal poisonings, the highest concentrations were found in the liver, lungs, blood, spleen and brain (NL Health Council 1999). However, the administration route also affects the distribution pattern (EPA 1992).

3.3 Metabolism

Metabolic inactivation of cyanide takes place mainly (about 80 %) via formation of thiocyanate, which is catalysed by rhodanese in the liver. The reverse reaction can be brought about by peroxidases. In addition, 2-imino-4-thiazolidine carboxylic acid or the tautomer 2-aminothiazoline-4-carboxylic acid can be formed by reaction of cyanide with cystine, and cyanocobalamin by reaction with hydroxocobalamin (vitamin B_{12a}). Further minor metabolites are cyanate, nitrite, CO_2 and formate. The limiting factor is not the rhodanese activity but the quantity of sulphanes available for sulphur transfer. Another crucial factor is transport of the sulphur compounds from the blood into the mitochondria. Binding of cyanide to methaemoglobin is used therapeutically (NL Health Council 1999). In humans (after i.v. injection), about 0.017 mg of cyanide per kg of body weight and minute (1.0 mg/kg body weight and hour) can be detoxified without therapeutic measures (EPA 1992). Another figure given in the literature for the detoxification capacity in man is 0.1 mg/kg body weight and hour (Dekant *et al.* 2001).

3.4 Excretion

Most cyanide is excreted in the urine as thiocyanate, but clear quantitative data are not available. After single doses (no further details), up to 95 % was excreted with the urine. In contrast, rats given KCN in the drinking water for 13 weeks in daily doses of 40, 80 or 160 mg/kg body weight excreted 11 % of the dose as thiocyanate each day. This is possibly a result of a decline in availability of thiosulphate during repeated exposure. A minor elimination route is exhalation of HCN (about 0.4–1.2 %) together with CO_2 (about 2.5–4 %) (NL Health Council 1999).

3.5 Endogenous formation

Cyanide is formed endogenously during vitamin B_{12} metabolism and is consumed in small quantities with food (e.g. soya beans) in the form of glycosidically bound cyanide (NL Health Council 1999).

4 Effects in Man

4.1 Single exposures

The acute toxicity of KCN and NaCN in humans is well documented. A large number of deliberate poisonings (suicide, murder) or accidents has been described. The following symptoms have been reported for cyanide poisoning: rapid breathing, weakness, headaches (pulsation), feeling of pressure in the chest, facial flushing, dyspnoea, nausea, vomiting, diarrhoea, dizziness, drowsiness, confusion, convulsions, coma, irregular breathing and death. A very early characteristic feature is tachypnoea (rapid breathing) and hyperpnoea (increased depth of breathing) (Ballantyne 1987). In addition, the following complications of acute cyanide poisoning are possible: rhabdomyolysis, diffuse cerebral oedema, degenerative changes in the central nervous system and pulmonary oedema (Messing 1991).

butyric acid and dopamine) and the levels of their precursors or metabolites (amino acids such as tyrosine and glutamic acid, L-dihydroxyphenylalanine, dihydroxyphenylacetic acid, hydroxyindoleacetic acid and homovanillic acid). Changes in the levels of cyclic guanosine monophosphate and intracellular calcium have also been found. Lipid peroxidation in the brain is regarded as clear evidence of the acute effects of cyanide. In addition, antioxidant enzymes in the brain are inhibited by cyanide. Cyanide treatment may cause an increase in the adrenaline and noradrenaline levels in the plasma. Changes in the electroencephalogram have been observed after acute cyanide exposure. Acute cyanide exposure leads to clear irreversible changes in the concentration of high-energy phosphate components in the rat brain, including a decrease in the ATP level. An increase in lactate and ADP and a large decrease in glycogen concentrations were observed in the brain of cyanide-exposed rats. The reaction of the cytochrome c oxidase in the brain is more pronounced than that of the cytochrome c oxidase in other tissues. Near lethal single doses of cyanide cause various histological changes in the brain. Acute exposure to cyanide may have adverse effects on movement and possibly on memory functions (NL Health Council 1999).

CNS depression was detected in mice exposed to HCN concentrations of 20 to 100 ml/m^3 (22–112 mg/m^3). The calculated DC_{50} was 63 ml/m^3 (Matijak-Schaper and Alarie 1982).

5.1.3 Cardiovascular effects

Acute cyanide exposure leads to changes in cardiac function, e.g. in heartbeat and ECG, to histological changes in the myocardium and to an increase in cardiospecific creatine phosphokinase in the blood (NL Health Council 1999).

5.2 Subacute, subchronic and chronic toxicity

5.2.1 Inhalation

No studies which can be used to establish a no observed adverse effect level (NOAEL) are available.

Histological changes were studied in the brains of dogs exposed for 30 minutes to an HCN concentration of 50 mg/m^3. Three of the dogs died during the study. Severe histological damage in the brain was found (Valade 1952). The study is inadequately documented.

In rabbits exposed to an HCN concentration of 0.5 mg/m^3 continuously for 4 weeks, no histopathological changes were found in the heart, lung or adjoining arteries (Hugod 1979). No other endpoints were investigated in this study.

5.2.2 Oral intake

The results of studies with histopathological examination and examination of the thyroid after repeated oral administration are listed in Table 8.

Table 8. Toxicity of cyanides after repeated oral exposure

Species number	Exposure	CN^- dose: effects	References
rat 6 ♂ 2 ♀	KCN: 2500 mg/kg feed (CN^-: about 36 mg/kg body weight and day) 84 days	**36 mg/kg body weight:** decreased feed consumption and weight gain; histopathology of liver, kidneys, thyroid and spleen unaffected	EPA 1992
rat F344/N 10 ♂ 10 ♀	NaCN: 3, 10, 30, 100, 300 mg/l drinking water, (CN^-: 0.16, 0.5, 1.6, 4.2, 12.5 mg/kg body weight per day) 13 weeks	**≥ 1.6 mg/kg body weight:** ♀: NOAEL ♂: dose-dependent, significant reduction in the weight of the cauda epididymis (7 %) **≥ 4.2 mg/kg body weight:** ♂/♀: decreased water consumption ♀: prooestrus and dioestrus longer and metoestrus and oestrus shorter ♂: NOAEL **about 12.5 mg/kg body weight:** ♂: decreased body weights (5 %), decreased spermatid head counts and spermatid counts (13 % each), decreased absolute testis weight (7 %) and absolute weight of *cauda epididymis* (13 %)	NTP 1993
rat 10 ♀	KCN: 50000, 100000 mg/kg feed (CN^-: about 500 and 1000 mg/kg body weight) [1] 14 weeks	**≥ 500 mg/kg body weight:** hair loss, apathy, decreased body weights, decreased haemoglobin, cell volumes, total serum protein, serum T4, relative spleen weights; increased relative thyroid weight; thymus atrophy, centrilobular liver cell necrosis	Olusi *et al.* 1979
rat 10 ♂	KCN: 1500 mg/kg feed (CN^-: about 30 mg/kg body weight and day) [2] feed with or without reduced levels of iodide, vitamin B_{12} and DL-methionine 11.5 months	**30 mg/kg body weight:** decreased body weights and plasma T4 levels after 4 months; slight primary myelin degeneration in the white matter of the spinal cord (more marked in rats with diets deficient in methionine and vitamin B_{12})	Philbrick *et al.* 1979
rat 10 ♂ 10 ♀	feed exposed to HCN gas (CN^-: about 1 or 3.5 mg/kg body weight) [3] 2 years	**3.5 mg/kg body weight:** NOAEL histopathology of 11 organs/tissues, haematology, mortality, feed consumption, body weights unaffected.	Howard and Hanzal 1955, NL Health Council 1999

Table 8. continued

Species number	Exposure	CN^- dose: effects	References
mouse B6C3F$_1$ 10 ♂ 10 ♀	NaCN: 3, 10, 30, 100, 300 mg/l drinking water, (CN^-: 0.3, 1, 3, 9, 26 mg/kg body weight and day) 13 weeks	**≥ 9 mg/kg body weight:** ♂/♀: decreased water intake; NOAEL **about 26 mg/kg body weight:** ♀: decreased body weight gain; menstrual cycle unchanged ♂: decreased absolute epididymis weight (10 %) and absolute weight of *cauda epididymis* (18 %); spermatid head density and sperm motility unchanged, decreased body weights (5 %) (not significant)	NTP 1993
dog 3 ♂ 2 ♀	NaCN: 2500 mg/kg feed (NaCN: 6 mg/kg body weight and day) [4] (CN^-: 3 mg/kg body weight) 32 days	**3 mg/kg body weight:** NOAEL feed consumption, haematology; autopsy, comprehensive histopathology (no further details)	EPA 1992
dog 3 (sex not specified)	NaCN: 0.5, 1–2 × 2 mg/kg body weight and day (CN^-: 0.27, 0.53–2 × 1.06 mg/kg body weight), (in capsules) 15 months	**≥ 0.27 mg/kg body weight:** degenerative changes in gangliocytes **> 0.53 mg/kg body weight:** acute toxicity	EPA 1992
pig 3 ♂ + ♀	CN^-: 0.4, 0.7, 1.2 mg/kg body weight and day, by gavage 24 weeks	**≥ 0.4 mg/kg body weight:** dose-dependent increase in T3 and T4, increased blood glucose levels in all treated groups, various behavioural parameters changed, prolonged narcosis **1.2 mg/kg body weight:** increased limping and stiffness, increased convulsions and trembling	Jackson 1988
rabbit 6 ♂	KCN: 1 755 mg/kg feed (CN^-: 702 mg/kg feed, about 15 mg/kg body weight and day) 10 months	**about 15 mg/kg body weight:** decreased body weight gain; LDH and AP increased in serum, kidney, liver; SDH, GPT and AP decreased in liver, increased in serum; AP in lung decreased; focal necrosis in liver, tubular and glomerular necrosis in kidney, pulmonary oedema and necrosis; pancreas and heart unaffected.	Okolie and Osagie 1999, 2000

[1] assumed feed consumption 25 g/kg body weight (normal 50 g/kg), since body weight loss up to 25 %
[2] assumed feed consumption 50 g/kg body weight
[3] according to NL Health Council 1999
[4] according to EPA 1992

Effects were found on the thyroid (Jackson 1988, Philbrick *et al.* 1979), the central nervous system and behaviour (EPA 1992, Jackson 1988, Philbrick *et al.* 1979), glucose metabolism (Jackson 1988), lungs, liver, kidneys (Okolie and Osagie 1999, 2000), the male reproductive organs (but without reduction in the sperm count) and the menstrual cycle of female rats (NTP 1993). In the NTP (1993) study, described in detail in Section

5.5, no changes in the thyroid were detected. The NOAEL for CN^- found in this study is 1.6 mg/kg body weight and day for rats and 9 mg/kg body weight for mice. The authors point out that the absence of effects on the CNS and the thyroid in the animal experiments is not consistent with the findings of human epidemiological studies but that this is possibly a result of a difference in the sensitivity of humans and rodents (NTP 1993).

The studies with administration of bolus doses to pigs (Jackson 1988) and of capsules to dogs (EPA 1992) are not used for the present assessment because the sudden rise in the cyanide level caused by these modes of administration is not equivalent to protracted workplace exposures. This difference is significant because of the steep dose-response curve for acute toxic effects. This is shown by the fact that on administration via the feed or drinking water a multiple of the acute oral LD_{50} is tolerated each day.

5.3 Effects on skin and mucous membranes

There are no studies of dermal irritation. Clear signs of eye irritation were observed on exposure of the animals' eyes to HCN, NaCN or KCN (Ballantyne 1983b, 1988).

Allergenic effects of cyanides have not been studied.

5.4 Reproductive and developmental toxicity

Pregnant golden hamsters (5–7 animals/dose group) received infusions of NaCN delivered in constant volumes by osmotic minipumps implanted subcutaneously in the neck. The pumps were implanted on day 6 of gestation and delivered calculated doses of 0, 0.126, 0.1275 or 0.1295 mmol/kg body weight and hour, equivalent to 6.17, 6.25 and 6.35 mg/kg body weight and hour. The calculation assumed a minipump delivery rate of 0.85 µl/hour. Preliminary tests had shown that ≤ 0.125 mmol/kg body weight and hour (6.12 mg/kg body weight and hour) had no teratogenic effects in 3 pregnant hamsters but that 0.133 mmol/kg body weight and hour (6.52 mg/kg body weight and hour) led to resorption of all the embryos and to high maternal toxicity. The tests were accordingly carried out in an extremely narrow dose range. The minipumps were removed 3 days after implantation on day 9 after conception, the hamsters were killed on day 10. More or less pronounced signs of maternal toxicity were found in the animals in all three cyanide-exposed groups: loss of weight, hypothermia, ataxia, salivation, dyspnoea. The animals recovered rapidly after removal of the minipumps. The cyanide doses were highly embryotoxic, producing early resorption rates of 63 %, 72 % and 83 % (control: 10 %). Except in the highest dose group, sufficient foetuses were still present, despite the high embryotoxicity, and 16/26 and 10/23 (control 4/75) showed malformations. The commonest anomalies were an open neural tube, exencephaly and encephalocele. The doses were thus also highly teratogenic, although, according to the authors, the malformations did not correlate with the intensity of maternal toxicity. It was possible to prevent both the maternal toxicity and the embryotoxic and teratogenic effects by simultaneous administration of sufficiently high thiosulphate doses (6.3 mg/kg body weight and hour). This indicates that the teratogenic effect is attributable to the cyanide itself and not to the metabolite thiocyanate. The authors of the study point out that no dose-effect relationships can be derived from the results because the delivery rate from the minipumps is very variable and the calculated delivery rates cannot be confirmed *in vivo*. Assuming an acute subcutaneous LD_{50} of 7.4 mg/kg body weight, the pregnant hamsters received from the minipumps in this study about 30 to 40 times the LD_{50} after bolus injection (Doherty *et al.* 1982). All these considerations must be taken into account in interpreting the study results. Accordingly, the study shows only that if highly maternally toxic cyanide doses are given during organogenesis they may cause malformations in hamsters like those produced by other cyanogenic substances such as acetonitrile (see "Acetonitrile" in the present volume). The available comparable investigations of the effects of cyanides or cyanogenic aliphatic nitriles on other species such as rats and rabbits suggest that the prenatal neural tube damage is probably specific to hamsters and is not a malformation seen in other species too (see "Acetonitrile" in the present volume).

Groups of 10 female rats were given KCN in concentrations of 50000 mg/kg diet or 100000 mg/kg diet for 13 weeks. None of the cyanide-treated rats, but 9 of 10 control animals, became pregnant (Olusi *et al.* 1979).

Female rats received a diet based on cassava meal (contains HCN in a concentration of 12 mg/kg meal) and containing 1250 mg of KCN per kg of feed, equivalent to about 125 mg of KCN per kg body weight. The animals were treated during mating (for up to 19 days), pregnancy, lactation and after weaning. Cyanide had no effect on the following parameters: body weight, weight gain during pregnancy, lactation and delivery, litter size, mortality of the pups during delivery, body weights of the pups at birth, weight gain of the pups during lactation, liver and kidney weights of the dams. Treatment of the pups after weaning resulted in a significant reduction in growth and feed consumption. There was likewise a reduction in the protein efficiency ratio (body weight gain/protein intake) for the treated pups (Tewe and Maner 1981a).

Groups of 6 pregnant sows received fresh cassava containing cyanide in a concentration of 277 and 521 mg/kg. The pigs were exposed from the day after mating until farrowing. No significant effects were found on reproduction parameters such as size of the piglets, size of the piglets during weaning, birth weights of the piglets, daily food intake of the sows and piglets, and body weight gain during pregnancy. Histopathological examination of the sows (n = 2/dose) revealed increased kidney cell hyperplasia. Histopathological changes were also found in the thyroid glands of the animals in the high-dose group (follicular cell hypoplasia) and increased relative thyroid weights. The foetuses in the high dose group showed reduced relative heart and spleen weights, whereas those in the low dose group were found to have reduced relative thyroid weights (Tewe and Maner 1981b).

In the 13-week study described in Section 5.2.2 (NTP 1993), sodium cyanide in the drinking water led to changes in some reproductive parameters in male rats and mice. In rats, the weight of the *cauda epididymis* was significantly reduced after CN^- doses of 1.6 mg/kg body weight and above, and after 12.5 mg/kg body weight there were significant reductions in the weights of the whole epididymis and of the testes, and in the number of (late) spermatids in the testes, together with reduced body weights. In male mice the weights of the whole epididymis and the *cauda epididymis* were significantly reduced only in the highest dose group. In both species there was no change relative to the control values in the sperm concentration in the epididymis after exposure. The authors therefore regard the observed reductions as not biologically relevant for the rodent species, but point out that humans are relatively more sensitive to such changes in reproductive parameters. In the female rats of the two higher dose groups there were merely slight shifts in the stages of the menstrual cycle, that is, prooestrus was longer and oestrus was shorter (see Table 8).

5.5 Genotoxicity

Tests with NaCN and KCN carried out in various *Salmonella typhimurium* strains (TA1535, TA1538, TA1537, TA98, TA100, TA97, TA102) and in the presence and absence of S9 mix yielded negative results. However, in one study positive results were obtained with HCN and the strain TA100 in the absence of metabolic activation. Negative results were obtained in tests for DNA damage with repair-competent and repair-deficient *Escherichia coli* strains WP67, CM871 and WP2, and in a test with the *Bacillus subtilis* strain N45. KCN did not inhibit DNA synthesis in the mouse testis (no further details; NL Health Council 1999, NTP 1993). NaCN did not cause DNA strand breaks in mouse lymphoma cells without metabolic activation (Garberg *et al.* 1988). KCN caused DNA double strand breaks in human lung epithelial cells only at concentrations which were cytotoxic, reducing cell survival by more than 40 % (Vock *et al.* 1998).

5.6 Carcinogenicity

There have been no studies of the carcinogenicity of HCN, KCN or NaCN. In a 14-week study with excessive doses, benign tumours of the coecum and colon were found in 5 of 10 rats in the 50 g/kg group and 7 of 10 rats in the 100 g/kg group. However, the MTD was clearly exceeded in these dose groups (25 % decrease in weight) (Olusi *et al.* 1979; Table 8). This study is unsuitable for assessment of carcinogenicity of these substances.

6 Manifesto (MAK value, classification)

Data available for human exposures show that long-term exposure to cyanide in concentrations of 4.7 to 13.9 mg/m^3 may lead to symptoms such as headaches, weakness, dizziness, throat irritation, dyspnoea, thyroid enlargement and an increase in thiocyanate excretion in the urine (El Ghawabi *et al.* 1975). No NOAEL was found in this study. Thiocyanate is the main metabolite of cyanide, and there is a causal association between exposure to thiocyanate and development of goitre. As stated in Section 4.2.1, occupational exposure to HCN in concentrations of 10 ml/m^3 for 8 hours must be expected to produce amounts of thiocyanate which are in the region of the LOEL for development of goitre. However, in the 13-week study on rats and mice, no effect on the thyroid was found with far higher doses. It is unclear whether or not humans are more sensitive to development of goitre caused by thiocyanate than are rats and mice. Irritant effects were seen in workers exposed to cyanide (alkali metal cyanides) in concentrations above 5 mg/m^3 (ACGIH 1996). In the 13-week study there were dose-dependent reductions in the weights of the reproductive organs of male rats and, in animals with reduced body weights, also a reduction in the spermatid concentration, which had, however, no effect on the sperm concentration. The NOAEL for cyanide in male rats is thus 4.6 mg/kg body weight. Changes in the phases of the menstrual cycle were found in female rats exposed at and above 4.6 mg/kg body weight. The NOAEL is 1.6 mg/kg body weight.

The cyanide detoxification capacity of humans is given as 0.1 to 1.0 mg/kg body weight and hour. On the basis of the lower figure, the amount of cyanide which can be detoxified per shift is calculated to be 56 mg, or 0.8 mg/kg body weight and day.

On the basis of the findings described above it was concluded that the MAK value of 10 ml/m^3 (11 mg/m^3) which was valid until 2001 was too high. A worker exposed to cyanide at a concentration of 2 mg/m^3 takes up a dose of 20 mg (worker weighing 70 kg, inhaling 10 m^3 in 8 hours, 100 % absorption). This is equivalent to 0.28 mg/kg body weight and day. Keeping the exposure concentration at this level or below would be sufficient to preclude all the effects mentioned above.

The MAK value for HCN, KCN and NaCN was therefore fixed uniformly in 2001 at 2 mg/m^3 (expressed as cyanide).

Because of the rapid onset of systemic effects, HCN is classified in Peak limitation category II. An excursion factor of 2 is appropriate because brief peaks at acutely toxic concentrations can be detoxified (see above: human detoxification capacity is given as 0.1 to 1.0 mg/kg body weight and hour). In the case of the salts it is not possible to decide whether the systemic or the local action is more important. As there are no data for the NOEL for local action of the substance, for safety the excursion factor is set at 1.

Scarcely any useful developmental toxicity studies are available; even the well-documented infusion study with pregnant hamsters showed only that the teratogenic effect of infused maternal-toxic to maternal-fatal doses can be prevented by thiosulphate treatment. This means that the cyanide ion is responsible for the teratogenic effects. The detoxification capacity mentioned above suggests that no prenatal toxicity is to be expected if the MAK value is observed, especially since it has been shown that the product of detoxification, thiocyanate, does not cause prenatal toxicity (Doherty *et al.* 1982). HCN, KCN and NaCN are classified in Pregnancy risk group C. This classification is also supported by data for the reproductive toxicity of acetonitrile, which is metabolized to cyanide and is likewise classified in Pregnancy risk group C with a far higher MAK value (see "Acetonitrile" in the present volume).

In an *in vitro* study, the permeability constant of human skin for aqueous solutions of cyanides was found to be 3.5×10^{-4} cm/hour. The permeability for an aqueous solution of HCN is almost 30 times higher, 10^{-2} cm/hour. A further study showed that toxic effects occur within 5 minutes after accidental exposure of a worker's hand to aqueous HCN solution. The designation "H" is therefore retained. The low dermal LD_{50} of HCN and cyanide salts also supports this designation.

No data are available for potential allergenic effects. Likewise, there are no useful studies of germ cell mutagenicity.

7 References

ACGIH (American Conference of Governmental Industrial Hygienists) (1996) Hydrogen cyanide and cyanide salts. in: *Documentation of TLVs and BEIs.* ACGIH, Cincinatti, OH, USA

Ahlmann H (1905) *Weitere Untersuchungen über die Giftigkeit der Blausäure* (Further studies of the toxicity of hydrogen cyanide) (German). Doctoral thesis, Bayerische Julius-Maximilians-Universität Würzburg

Ballantyne B (1983a) The influence of exposure route and species on the acute lethal toxicity and tissue concentrations of cyanide. in: Hayes AW, Schnell RC, Miya TS (Eds) *Developments in the Science and Practice of Toxicology*, New York, Elseviers Science Publishers BV

Ballantyne B (1983b) Acute systemic toxicity of cyanides by topical application to the eye. *J Toxicol Cut Ocular Toxicol 2*: 119–129

Ballantyne B (1987) Toxicology of cyanides. in: Ballantyne B, Marrs TC (Eds) *Clinical and Experimental Toxicology of Cyanides*, Wright, Bristol, UK, 41–126

Ballantyne B, Marrs TC (Eds) (1987) *Clinical and experimental toxicology of cyanides*, Wright, Bristol, UK

Ballantyne B (1988) Toxicology and hazard evaluation of cyanide fumigation powders. *J Toxicol Clin Toxicol 26*: 325–335

Ballantyne B (1994) Acute percutaneous systemic toxicity of cyanides. *J Toxicol Cut Ocular Toxicol 13*: 249–262

Banerjee KK, Bishayee A, Marimuthu P (1997) Evaluation of cyanide exposure and its effect on thyroid function in workers in a cable industry. *J Occup Environ Med 39*: 258–260

Blanc P, Hogan M, Mallin K, Hryhorczuk D, Hessl S, Bernard B (1985) Cyanide intoxication among silver-reclaiming workers. *J Am Med Assoc 253*: 367–371

Borgohain R, Singh AK, Radhakrishna H, Rao C, Mohandas S (1995) Delayed onset generalised dystonia after cyanide poisoning. *Clin Neurol Neurosurg 97*: 213–215

Carella F, Grassi MP, Savoiardo M, Contri P, Rapuzzi B, Mangoni A (1988) Dystonic-parkinsonian syndrome after cyanide poisoning: clinical and MRI findings. *J Neurol Neurosurg Psychiatry 51*: 1345–1348

Chandra H, Gupta BN, Bhargava SK, Clerk SH, Mahendra PN (1980) Chronic cyanide exposure, a biochemical and industrial hygiene study. *J Anal Toxicol 4*: 161–165

Chen KK, Rose CL (1952) Nitrite and thiosulfate therapy in cyanide poisoning. *J Am Med Assoc 149*: 113–119

Dekant W, Vamvakas S, Henschler D (2001) Wichtige Gifte und Vergiftungen (Prominent poisons and poisonings) (German). in: Forth W, Henschler D, Rummel W, Förstermann U, Starke K (Eds) *Allgemeine und spezielle Pharmakologie und Toxikologie*, Urban & Fischer, München, 1034

Doherty PA, Ferm VH, Smith RP (1982) Congenital malformations induced by infusion of sodium cyanide in the golden hamster. *Toxicol Appl Pharmacol 64*: 456–464

Dschang KY (1928) *Entgiftung eingeatmeter Blausäure durch Schwefelpräparate* (Detoxification of inhaled hydrogen cyanide by sulfur compounds) (German). Doctoral thesis, Bayerische Julius-Maximilians-Universität Würzburg

Dudley HC, Sweeny TR, Miller JW (1942) Toxicology of acrylonitrile (vinyl cyanide) II. Studies of effects of daily inhalation. *J Ind Hyg 24*: 255–258

Dugard PH (1987) The absorption of cyanide through human skin *in vitro* from solutions of sodium cyanide and gaseous HCN. in: Ballantyne B, Marrs TC (Eds) (1987) *Clinical and Experimental Toxicology of Cyanides, Wright*, Bristol, UK, 127–137

El Ghawabi SH, Gaafar MA, El Saharti AA, Ahmed SH, Malash KK, Fares R (1975) Chronic cyanide exposure: a clinical, radioisotope, and laboratory study. *Br J Ind Med 32*: 215–219

EPA (US Environmental Protection Agency) (1992) *Drinking water criteria document for cyanide.* prepared by the Office of Health and Environmental Assessment, Environmental Criteria and Assessment Office, Cincinatti, OH, for US EPA, Washington DC, External Review Draft

Fassett DW (1962) Cyanides and Nitriles. in Patty FA (Ed.) *Industrial Hygiene and Toxicology Vol. II*, 2nd revised edition, Interscience Publishers, John Wiley & Sons, New York, London, p 1996

Flury F, Heubner W (1919) Über Wirkung und Entgiftung eingeatmeter Blausäure (Action and detoxification of inhaled hydrogen cyanide) (German). *Biochem Z 95*: 249–256

Flury F, Zernik F (1931) *Schädliche Gase* (Hazardous gases) (German). Springer, Berlin, p 400

Forst JA (1928) Zur Entgiftung der Blausäure (Detoxification of hydrogen cyanide) (German). *Naunyn-Schmiedebergs Arch Pharmakol 128*: 1–66

Friedberg KD, Schwarzkopf HA (1969) Blausäureexhalation bei der Cyanidvergiftung (Exhalation of hydrogen cyanide in cyanide poisoning) (German). *Arch Toxikol 24*: 235–248

Garberg P, Akerblom E-L, Bolcsfoldi G (1988) Evaluation of a genotoxicity test measuring DNA-strand breaks in mouse lymphoma cells by alkaline unwinding and hydroxyapatite elution. *Mutat Res 203*: 155–176

Grandas F, Artieda J, Obeso JA (1989) Clinical and CT scan findings in a case of cyanide intoxication. *Mov Disord 4*: 188–193

Hardy HL, Jeffries WMcK, Wasserman MM, Wadell WR (1950) Thiocyanate effect following industrial cyanide exposure. Report of two cases. *N Engl J Med 242*: 968–972

Henschler D (1965) unpublished data from the Department of Toxicology, University of Würzburg

Henschler D (Ed.) (1971) Cyanwasserstoff (Hydrogen cyanide) (German). in: *Toxikologisch-arbeitsmedizinische Begründungen von MAK-Werten*, 1st issue, 1972. VCH-Verlagsgesellschaft, Weinheim

Henschler D (Ed.) (1981) Stickstoffwasserstoffsäure, Natriumazid. in: *Toxikologisch-arbeitsmedizinische Begründungen von MAK-Werten*, 8th issue. VCH-Verlagsgesellschaft, Weinheim

Hlynczak JA, Kersten E, Wysocki K, Stamm E, Fokt M, Raczynski A (1980) Untersuchungen zur Aktivität einiger Enzyme im Serum HCN-exponierter Frauen. *Z Aerztl Fortbild 74*: 591–593

Hofmann HTh (1971) personal communication of data from the toxicology laboratory of the BASF AG, Ludwigshafen

Howard JW, Hanzal RF (1955) Chronic toxicity for rats of food treated with hydrogen cyanide. *Agric Food Chem 3*: 325–329

Hugod C (1979) Effect of exposure to 0.5 ml/m^3 hydrogen cyanide singly or combined with 200 ml/m^3 carbon monoxide and/or 5 ml/m^3 nitric oxide on coronary arteries, aorta, pulmonary artery, and lungs in the rabbit. *Int Arch Occup Environ Health 44*: 13–23

Hunt R (1923) Cyanwasserstoff, Nitrilglukoside, Nitrile, Rhodanwasserstoff, Isocyanide (Hydrogen cyanide, nitrile glycosides, nitriles, thiocyanate, isocyanides) (German). in Heffter A (Ed.) *Handbuch der experimentellen Pharmakologie, Vol. 1*, Springer, Berlin, 771–774

Isom GE, Liu DHW, Way JL (1975) Effect of sublethal doses of cyanide on glucose catabolism. *Biochem Pharmacol 24*: 871–875

Jackson LC (1988) Behavioral effects of chronic sublethal dietary cyanide in an animal model: Implications for humans consuming cassava (*Manihot esculenta*). *Hum Biol 60*: 597–614

Katsumata Y, Sato K, Oya M, Suzuki O, Yoshino M (1980) Kinetic analysis of the shift of aerobic to anaerobic metabolism in rats during acute cyanide poisoning. *Life Sci 27*: 1509–1512

Katsumata Y, Sato K, Yada S, Suzuki O, Yoshino M (1983) Kinetic analysis of anaerobic metabolism in rats during acute cyanide poisoning. *Life Sci 33*: 151–155

Maehly AC, Swensson A (1970) Cyanide and thiocyanate levels in blood and urine of workers with low-grade exposure to cyanide. *Int Arch Arbeitsmed 27*: 195–209

Matijak-Schaper M, Alarie Y (1982) Toxicity of carbon monoxide, hydrogen cyanide and low oxygen. *J Combustion Toxicol 9*: 21–61

Mercker H, Lochner W, Gestenberg E (1958) Untersuchungen über das Verhalten des Kreislaufes, der Atmung und der Sauerstoffsättigung des Blutes bei Cyanidvergiftung (Effects of cyanide poisoning on circulation, respiration and oxygen saturation of the blood) (German). *Naunyn-Schmiedebergs Arch Pharmakol 232*: 459–469

Messing B (1991) Extrapyramidal disturbances after cyanide poisoning (first MRT-investigation of the brain). *J Neural Transm 33, Suppl*: 141–147

NL Health Council (1999) *Hydrogen cyanide, sodium cyanide, and potassium cyanide, Draft.* Dutch Expert Committee on Occupational Standards, Health Council of the Netherlands, Den Haag

NTP (1993) *Technical report on toxicity studies of sodium cyanide, 37, NIH Publication 94-3386*, US Department of Health and Human Services, National Institutes of Health, Bethesda, MD, USA

O'Donnell JE, Mundt HW, Knudsen WN, Delano PH (1940) Hydrogen cyanide gas fumigation. *J Ind Hyg 22*: 253–275

Offterdinger H, Weger N (1969) Kreislauf und Atmung bei Blausäurevergiftung und Therapie mit Ferrihämoglobinbildnern und Kobaltverbindungen (Circulation and respiration in cyanide poisoning and therapy with ferrihemoglobin-forming and cobalt compounds) (German). *Naunyn-Schmiedebergs Arch Pharmakol 264*: 289

Okolie NP, Osagie AU (1999) Liver and kidney lesions and associated enzyme changes induced in rabbits by chronic cyanide exposure. *Food Chem Toxicol 37*: 745–750

Okolie NP, Osagie AU (2000) Differential effect of chronic cyanide intoxication on heart, lung and pancreatic tissue. *Food Chem Toxicol 38*: 543–548

Olusi SO, Oke OL, Odusote A (1979) Effects of cyanogenic agents on reproduction and neonatal development in rats. *Biol Neonat 36*: 233–243

Philbrick DJ, Hopkins JB, Hill DC, Alexander JC, Thomson RG (1979) Effects of prolonged cyanide and thiocyanate feeding in rats. *J Toxicol Environ Health 5*: 579–592

Rosenberg NL, Myers JA, Martin WRW (1989) Cyanide-induced parkinsonism: clinical, MRI, and 6 fluorodopa (PET) studies. *Neurology 39*: 142–144.

Sandberg CG (1967) A case of chronic poisoning with potassium cyanide? *Acta Med Scand 181*: 233–236

Sheehy M, Way JL (1968) Effect of oxygen on cyanide intoxication. 3. Mithridate. *J Pharm Exp Ther 161*: 163–168

Schütze W (1927) Über die Gefährdung von Mensch und Tier durch große Konzentrationen einiger giftiger Gase von der Haut aus (Danger from exposure of the skin of man and animals to high concentrations of poisonous gases) (German). *Arch Hyg 98*: 70–83

Schwab R (1929) Der Einfluß von Traubenzucker auf den Verlauf von Giftwirkungen (The effect of glucose on the course of intoxication) (German). *Z Ges Exp Med 67*: 513–537

Szinicz L (1996) Toxische Gase. in: Greim H, Deml E (Eds) *Toxikologie: Eine Einführung für Naturwissenschaftler und Mediziner*, VCH-Verlagsgesellschaft, Weinheim

Tewe OO, Maner JH (1981a) Long-term and carry-over effect of dietary inorganic cyanide (KCN) in the life cycle performance and metabolism of rats. *Toxicol Appl Pharmacol 58*: 1–7

Tewe OO, Maner JH (1981b) Performance and pathophysiological changes in pregnant pigs fed cassava diets containing different levels of cyanide. *Res Vet Sci 30*: 147–151

Uitti RJ, Rajput AH, Ashenhurst EM, Rozdilsky B (1985) Cyanide-induced parkinsonism: a clinicopathological report. *Neurology 35*: 921–925

Valade MP (1952) Lésions du système nerveux central dans le intoxications chroniques expérimentales par l'acide cyanhydrique gazeuse. *Bull Acad Natl Med, Paris 136*: 280–285

Valenzuela R, Court J, Godoy J (1992) Delayed cyanide induced dystonia. *J Neurol Neurosurg Psychiatry 55*: 198–199

Vock EH, Lutz WK, Hormes P, Hoffmann HD, Vamvakas S (1998) Discrimination between genotoxicity and cytotoxicity in the induction of DNA double-strand breaks in cells treated with etoposide, melphalan, cisplatin, potassium cyanide, Triton X-100, and gamma-irradiation. *Mutat Res 413*: 83–94

Wagschal F (1903) *Quantitative Studien über die Giftigkeit der Blausäure-Dämpfe* (Quantitative studies of the toxicity of hydrogen cyanide) (German). Doctoral thesis, Bayerische Julius-Maximilians-Universität Würzburg

Way JL, Gibbon SL, Sheehy M (1966) Effect of oxygen on cyanide intoxication I. Prophylactic protection. *J Pharm Exp Ther 153*: 381–385

completed 01.03.2001

Isobutyl acetate

MAK value (1997)	**100 ml/m³ (ppm) ≙ 480 mg/m³**
Peak limitation (2000)	**Category I, excursion factor 2**
Absorption through the skin	**–**
Sensitization	**–**
Carcinogenicity	**–**
Prenatal toxicity (1999)	**Pregnancy risk group C**
Germ cell mutagenicity	**–**
BAT value	**–**
Synonyms	acetic acid isobutyl ester 2-methyl-1-propyl acetate 2-methylpropyl acetate *beta*-methylpropyl ethanoate natural isobutyl acetate
Chemical name (CAS)	acetic acid 2-methylpropyl ester
CAS number	110-19-0
Structural formula	$CH_3COO–CH_2–CH(CH_3)–CH_3$
Molecular formula	$C_6H_{12}O_2$
Molecular weight	116.16
Melting point	–99°C
Boiling point	111–118°C
Density at 20°C	0.87 g/cm³
Vapour pressure at 20°C	18 hPa
log P_{ow}*	1.72
1 ml/m³ (ppm) ≙ 4.82 mg/m³	**1 mg/m³ ≙ 0.207 ml/m³ (ppm)**

* *n*-octanol/water distribution coefficient

1 Toxic Effects and Mode of Action

Once absorbed by the organism or in contact with mucous membranes, isobutyl acetate is probably hydrolysed to isobutyl alcohol and acetic acid. Quantitative data for this reaction are not available. According to unpublished results, the critical effects in man, the irritative effects on the eyes and mucous membranes of the upper respiratory tract, are weaker than those of *n*-butyl acetate. The acute toxicity of isobutyl acetate is low after inhalation, oral and dermal exposure. The substance does not affect rabbit skin, and has only weak irritative effects on the rabbit eye. Neither in man nor in experiments with animals have sensitizing effects been demonstrated. Isobutyl acetate was not found to be mutagenic in various *Salmonella typhimurium* strains nor clastogenic in V79 cells.

2 Mechanism of Action

There are no data available for the mechanism of action of isobutyl acetate. The irritative effects may be caused by the acetic acid formed during hydrolysis.

3 Toxicokinetics and Metabolism

After contact with mucous membranes and absorption by the organism, isobutyl acetate may be assumed to be hydrolysed to the corresponding butyl alcohol and acetic acid in reactions catalysed by various esterases; this has been demonstrated for all butyl acetates with isolated nasal tissue from the rat. The rate of hydrolysis decreased with increasing branching of the butyl residue and was between 42 and 77 nmol/mg protein and minute. This can be explained by increasing steric hindrance (Dahl *et al.* 1987). There is no other information available.

4 Effects in Man

Unpublished results indicate that the irritative effects of isobutyl acetate are weaker than those of *n*-butyl acetate (ACGIH 1991). A maximization test with 2 % isobutyl acetate in petrolatum applied to 28 persons yielded negative results (no other details) (Opdyke 1978).

5 Animal Experiments and *in vitro* Studies

There are no data available for the subacute, subchronic or chronic toxicity of isobutyl acetate, or its carcinogenicity.

5.1 Acute toxicity

LC_{50} values are not available for isobutyl acetate. At concentrations of 8000 ml/m^3 4 of 6 rats died, and 16000 ml/m^3 was lethal for all 6 animals (Smyth *et al.* 1962). The oral LD_{50} for rats was found to be 15400 mg/kg body weight (Smyth *et al.* 1962) and for rabbits 4763 mg/kg body weight (Bayer 1997). The dermal LD_{50} for rabbits was more than 20 ml/kg body weight (17400 mg/kg body weight) (Smyth *et al.* 1962). The RD_{50} for mice was found to be 3890 mg/m^3 (805 ml/m^3). The concentration–effect relationship was not described (Muller and Greff 1984).

5.2 Local effects on skin and mucous membranes

Non-occlusive dermal application of 0.01 ml isobutyl acetate did not lead to irritation of the rabbit skin. The instillation of up to 0.5 ml into the conjunctival sac led to slight eye irritation in rabbits (Smyth *et al.* 1962).

5.3 Allergenic effects

In a maximization test with guinea pigs carried out in accordance with OECD guideline 406, isobutyl acetate was not found to have sensitizing effects (Bayer 1997).

5.4 Reproductive and developmental toxicity

There are no data available for the reproductive and developmental toxicity of isobutyl acetate. The toxic potential of the metabolites can, however, be evaluated. Isobutyl alcohol, with a MAK value of 100 ml/m^3, is classified in Pregnancy risk group C (see "Isobutyl alcohol" in this volume). The other metabolite, acetic acid, is at present not classified in any Pregnancy risk group. The effective principle of a potential prenatal toxicity in this case could be acidosis. However, even exposure of rabbits for over 2 hours to very high concentrations of various alkyl acetate esters (e.g. *n*-propyl acetate concentrations of 54000 mg/m^3, corresponding to 13000 ml/m^3) did not produce such effects (Flury and Wirth 1933), so that during exposure to concentrations in the range of the MAK value, prenatal toxicity resulting from acetic acid is not to be expected for isobutyl acetate. This is confirmed by a study with rabbits given cider vinegar in doses of 1600 mg/kg body weight from days 6 to 18 of gestation. Assuming an acetic acid content of 5 %, this corresponds to acetic acid doses of 80 mg/kg body weight. Neither maternal toxicity nor foetal toxicity, nor an increased incidence of malformations in the foetuses were observed (FDA 1977).

5.5 Genotoxicity

A *Salmonella* mutagenicity test with the strains TA98, TA100, TA1535, TA1537 and TA1538 with and without an added metabolic activation system, yielded negative results up to the highest concentration tested of 5 mg/plate (Bayer 1997). In V79 cells the substance did not cause structural chromosomal aberrations. Concentrations up to 2500 µg/ml were tested with and without the addition of a metabolic activation system. At this concentration the mitotic index was reduced by as much as 50 % (BAU 1996).

6 Manifesto (MAK value/classification)

For the workplace situation, the critical effects of isobutyl acetate are its irritative effects. There are no data for the irritation threshold of isobutyl acetate. It can be assumed, however, that this is not very different from the irritation threshold of the better investigated *n*-butyl acetate; this has a MAK value of 100 ml/m^3 (see "*n*-Butyl acetate" in this volume). The MAK value for isobutyl acetate has therefore been set at 100 ml/m^3 in analogy to *n*-butyl acetate. Also in analogy to *n*-butyl acetate, for Peak limitation category I an excursion factor of 2 has been established provisionally.

Although studies of the systemic toxicity of isobutyl acetate itself are not available, this can be estimated by evaluating the systemic toxicity of the metabolites. The amount of acetic acid taken up daily if the MAK value is observed, at most about 2.5 g, is of the order of magnitude of the average amount taken up with food (Guest *et al.* 1982). In addition, acetic acid is formed endogenously and is produced in numerous pathways of intermediary metabolism, so that no adverse effects are expected as a result of the additional occupational exposure. For isobutyl alcohol, the MAK value has also been set at 100 ml/m^3, a concentration at which no systemic effects occur (see "Isobutyl alcohol" in this volume).

Studies of the reproductive toxicity of isobutyl acetate itself are not available. Evaluation of the toxic effects of the metabolites on reproduction can, however, provide information about the effects of the original substance. Isobutyl alcohol, with a MAK value of 100 ml/m^3, is classified in Pregnancy risk group C (see "Isobutyl alcohol" in this volume). Given observation of the MAK value for isobutyl acetate, prenatal toxicity is not to be expected from acetic acid, the other metabolite (see Section 5.4). There is thus no reason to fear a risk of damage to the embryo or foetus when the MAK value is observed, and isobutyl acetate has therefore been classified in Pregnancy risk group C.

Because of the lack of data, the substance is not designated with an "H" (for substances for which there is danger from cutaneous absorption).

Sensitizing effects were not detected in either man or experiments with animals; isobutyl acetate is therefore not designated with an "S".

7 References

ACGIH (American Conference of Governmental Industrial Hygienists) (1991) Isobutyl acetate, *sec*-butyl acetate, *tert*-butyl acetate. in: *Documentation of the threshold limit values and biological exposure indices*, ACGIH, Cincinnati, OH, USA

BAU (Bundesanstalt für Arbeitsschutz) (Eds) (1996) *Prüfung exemplarisch ausgewählter Altstoffe auf eine cytogenetische Wirkung in vitro: Induktion von Chromosomenaberrationen* (Testing of a selection of existing chemicals for cytogenetic effects *in vitro*: induction of chromosomal aberrations) (German). Fraunhofer-Institut für Toxikologie und Aerosolforschung, Report No. F 1250

Bayer (1997) *IUCLID data sheet, isobutyl acetate, 28.01.1997*, unpublished

Dahl AR, Miller SC, Petridou-Fischer J (1987) Carboxylesterases in the respiratory tracts of rabbits, rats and Syrian hamsters. *Toxicol Lett 36*: 129–136

Flury F, Wirth W (1933) Zur Toxikologie der Lösungsmittel (Toxicology of solvents) (German). *Arch Gewerbepathol Gewerbehyg 5*: 1–90

FDA (Food and Drug Administration) (1977) *Evaluation of the health aspects of acetic acid, sodium acetate, and sodium diacetate as food ingredients*. Report No. FDA/BF-78/16; NTIS Doc No. PB-274670, NTIS, Springfield, VA, USA

Guest D, Katz GV, Astill BD (1982) Aliphatic carboxylic acids. in: Clayton GD, Clayton FE (Eds) *Patty's Industrial Hygiene and Toxiology, Vol. 2C*, John Wiley & Sons, New York, 4910

Muller J, Greff G (1984) Recherche de relations entre toxicité de molécules d'intérêt industriel et propriétés physico-chimiques: test d'irritation des voies aériennes supérieures appliqué à quatre familles chimiques (Relation between the toxicity of molecules of industrial value and their physico-chemical properties: test of upper airway irritation applied to 4 chemical groups) (French). *Food Chem Toxicol 22*: 661–664

Opdyke DLJ (1978) Fragrance raw material monographs, isobutyl acetate. *Food Cosmet Toxicol 16*: 795–796

Smyth Jr HF, Carpenter CP, Weil CS, Pozzani UC, Striegel JA (1962) Range-finding toxicity data: List VI. *Am Ind Hyg Assoc J 23*: 95–107

completed 06.04.1999; peak limitation category 01.05.2000

Isobutyl alcohol

MAK value (1969)	**100 ml/m^3 (ppm) ≙ 310 mg/m^3**
Peak limitation (2000)	**Category I, excursion factor 1**
Absorption through the skin	**–**
Sensitization	**–**
Carcinogenicity	**–**
Prenatal toxicity (1992)	**Pregnancy risk group C**
Germ cell mutagenicity	**–**
BAT value	**–**
Synonyms	fermentation butyl alcohol 1-hydroxymethylpropane isopropylcarbinol
Chemical name (CAS)	2-methyl-1-propanol
CAS number	78-83-1
Structural formula	$(CH_3)_2CH{-}CH_2OH$
Molecular formula	$C_4H_{10}O$
Molecular weight	74.12
Melting point	–108°C
Boiling point	107.9°C
Density at 20°C	0.80 g/cm^3
Vapour pressure at 20°C	11.7 hPa
log P_{ow}*	0.83
1 ml/m^3 (ppm) ≙ 3.03 mg/m^3	**1 mg/m^3 ≙ 0.330 ml/m^3 (ppm)**

The MAK value for isobutyl alcohol of 100 ml/m^3 was established in 1969 in analogy to the TLV value at the time. The present documentation was based on reviews of toxicological data for isobutyl alcohol (ECB 1995, WHO 1987) and supplemented with more recent results and includes the 1992 MAK documentation of the toxic effects of isobutyl alcohol on reproduction.

* *n*-octanol/water distribution coefficient

1 Toxic Effects and Mode of Action

Absorbed isobutyl alcohol is distributed evenly in the organism. The substance is probably oxidized via alcohol dehydrogenase and aldehyde dehydrogenase to form isobutyric acid, which after further metabolism enters the citric acid cycle as propionyl-CoA. In man, isobutyl alcohol vapour has similar effects to *n*-butyl alcohol vapour; but there are no data for effect thresholds. In addition, keratopathy with vacuolization of the cornea has been described. Allergic reactions of the skin to isobutyl alcohol are rare and usually interpreted as cross-reactions. The acute toxicity of the substance in experimental animals is low for all absorption routes. Even after repeated inhalation exposure of rats and mice, apart from reversible impairment of central nervous functions no other adverse effects were observed up to the highest concentration tested of 2500 ml/m^3. For oral administration of the substance to rats, the NOEL (no observed effect level) was found to be over 1450 mg/kg body weight. On rabbit skin isobutyl alcohol has moderately irritative effects. In the rabbit eye it is irritative to highly irritative. Isobutyl alcohol was not found to be mutagenic in prokaryotes. In neither rats nor rabbits exposed to concentrations of up to 3300 ml/m^3 were effects on the offspring detected during organogenesis.

2 Mechanism of Action

The irritative effects of the butyl alcohol isomers are probably the result of non-covalent interaction with the receptors of the sensory nerve endings in the mucous membranes of the respiratory tract and are a function of the physico-chemical properties (e.g. vapour pressure, lipophilia) of the substances (Abraham *et al.* 1996, Alarie *et al.* 1998, Hansen and Nielsen 1994). The central nervous effects which occur at higher concentrations are probably the result of the interaction of the relatively hydrophobic substances with neuronal membranes (Fang *et al.* 1997, Frantik *et al.* 1994).

3 Toxicokinetics and Metabolism

There are no studies available of the rate of absorption of isobutyl alcohol via various routes. On the basis of the physico-chemical properties of isobutyl alcohol, a penetration rate through human skin of 1.78 mg/cm^2 and hour (Fiserova-Bergerova *et al.* 1990) or 0.2 mg/cm^2 and hour (Guy and Potts 1993) has been estimated for a saturated aqueous isobutyl alcohol solution. Assuming both hands and forearms were wetted for an hour (skin area of about 2000 cm^2), with a flux of about 1 mg/cm^2 and hour the amount of isobutyl alcohol absorbed would be 2000 mg or a dose of about 30 mg/kg body weight.

Six volunteers drank a drink containing ethanol to which isobutyl alcohol had been added. The isobutyl alcohol dose was 5 mg/kg body weight, the ethanol dose 800 mg/kg body weight. The isobutyl alcohol concentration in serum increased in the first 45 minutes after administration to a maximum of 5.5 mg/kg; elimination then followed an exponential course; 4 hours after administration 1.1 mg/kg was still detectable. The authors calculated a half-time for elimination from the plasma of 1.45 hours. The potential mutual influence of the substances on their elimination behaviour was not investigated (Bilzer *et al.* 1990). Also in rabbits given oral isobutyl alcohol doses of 2 ml/kg body weight, the maximum concentrations in blood were reached 1 hour after administration; 6 hours after administration, the substance could no longer be detected (WHO 1987). There are no studies of the distribution of isobutyl alcohol in the human organism. A blood/air distribution coefficient of 453 ± 37 was determined with human blood for isobutyl alcohol at 37°C, and tissue/air distribution coefficients with human tissue homogenates of 343 ± 46 for muscle tissue, 371 ± 28 for kidney tissue, 400 ± 42 for lung tissue, 337 ± 57 for white brain matter, 387 ± 29 for grey brain matter and 388 ± 33 for adipose tissue (Fiserova-Bergerova and Diaz 1986). The substance can therefore be assumed to be distributed evenly in the organism. Also from the distribution coefficients liver/air, muscle/air and fat/air for isobutyl alcohol of 880, 850 and 720 (Kaneko *et al.* 1994) determined at 37°C with rat tissue, it can be concluded that the inhaled substance is distributed evenly in the whole body.

Isobutyl alcohol can be oxidized to isobutyric acid, which after β-oxidation enters the citric acid cycle as propionyl-CoA. Acetaldehyde, acetic acid, isobutyraldehyde, isobutyl alcohol and isovaleric acid were detected in the urine of rabbits given single oral isobutyl alcohol doses of 2 ml/kg body weight and then offered drinking water saturated with isobutyl alcohol. The source of the isovaleric acid was not discussed by the authors. It is not a plausible isobutyl alcohol metabolite. In experiments with rat liver homogenate and with *in situ* perfused rat liver it was demonstrated that the rate of oxidation of isobutyl alcohol is higher than that of ethanol. From their experiments with pyrazole, an alcohol dehydrogenase inhibitor, the authors concluded that isobutyl alcohol is oxidized by this enzyme.

The elimination of isobutyl alcohol and its metabolites has not been sufficiently investigated. After ingestion of 2 g/kg body weight, rats excreted 0.27 % of the dose with the urine within 8 hours. Rabbits given oral doses of about 2 ml excreted 4.4 % of the administered substance as the glucuronide conjugate with the urine within 24 hours. In the exhaled air of rabbits given 6 ml isobutyl alcohol, no aldehydes or ketones were detected during the 6 hours after administration (ECB 1995, WHO 1987).

4 Effects in Man

In the few reports of exposures to isobutyl alcohol there are no data for the levels of exposure, and in all cases the persons were exposed to mixtures of substances (ECB 1995, WHO 1987). The symptoms described were like those after exposure to *n*-butyl alcohol. Also the corneal changes (keratopathy with vacuolization of the corneal epithelium) characteristic of exposure to *n*-butyl alcohol vapour (see the chapter on *n*-Butyl alcohol in this volume) were observed after exposure to isobutyl alcohol (Rowe and McCollister 1982).

4.1 Allergenic effects

There are only a few reports of positive results obtained in patch tests with isobutyl alcohol. It seems probable that they were the result of cross-reactions because in these patients ethanol was very likely the sensitizing agent. In patch tests with such patients, positive results are frequently obtained with various primary and secondary alcohols and with acetone. In some of the persons, skin reactions also developed after the consumption of alcoholic drinks, and in addition erythema, aphthous lesions and a burning sensation in the oral mucosa (Fisher 1983, Fregert *et al.* 1969, Ophaswongse and Maibach 1994, Stotts and Ely 1977, Van Ketel and Tan-Lim 1975). From these results, it cannot be concluded that isobutyl alcohol has a significant sensitizing potential, if any.

5 Animal Experiments and *in vitro* Studies

There are no data available for allergenicity of isobutyl alcohol in animals.

5.1 Acute toxicity

Like all butyl alcohol isomers, isobutyl alcohol leads via all routes of absorption to the typical symptoms of acute alcohol intoxication, from central nervous depression to narcosis. The severity of the effects of the individual isomers is about the same and overall greater than that of ethanol (WHO 1987).

5.1.1 Inhalation

All rats survived exposure for 2 hours to an atmosphere saturated with isobutyl alcohol (about 16000 ml/m^3). Exposure to concentrations of 8000 ml/m^3 for 4 hours was lethal for 2/6 animals. In a Russian study LC_{50} values in the range from 15500 to 26250 mg/m^3 were given for the species rat, mouse, rabbit and guinea pig. In rats and rabbits, irritation of the respiratory tract was described after exposure for 4 hours to concentrations of 15700 mg/m^3 (about 5000 ml/m^3) and more. Three days after the end of exposure, central nervous effects, a reduced number of lymphocytes in the bone marrow, and morphological changes such as dystrophy of the hepatocytes and the olfactory neurons in the brain were observed in the animals. The same effects were also seen at concentrations of

8000 mg/m^3, but were less marked. At 1300 mg/m^3 only the lymphocyte count in bone marrow was decreased, while 100 mg/m^3 led to a changed respiration rate (WHO 1987). The RD_{50} value in Swiss-OF_1 mice was given as 1818 ml/m^3 for 5 minutes exposure (Bos *et al.* 1992).

5.1.2 Ingestion and dermal absorption

For the species mouse, rat and rabbit the oral LD_{50} values were between 2500 and 3500 mg/kg body weight. For the rabbit, dermal LD_{50} values in the range from 2500 to 4200 mg/kg body weight were found. More details may be found in other reviews (ECB 1995, WHO 1987).

5.2 Subacute, subchronic and chronic toxicity

5.2.1 Inhalation

Groups of five 8-week-old Sprague-Dawley rats per sex given food *ad libitum* and groups of five 16-week-old male rats whose calorie uptake was limited were exposed to isobutyl alcohol concentrations of 750, 1500 or 3000 ml/m^3 for 6 hours a day on 5 days a week for 2 weeks. During the exposure periods the animals of the 750 ml/m^3 group were less sensitive to external stimuli, while the breathing of the animals of the two high concentration groups was more laboured and clear signs of central nervous depression were observed. The haematological and ophthalmological examinations and autopsy did not reveal pathological findings. A histological examination of the organs was not carried out. The results of a battery of behavioural tests carried out before the beginning of and after the 2-week exposure did not differ (CMA 1996a). A medium-term inhalation study with groups of 10 to 20 Sprague-Dawley rats per concentration and sex investigated the influence of the exposure on learned operant behaviour patterns with food rewards as positive reinforcement. For this purpose the animals were offered only a limited amount of food before the experiment to keep their body weights between 300 and 350 g. Exposure to isobutyl alcohol concentrations of 250, 1000 or 2500 ml/m^3 for 6 hours a day on 5 days a week for 90 days had no effect on the results of the daily behavioural tests. During the exposure periods the animals of all concentration groups reacted less sensitively to external stimuli. In the exposure-free intervals no differences from the controls were detected in this respect. Body weight gains were the same in all groups. The ophthalmological examination did not reveal any differences between the 2500 ml/m^3 group and the control group. Also the results of a battery of behavioural tests and motor activity tests were normal. The haematological examination revealed a significant increase in the erythrocyte count of the female animals of the high concentration group, a finding that did not seem to be of toxicological significance. In half of the 10 to 20 animals per group the usual gross pathological and histological examinations were carried out, in the other half only the nervous system was examined. Pathological findings related to the exposure were not detected. Thus, up to the highest concentration tested of 2500 ml/m^3, isobutyl alcohol led merely to reversible impairment of central nervous functions during exposure. Specific neurotoxic effects were not detected, even using specific methods (CMA 1996b).

5.2.2 Ingestion

In a study of medium-term oral toxicity carried out according to OECD guideline 408, a NOAEL (no observed adverse effect level) of at least 1450 mg/kg body weight was determined for isobutyl alcohol in the rat. Groups of 10 male and 10 female Wistar rats were given the substance with the drinking water in concentrations of 1000, 4000 and 16000 mg/l for 90 days. The resulting doses were about 80, 340 and 1450 mg/kg body weight and day. The high dose was determined in pilot studies as the highest tolerable dose. At higher doses water consumption was significantly decreased, probably as a result of reduced palatability. In the extensive clinico-chemical, haematological, gross pathological and histopathological examinations, substance-related adverse effects were not detected in any dose group (Schilling *et al.* 1997). In a study from 1974, 5 to 6 male Wistar rats were given isobutyl alcohol in concentrations of 74000 mg/l in the drinking water for 4 months or 148000 mg/l for 2 months. At the low dose, bleeding was observed in the small intestines of some animals; liver damage was not detected. At the high dose, the levels of fat, glycogen and DNA in the liver were reduced, and the liver cells were smaller (no other details) (ECB 1995, WHO 1987).

5.3 Local effects on skin and mucous membranes

The irritative effects of isobutyl alcohol on the skin and eyes of the rabbit were investigated in several studies, some of which, however, are unpublished. According to a review, using various methods (e.g. Draize test, OECD guideline 404, BASF test) the substance was either not found to be irritative to the skin, or only moderately irritative. In one study, necrosis was observed in 2 of 6 rabbits; isobutyl alcohol was therefore regarded as caustic (ECB 1995). In most studies—including 2 carried out in accordance with OECD guideline 405—isobutyl alcohol was found to be irritative to strongly irritative in the rabbit eye. In contrast, in a Draize test and a BASF test, no irritating effects were recorded (ECB 1995).

5.4 Reproductive and developmental toxicity

There are no data available for the effects of isobutyl alcohol on fertility.

5.4.1 Developmental toxicity

Female Wistar rats were exposed to isobutyl alcohol concentrations of 500, 2500 and 10000 mg/m^3 air for 6 hours a day on days 6 to 15 of gestation. Neither in the dams nor in the foetuses were substance-related effects observed (BASF 1990a, Klimisch and Hellwig 1995).

Female Russian rabbits were exposed to isobutyl alcohol concentrations of 500, 2500 and 10000 mg/m^3 air for 6 hours a day on days 7 to 19 of gestation. Apart from marginally retarded body weight gains in the dams of the high concentration group, substance-related effects were not observed in either the dams or the foetuses (BASF 1990b, Klimisch and Hellwig 1995).

Dose-finding studies were carried out before the two main studies. No maternally toxic effects were found up to concentrations of 5000 mg/m^3 (limit test concentration, according to EPA-TSCA guideline). As foetotoxic effects were described at slightly higher concentrations in studies of other alcohols, 10000 mg/m^3 was selected as the highest concentration.

5.5 Genotoxicity

5.5.1 *In vitro*

In *Salmonella* mutagenicity tests with the strains TA97, TA98, TA100, TA1535, TA1537 and TA1538, isobutyl alcohol was not found to be mutagenic either with or without an added metabolic activation system (ECB 1995, Zeiger *et al.* 1988). A gene mutation test with *Escherichia coli* WP2uvrA also yielded negative results (ECB 1995). In a study from 1969, an increase in the frequency of reversions in *E. coli* CA 274 after incubation with 0.7 % isobutyl alcohol is described. The test was carried out only without metabolic activation, and the test concentration chosen was cytotoxic (ECB 1995, WHO 1987).

5.5.2 *In vivo*

In a Russian study, all butyl alcohol isomers were tested with groups of 8 rats in a cytogenetic *in vivo* test. The dose in each case was one fifth of the LD_{50}. All isomers led to increased chromosomal aberrations. It is unusual that no aberrations were found in the controls. The incidence of polyploid bone marrow cells was also increased in the animals treated with *n*-butyl alcohol, *sec*-butyl alcohol and isobutyl alcohol. No chromosome breaks were found. As the study did not investigate whether the effects were dose-dependent, they cannot be used as evidence of genotoxic effects of the butyl alcohol isomers (Barilyak and Kozachnuk 1988).

5.6 Carcinogenicity

In a study of long-term administration of isobutyl alcohol, a substance-related increase in the tumour incidence in Wistar rats was reported. One group of 19 animals were given gavage doses of isobutyl alcohol of 0.2 ml/kg body weight (about 150 mg/kg body weight) twice a week until their spontaneous death, another group of 24 animals subcutaneous doses of 0.05 ml/kg body weight twice a week. The average survival of the orally treated animals was 495 days, that of the subcutaneously treated animals 544 days. The control animals treated with physiological saline (two groups of 25) were killed at the age of 643 days. The only benign tumours in the control group were papillomas of the forestomach or fibroadenomas of the mamma, and they occurred only three times in the orally treated group and twice in the subcutaneously treated group. Malignant tumours were not found in the control animals. In the animals treated orally with isobutyl alcohol, 3 malignant tumours were found in various locations (forestomach carcinoma, liver cell carcinoma, myeloid leukaemia), in the animals treated subcutaneously with isobutyl alcohol there were 8 (forestomach carcinomas, liver cell sarcomas, a spleen sarcoma, a mesothelioma, retroperitoneal sarcomas). The numbers of benign tumours in the groups treated with isobutyl alcohol were 9 (oral group) and 3 (subcutaneous group). The signs of toxic effects described without further details included hyperplasia of the haematopoietic parenchyma in the bone marrow, metaplasia in the liver and spleen, and toxic liver damage including cirrhosis (Gibel *et al.* 1974, 1975). This and the markedly reduced survival of the animals treated with isobutyl alcohol indicates that the only dose tested in this study was far above the maximum tolerated dose. Therefore, and as a result of other shortcomings such as the lack of data for the numbers of animals with tumours and the lack of statistical analysis, this study cannot be included in the evaluation of the carcinogenic potential of isobutyl alcohol.

6 Manifesto (MAK value/classification)

Although there are no relevant data available for the effects in man of exposure to isobutyl alcohol, it can be assumed that the local irritation and central nervous depression are similar to that caused by *n*-butyl alcohol, as the two structurally closely related compounds are both primary alcohols. Therefore, on the basis of the analogy to *n*-butyl alcohol ("*n*-Butyl alcohol", this volume), the MAK value for isobutyl alcohol (100 ml/m^3) has been retained provisionally. When valid data for the threshold concentration in man become available, this MAK value must be re-assessed. Peak limitation category I, excursion factor 1 has also been established in analogy to *n*-butyl alcohol. Systemic toxicity is not to be expected at this MAK value, as repeated inhalation exposure of rats to concentrations up to 2500 ml/m^3 did not produce such effects (CMA 1996a, 1996b). Merely transient central nervous depression was reported from 250 ml/m^3. In a valid medium-term drinking water study with rats, no adverse effects were observed up to the highest dose of 1450 mg/kg body weight (Schilling *et al.* 1997).

Under the exposure conditions expected at the workplace, the amount of isobutyl alcohol taken up by the skin cannot have adverse effects (see Section 3). Designation with an "H" is therefore not necessary.

Despite the described increase in tumour incidences, the only study of the carcinogenic effects of isobutyl alcohol cannot be used as evidence of such effects because of its various methodological shortcomings. The germ cell mutagenicity of the substance has not been investigated.

In view of the results of studies with two species after inhalation exposure, prenatal toxicity is not to be expected given observance of the MAK value of 310 mg/m^3 (100 ml/m^3). Isobutyl alcohol is therefore classified in Pregnancy risk group C.

The available data indicate that designation with an "S" (for substances which cause sensitization) is not necessary.

7 References

Abraham MH, Andonian-Haftvan J, Cometto-Muñiz JE, Cain WS (1996) An analysis of nasal irritation thresholds using a new solvation equation. *Fundam Appl Toxicol 31*: 71–76

Alarie Y, Schaper M, Nielsen GD, Abraham MH (1998) Structure-activity relationships of volatile organic chemicals as sensory irritants. *Arch Toxicol 72*: 125–140

Barilyak IR, Kozachnuk YS (1988) Studies in cytogenetic action of some monohydric alcohols on the rat bone marrow cells (Russian). *Tsitol Genet 22*: 49–52

BASF Aktiengesellschaft, Abteilung Toxikologie (1990a) *Prenatal toxicity of 2-methyl-1-propanol-1 in rats after inhalation*, BG No. 96, Project No. 37R0057/88047, for BG Chemie, Heidelberg, 17th December 1990

BASF Aktiengesellschaft, Abteilung Toxikologie (1990b) *Prenatal toxicity of 2-methyl-1-propanol in rabbits after inhalation*, BG No. 96, Project No. 90R0057/88047, for BG Chemie, Heidelberg, 14th December 1990

Bilzer N, Schmutte P, Jehs M, Penners BM (1990) Kinetik aliphatischer Alkohole (Methanol, Propanol-1 und Isobutanol) bei Anwesenheit von Äthanol im menschlichen Körper (Kinetics of aliphatic alcohols (methanol, propanol-1 and isobutanol) in the presence of alcohol in the human body) (German). *Blutalkohol 27*: 385–409

Bos PMJ, Zwart A, Reuzel PGJ, Bragt PC (1992) Evaluation of the sensory irritation test for the assessment of occupational health risk. *Crit Rev Toxicol 21*: 423–450

CMA (Chemical Manufacturers Association) (1996a) *Exposure concentration range-finding study for a subchronic inhalation neurotoxicity study of isobutanol administered by whole-body inhalation to CD rats, with cover letter dated 3/14/96.* NTIS/OTS 0558856, EPA/OTS 44624, NTIS, Springfield, VA, USA

CMA (Chemical Manufacturers Association) (1996b) *Three month schedule controlled operant behavior study of isobutanol by whole body inhalation to CD rats, with cover letter dated 3/14/96.* NTIS/OTS 0558855, EPA/OTS 44624, NTIS, Springfield, VA, USA

ECB (European Chemicals Bureau) (1995) *IUCLID data sheet, 2-methylpropan-1-ol,* 23.10.95, Ispra, Italy

Fang Z, Ionescu P, Chortkoff BS, Kandel L, Sonner J, Laster MJ, Eger II EI (1997) Anesthetic potencies of *n*-alkanols: results of additivity and solubility studies suggest a mechanism of action similar to that for conventional inhaled anesthetics. *Anesthesia Analgesia 84*: 1042–1048

Fiserova-Bergerova V, Diaz ML (1986) Determination and prediction of tissue-gas partition coefficients. *Int Arch Occup Environ Health 58*: 75–87

Fiserova-Bergerova V, Pierce JT, Droz PO (1990) Dermal absorption potential of industrial chemicals: criteria for skin notation. *Am J Ind Med 17*: 617–635

Fisher AA (1983) Topically applied alcohol as a cause of contact dermatitis. *Cutis 31*: 588–600

Frantik E, Hornychova M, Horvath M (1994) Relative acute neurotoxicity of solvents: isoeffective air concentrations of 48 compounds evaluated in rats and mice. *Environ Res 66*: 173–185

Fregert S, Groth O, Hjorth N, Magnusson B, Rorsman H, Övrum P (1969) Alcohol dermatitis. *Acta Derm Venereol 49*: 493–497

Gibel W, Lohs KH, Wildner GP, Schramm T (1974) Experimentelle Untersuchungen zur kanzerogenen Wirkung höherer Alkohole am Beispiel von 3-Methyl-1-butanol, 1-Propanol und 2-Methyl-1-propanol (Experimental studies on the carcinogenic effect of higher alcohols using 3-methyl-1-butanol, 1-propanol and 2-methyl-1-propanol as examples) (German). *Z Exp Chirurg 7*: 235–239

Gibel W, Lohs KH, Wildner GP (1975) Experimentelle Untersuchung zur kanzerogenen Wirkung von Lösungsmitteln am Beispiel von Propanol-1, 2-Methylpropanol-1 und 3-Methylbutanol-1 (Experimental study on the carcinogenic activity of propanol-1, 2-methylpropanol-1 and 3-methylbutanol-1) (German). *Arch Geschwulstforsch 45*: 19–24

Guy RH, Potts RO (1993) Penetration of industrial chemicals across the skin: a predictive model. *Am J Ind Med 23*: 711–719

Hansen LF, Nielsen GD (1994) Sensory irritation, pulmonary irritation and structure-activity relationships of alcohols. *Toxicology 80*: 81–99

Kaneko T, Wang PY, Sato A (1994) Partition coefficients of some acetate esters and alcohols in water, blood, olive oil, and rat tissues. *Occup Environ Med 51*: 68–72

Klimisch HJ, Hellwig J (1995) Studies on the prenatal toxicity of 3-methyl-1-butanol and 2-methyl-1-propanol in rats and rabbits following inhalation exposure. *Fundam Appl Toxicol 27*: 77–89

Ophaswongse S, Maibach HI (1994) Alcohol dermatitis: allergic contact dermatitis and contact urticaria syndrome. A review. *Contact Dermatitis 30*: 1–6

Rowe VK, McCollister SB (1982) Alcohols. in: Clayton GD, Clayton FE (Eds) *Patty's Industrial Hygiene and Toxiology*, Vol. 2C, John Wiley & Sons, New York, 4578–4582

Schilling K, Kayser M, Deckardt K, Kuettler K, Klimisch HJ (1997) Subchronic toxicity studies of 3-methyl-1-butanol and 2-methyl-1-propanol in rats. *Hum Exp Toxicol 16*: 722–726

Stotts J, Ely WJ (1977) Induction of human skin sensitization to ethanol. *J Invest Dermatol 69*: 219–222

Van Ketel WG, Tan-Lim KN (1975) Contact dermatitis from ethanol. *Contact Dermatitis 1*: 7–10

WHO (World Health Organization) (1987) *Butanols – four isomers: 1-butanol, 2-butanol, tert-butanol, isobutanol. IPCS – Environmental health criteria 65*, WHO, Genf

Zeiger E, Anderson B, Haworth S, Lawlor T, Mortelmans K (1988) *Salmonella* mutagenicity tests: IV. Results from the testing of 300 chemicals. *Environ Mol Mutagen 11, Suppl 12*: 1–158

completed 15.01.1999, peak limitation category 25.04.2000, pregnancy risk group 1992

Nitrobenzene

MAK value	**–**
Peak limitation	**–**
Absorption through the skin (1958)	**H**
Sensitization	**–**
Carcinogenicity (1998)	**Category 3**
Prenatal toxicity	**–**
Germ cell mutagenicity	**–**
BAT value (1989)	**aniline released from aniline-haemoglobin conjugate: 100 µg/l blood**
Synonyms	essence of mirbane nitrobenzol oil of mirbane
Chemical name (CAS)	nitrobenzene
CAS number	98-95-3
Structural formula	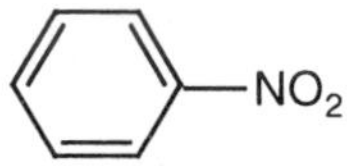
Molecular formula	$C_6H_5NO_2$
Molecular weight	123.11 g/mol
Melting point	5.6°C
Boiling point	210°C at 1013 hPa
Density at 20°C	1.2 g/cm^3
Vapour pressure at 20°C	0.26 hPa
log P_{ow}*	1.86

The MAK value for nitrobenzene which was valid until 1998 (1 ml/m^3, 5.1 mg/m^3) was established in 1958. The present documentation is based on reviews of the toxicological data for nitrobenzene (ACGIH 1996, BUA 1991, 1994, IARC 1995). More recent studies have also been included.

---

* *n*-octanol/water distribution coefficient

1 Toxic Effects and Mode of Action

Nitrobenzene poisoning in man causes marked methaemoglobin formation, cyanosis, neurotoxic effects and unconsciousness.

Nitrobenzene is taken up rapidly via the airways and the skin but eliminated only slowly. In man, nitrobenzene can be metabolized to *p*-aminophenol and *p*-nitrophenol.

In experimental animals, nitrobenzene is acutely toxic. For the rat the oral LD_{50} is 640 mg/kg body weight and the dermal LD_{50} 2100 mg/kg body weight. High doses cause neurological symptoms. In rats exposed by inhalation for 90 days to nitrobenzene concentrations of 5 ml/m^3 or more, dose-dependent methaemoglobin formation was seen. In animals exposed for 2 years to concentrations of 1 to 50 ml/m^3 (rats) or 250 mg/m^3 (mice) the target organs were found to be spleen, liver, kidneys, testes and nose.

The substance causes only slight irritation of the skin and eyes. There are no studies of its sensitization potential.

Nitrobenzene causes degeneration of the germinal epithelium of the testis; reduced fertility is the result. Embryotoxic or teratogenic effects have not been found.

Inhaled by rats or mice, nitrobenzene can induce tumours in the lungs, liver, kidneys, thyroid gland, mammary gland and uterus; in these organs concurrent chronic toxic effects are generally seen as well. In the studies carried out to date, a genotoxic potential has not been detected.

2 Mechanism of Action

Nitrobenzene is metabolized to nitrosobenzene and phenylhydroxylamine, metabolites which are mainly produced by intestinal bacteria and are responsible for the marked methaemoglobin formation.

The mechanism by which the tumours are formed is not yet known. It is unlikely that nitrobenzene has mutagenic potential because tests for mutagenicity *in vitro* (except for the Ames test in the presence of norharman) and *in vivo* have yielded negative results. However, a metabolite of nitrobenzene, phenylhydroxylamine, has been shown to be mutagenic. As proliferative, degenerative or inflammatory changes have been found in most target organs in which tumours have developed, it seems likely that the mechanism of tumour induction involves chronic toxic effects.

3 Toxicokinetics and Metabolism

Nitrobenzene is taken up rapidly via the airways and the skin but eliminated only slowly (ACGIH 1996).

Nitrobenzene is taken up through the skin. When nitrobenzene (dose not specified) was applied one or three times to about one tenth of the surface of the unshaved or shaved skin of mice, the animals died within 2 to 3 days as a result of reduction of the respiration rate to zero. The main symptoms of intoxication were cyanosis, restlessness, narcosis. Histological examination revealed necrosis in the liver, and swelling of the glomerulus and tubule epithelium in the kidneys. No changes were found in the spleen, lungs or testes. When nitrobenzene was applied in doses of 0.25 ml per animal to the shaved skin of rats, one dog, guinea pigs, rabbits or cats (about 1.0, 0.025, 1.0, 0.1 or 0.075 ml/kg body weight, respectively) or administered by subcutaneous injection, the treatment caused dyspnoea and weakness within 6 to 12 hours. Cyanosis was seen in the cats, rats, rabbits and the dog. All animals died within one day (BUA 1991).

5.3 Local effects on skin and mucous membranes

The substance is only slightly irritating for the mucous membranes of the rabbit eye and skin.

There are no data available for allergenic effects of nitrobenzene.

5.4 Reproductive and developmental toxicity

Groups of 26 pregnant CD (Sprague-Dawley) rats were exposed to nitrobenzene concentrations of 0, 1, 10 or 40 ml/m^3 (concentrations found by analysis: 0, 1.06, 9.8 and 39.4 ml/m^3) in whole animal exposure chambers for 6 hours daily from day 6 to day 19 of gestation. On day 21 of gestation the dams underwent caesarean section. During the exposure period the body weight gains of the dams in the 40 ml/m^3 group were significantly lower than those in the control group. Absolute and relative spleen weights were significantly increased in the dams exposed to 10 or 40 ml/m^3. Inhalation of nitrobenzene had no effects on the weight of the gravid uterus, on preimplantation or postimplantation losses including resorptions, or on the numbers of dead foetuses or foetal body weights. There were no treatment-related increases in the incidence of foetal malformations or variations (Tyl *et al.* 1987).

Groups of 30 male and 30 female CD rats were exposed to nitrobenzene in concentrations of 0, 1, 10, or 40 ml/m^3, 6 hours/day, 5 days/week for 10 weeks. The animals were mated with animals from the same exposure group. The F_0 females were exposed until day 19 of gestation. The pups were weaned when they were five days old and the dams then exposed until day 20 after the birth. After 2 weeks, 30 F_1 rats per sex and exposure group were exposed in the same way as the F_0 generation. The F_2 pups were killed at the age of 21 days (two-generation reproduction study). Exposure to nitrobenzene in concentrations of 1 or 10 ml/m^3 had no effects on reproduction. Exposure to 40 ml/m^3 resulted in reduced fertility in the F_0 and F_1 generations. Testis and epididymis weights were reduced; histological examination revealed tubule atrophy in the testes, spermatocyte degeneration and the presence of giant spermatocytes. In the 40 ml/m^3 group, the body weights of the F_1 pups at age 21 days were reduced by about 12 %. F_1 males of the 40 ml/m^3 group were mated with untreated females after a 9-week exposure-free interval. The results showed that the damage to the male reproductive organs was clearly reversible (Dodd *et al.* 1987).

In the seminiferous tubules of exposed SD rats (one oral dose of 300 mg/kg body weight in 1 ml/kg corn oil)—and in those exposed *in vitro* (to nitrobenzene in concentrations up to 10^{-3} M for 24 hours)—secretion of the specific immunoactive protein inhibin was increased (Allenby *et al.* 1990). This protein can be seen as a marker for the toxic effects of nitrobenzene on spermatogenesis. Another study identified six proteins as markers of toxic effects on protein secretion by the seminiferous tubules during the stages of spermatogenesis. In Wistar rats given a single oral nitrobenzene dose of 300 mg/kg body weight, the average testis weights on days 1 and 3 *post applicationem* were significantly lower than in the control animals. Histological examination was not carried out. In addition, the level of secretion of two proteins was reduced during stages VI to VIII of spermatogenesis (McLaren *et al.* 1993). In a 2-year study (see Section 5.6), bilateral testis atrophy and hypospermia in the epididymis was seen in CD rats exposed to 25 mg/m^3 (Cattley *et al.* 1994).

In the teratogenicity study, no embryotoxic, foetotoxic or teratogenic effects were seen even after exposure of the dams to the clearly maternally toxic concentration of 40 ml/m^3. In the two-generation study, this high concentration caused reversible testis damage in the male parents, but the only prenatal toxic effect was that the F_1 generation pups gained weight slightly more slowly than the controls. These inhalation studies with the rat demonstrate that prenatal toxic effects of nitrobenzene are unlikely after occupational exposures during which the MAK value of 1 ml/m^3 (5 mg/m^3), which was valid until 1998, is observed.

5.5 Genotoxicity

In vitro

Mutagenicity tests with various *Salmonella typhimurium* strains in the presence and absence of S9 mix yielded negative results with nitrobenzene concentrations up to 2500 ng per plate (Anderson and Styles 1978, Benkendorf 1978, Chiu *et al.* 1978, Fu *et al.* 1985, Garner and Nutman 1977, Nohmi *et al.* 1984, Purchase *et al.* 1978, Suzuki *et al.* 1983, Vance and Levin 1984). In the presence of both norharman and S9 mix, nitrobenzene was mutagenic in the *S. typhimurium* strain TA98 (Suzuki *et al.* 1983; see BUA 1991).

In concentrations of 0.01 to 0.02 M (1.2–2.4 mg/ml) which were markedly toxic, nitrobenzene caused changes in the form of yeast cells which could, however, also result from non-mutagenic effects. A concentration without toxic effects was not determined (Levan 1947). No mutagenic activity was found in tests with *Actinomyces* strains. Cytological and morphological changes (mitotic disturbances like those caused by the spindle toxin colchicine) have been seen in the seeds of barley (Favorsky 1939), wheat and flax (Simonet and Guinochet 1939). A concentration without toxic effects was apparently not determined (see BUA 1991).

Unlike nitrobenzene itself, one of its metabolites, phenylhydroxylamine, proved to be mutagenic in the *S. typhimurium* strain TA100 in the presence of S9 mix and caused sister chromatid exchange (SCE) in human fibroblasts (see 1992 MAK documentation for Aniline in Volume 6 of the present series).

In vivo

Male Fischer 344 rats were exposed 21 times within 29 days to nitrobenzene in concentrations of 0, 5, 16 or 50 ml/m^3 for 6 hours daily. Lymphocytes were isolated from blood and spleen and cultured in the presence of 5-bromodeoxyuridine. No effects on the incidence of SCE or chromosome breaks were detected. Reduced mitotic activity of lymphocytes and cell cycle delays were observed. The authors concluded from these results that nitrobenzene has no genotoxic potential (Kligerman *et al.* 1982; see also BUA 1991).

Nitrobenzene was administered to groups of 5 male and 5 female B6C3F$_1$ mice by intraperitoneal injection of doses of 62.5, 125 and 250 mg/kg body weight. Examination after 24 and 48 hours revealed neither inhibition of erythropoiesis nor micronucleus induction (BASF 1996).

Genotoxic effects of nitrobenzene were not detected in the *in vivo–in vitro* hepatocyte DNA repair test. For this test, rats were given nitrobenzene doses of 200 mg/kg body weight in corn oil by gavage, killed 12 hours later and their livers removed; hepatocyte cultures were set up and used for the study (Mirsalis *et al.* 1982; see also BUA 1991).

In female rats given oral doses of ^{14}C-nitrobenzene (8.5 mCi/µmol), a binding index of 73 ± 10 [binding (µmol/mol haemoglobin) per dose (µmol/kg body weight)] was determined (Albrecht and Neumann 1985). Nitrobenzene causes less methaemoglobin formation than does aniline but the binding index obtained for nitrobenzene was higher.

Single subcutaneous nitrobenzene doses of 4 mg/kg body weight were administered to a group of six Fischer 344 rats of both sexes and one of eight B6C3F$_1$ mice of both sexes; 24 hours later, the animals were killed and DNA and chromatin proteins from rat liver and kidneys and mouse liver and lungs were isolated and purified. Of the applied dose, 37 % to 51 % was excreted in the 24-hour urine. The covalent binding indices (CBI) (µmol adduct/mol DNA-nucleotide/mmol substance/kg body weight) were determined for rat liver (0.11), rat kidney (0.11), mouse liver (0.15), male mouse lung (0.73), and female mouse lung (0.22). These values are about one tenth of those found for weakly genotoxic carcinogens. Binding to chromatin proteins was also demonstrated (BASF 1997).

5.6 Carcinogenicity

In a carcinogenicity study, Fischer 344 rats and male CD rats were exposed to nitrobenzene concentrations of 0, 1, 5 or 25 ml/m^3 and B6C3F$_1$ mice to 0, 5, 24 or 50 ml/m^3 for a period of 24 months (Cattley *et al.* 1994). The study is of limited validity because of shortcomings reported in the publication. High losses were reported, especially of mice (e.g. of 70 female mice 27 were found dead and 13 went missing), and wrongly assigned animals (CIIT 1993). The report contains no GLP "Quality Assurance Statement".

Both the tumour incidences (Table 3) and the toxic effects (Tables 1 and 2) are described below for each target organ.

Liver

In the male Fischer 344 rats, the incidence of liver cell adenomas or liver cell adenomas or carcinomas (kind of tumour not specified) was 1/69 (1 %) in the control group, 4/69 (6 %) at the low concentration, 5/70 (4 %) at the medium concentration and 6/70 (23 %, significant) at the high concentration. The historical control incidence was given as 2 % (Tarone *et al.* 1981). The relative liver weights were increased slightly but significantly only at 1 and 25 ml/m^3. However, the incidences of centrilobular liver cell enlargement, of foci of eosinophilic cells and of focal cystic degeneration were increased (Table 1).

As in the Fischer 344 rats, in the CD rats the incidence of liver cell adenomas or carcinomas was increased from 2/63 (3 %) in the control group and 1/67 (1 %) at the low concentration to 4/70 (6 %) at the medium concentration and significantly to 9/65 (14 %) at the high concentration. The liver cell adenomas in the CD rats were also associated with chronic toxic liver damage (centrilobular hypertrophy) (Table 1). In the livers of male mice, the incidences of centrilobular liver cell enlargement and of multinucleate liver cells were increased (Table 2). The number of hepatocellular carcinomas or adenomas was not significantly different from the control value. Female mice developed centrilobular hepatocytomegaly but tumour incidences were not significantly increased.

Lung

In the lungs of the male mice, the incidence of alveolar/bronchiolar adenomas or carcinomas was 9/68 (13 %) in the control group, 21/67 (31 %) at the low concentration, 21/65 (32 %, significant) at the medium concentration and 23/66 (35 %, significant) at the high concentration. The historical control incidence was given as 17 % (Tarone *et al.* 1981). In addition to alveolar/bronchiolar tumours, alveolar/bronchiolar hyperplasia was observed in male mice and alveolar bronchiolization was also frequently detected (Table 2). In the female mice, no tumour incidences were increased but an increased incidence of alveolar bronchiolization was found (Table 2).

Kidney

In the male Fischer 344 rats, the incidence of renal tubule adenomas was 0/69 (0 %) in the control group, 0/68 (0 %) at the low concentration, 0/70 (0 %) at the medium concentration and 5/70 (7 %, significant) at the high concentration. The historical control incidence was given as 0.2 % (Solleveld *et al.* 1984). Chronic toxic effects on the renal tubules (hyperplasia, eosinophilic droplets, chronic nephropathy) were also observed (Table 1). Male Sprague-Dawley (CD) rats did not develop kidney tumours and the incidence of chronic toxic effects was not increased significantly. In the female Fischer 344 rats of the highest concentration group, the incidence of intratubular eosinophilic droplets was significantly increased (Table 1) but there were no tumours. In the male mice there were no kidney tumours but an increased incidence of cysts was found in the animals of the highest concentration group (Table 2).

Thyroid gland

The incidence of follicle cell adenomas of the thyroid was increased in male mice. The incidence was 0/65 (0 %) in the control group, 4/65 (6 %) at the low concentration, 1/65 (2 %) at the medium concentration and 7/64 (11 %, significant) at the high concentration. The historical control incidence was given as 0.5 % to 1 % (Chandra and Frith 1992). In the male mice with follicle cell adenomas, hyperplasia was also detected. In the female mice of the highest concentration group, the incidence of follicle cell hyperplasia was also increased but no thyroid tumours developed (Table 2).

In the male Fischer 344 rats, the incidence of follicle cell neoplasms (adenomas or adenocarcinomas) in the thyroid was 2/69 (3 %) in the control group, 1/69 (1 %) at the low concentration, 5/70 (7 %) at the medium concentration, and 8/70 (11 %) at the high concentration. The historical control incidence for C cell carcinomas is 7.1 % (Tarone *et al.* 1981). In these rats follicle cell hyperplasia was also detected (Table 1).

Mammary gland

In the female mice of the highest concentration group, the incidence of mammary adenocarcinomas (5/60, 8 %) was increased significantly above that for the control group (0/48, 0 %). The historical control incidence is 1 % to 2 % (Chandra and Frith 1992).

Endometrium

In the female Fischer 344 rats, the incidence of uterine endometrial polyps was 11/69 (16 %) in the control group, 17/65 (26 %) at the low concentration, 15/65 (23 %) at the medium concentration and 25/69 (36 %, significant) at the high concentration. The historical control incidence is 15 % (Tarone *et al.* 1981).

6 Manifesto (MAK value/classification)

In persons exposed long-term to nitrobenzene, hypoxic damage in the internal organs, methaemoglobin formation, cyanosis and haemolytic anaemia can develop. No epidemiological studies have been published. Nitrobenzene induces neoplastic changes in animals exposed by inhalation. The substance induces statistically significant increases in the incidences of lung and thyroid gland tumours in male B6C3F$_1$ mice, of tumours of the mammary gland in female mice, liver and kidney in male Fischer 344 rats, uterus in female rats, and liver in male CD rats. In the Ames test, nitrobenzene was mutagenic only after addition of norharman in the presence of S9 mix. In *in vivo* studies, nitrobenzene had no effect on the incidence of SCE or on DNA repair in spleen lymphocytes and did not induce micronucleus formation in bone marrow cells. The results of the long-term studies reveal that there were mostly also chronic toxic inflammatory or proliferative changes in the tissues in which tumours developed. It is conceivable that the tumours develop primarily as a result of chronic tissue damage. A NOAEL, however, could not be established in the long-term inhalation studies.

Table 3. Carcinogenicity studies with nitrobenzene

Author:	Cattley *et al.* 1994
Substance:	nitrobenzene (purity > 99.8 %)
Species:	rat, Fischer 344, 70 ♂, 70 ♀ rat, CD 70 ♂
Administration route:	inhalation
Concentration:	0, 1, 5, 25 ml/m^3
Duration:	24 months (6 h/day, 5 days/week) (10 animals per sex and strain killed after 15 months)
Toxicity:	1 ml/m^3 or more: significant reductions in body weights (Fischer 344 rat)

Tumours:				Concentration (ml/m^3)			
				0	1	5	25
liver	adenomas	F344	♂	1/69	3/69	3/70	15/70*
			♀	0/70	2/66	0/66	3/70
		CD	♂	1/63	1/67	2/70	7/67*
	carcinomas	F344	♂	0/69^T	1/69	2/70	4/70
			♀	0/70^T	0/66	0/66	2/70
		CD	♂	2/63	0/67	2/70	2/65
	adenomas or carcinomas	F344	♂	1/69^T	4/69	5/70	16/70*
			♀	0/70^T	2/66	0/66	4/70
		CD	♂	2/63	1/67	4/70	9/65*
kidney	adenomas (tubule)	F344	♂	0/69	0/68	0/70	5/70
			♀	0/70	0/66	0/66	0/70
		CD	♂	2/63	0/67	2/70	0/65
	carcinomas (tubule)	F344	♂	0/69^T	0/68	0/70	5/70*
			♀	0/70	0/66	0/66	0/70
		CD	♂	0/63	1/67	0/70	0/65
	adenomas or carcinomas	F344	♂	0/69	0/68	0/70	6/70
	(tubule)		♀	0/70	0/66	0/66	0/70
		CD	♂	2/63	0/67	2/70	0/65
thyroid gland	adenomas (follicle)	F344	♂	0/69	0/69	2/70	2/70
			♀	0/69	–	–	2/68
		CD	♂	2/63	4/64	2/68	3/64
	adenocarcinomas	F344	♂	2/69	1/69	3/70	6/70
	(follicle)		♀	0/69	–	–	1/68
		CD	♂	4/63	1/64	1/68	2/64
	follicle adenomas or	F344	♂	2/69^T	1/69	5/70	8/70
	adenocarcinomas		♀	0/69	–	–	3/68
		CD	♂	5/63	5/64	3/68	5/64

Table 3. continued

uterus	endometrial stroma polyps	F344	♀	11/69 [T]	17/65	15/65	25/69*
testis	interstitial tumours	F344	♂	61/69	52/56	58/61	65/70
		CD	♂	3/62	6/66	7/70	4/61
leukaemia	multiple mononuclear leukaemia	F344	♂	12/69	5/69	4/70	0/70
			♀	11/70	5/66	4/66	0/70
		CD	♂	1/63	4/67	0/70	0/65

* $p < 0.05$
[T] significant positive trend, $p < 0.05$

Author:	Cattley *et al.* 1994
Substance:	nitrobenzene (purity > 99.8 %)
Species:	mouse, B6C3F$_1$ 70 ♂, 70♀
Administration route:	inhalation
Concentration:	0, 5, 25, 50 ml/m^3
Duration:	24 months (6 h/day, 5 days/week)
Toxicity:	50 ml/m^3 or more: significant reductions in body weights

Tumours:			Concentration (ml/m^3)			
			0	5	25	50
lung	adenomas (alveolar/bronchiolar)	♂	7/68 [T]	12/67	15/65*	18/66*
		♀	4/53	11/60	3/64	2/62
	carcinomas (alveolar/bronchiolar)	♂	4/68	10/67	8/65	8/66
		♀	2/53	0/60	4/64	4/62
	adenomas or carcinomas	♂	9/68 [T]	21/67*	21/65*	23/66*
		♀	6/53	11/60	6/64	6/62
thyroid gland	follicle adenomas	♂	0/65 [T]	4/65	1/65	7/64*
		♀	2/49	0/59	3/61	2/61
mammary gland	adenocarcinomas	♀	0/48	–	–	5/60 P
liver	adenomas	♂	14/68	18/65	15/65	14/64
		♀	6/51 [T]	5/61	5/64	13/62
	carcinomas	♂	12/68	13/65	12/65	8/64
		♀	1/51	2/61	3/64	1/62
	adenomas or carcinomas	♂	25/68	30/65	22/65	21/64
		♀	7/51	7/61	7/64	14/62

* significant, $p < 0.05$
[T] significant positive trend, $p < 0.05$

Because of the minor significance of genotoxicity and the presumably non-linear dose-effect relationship, nitrobenzene could be a candidate for classification in Carcinogenicity category 4. However, data suitable for the derivation of a MAK value are not available. Therefore, nitrobenzene is classified for the present in Carcinogen category 3 in Section II of the *List of MAK and BAT Values*. The MAK value of 1 ml/m^3 and the classification in Pregnancy risk group D which applied until 1998 have been withdrawn.

Toxic amounts of nitrobenzene can be absorbed through the skin. The designation with "H" has therefore been retained.

Studies which suggest that the substance should be designated with an "S" are not available.

7 References

ACGIH (1996) American Conference of Governmental Industrial Hygienists (1996) *Documentation of the threshold limit values and biological exposure indices, nitrobenzene*, ACGIH, Cincinnati, Ohio

Albrecht W, Neumann H-G (1985) Biomonitoring of aniline and nitrobenzene. *Arch Toxicol 57*: 1–5

Allenby G, Foster PMD, Sharpe RM (1990) Evaluation of changes in the secretion of immunoactive inhibin by adult rat seminiferous tubules *in vitro* as an indicator of early toxicant action on spermatogenesis. *Fundam Appl Toxicol 16*: 710–724

Anderson D, Styles JA (1978) The bacterial mutation test. *Br J Cancer 37*: 924–930

BASF (1996) *Cytogenetic study in vivo of nitrobenzene in mice*. Project No. 26M0116/949010, unpublished report

BASF (1997) *Investigation of the potential for DNA binding of nitrobenzene*. Final Report No. CB94/16, unpublished report

Benkendorf DAR (1978) Metabolic pathways of the *N*-substituted naphthalenes: Their relationship to mutagenicity and carcinogenicity. *Diss Abstr Int B 39*: 2723-B

BUA (Beratergremium für umweltrelevante Altstoffe der Gesellschaft Deutscher Chemiker) (1991) *Nitrobenzol, BUA-Stoffbericht 59* (Also available in English translation). VCH Verlagsgesellschaft, Weinheim

BUA (Beratergremium für umweltrelevante Altstoffe der Gesellschaft Deutscher Chemiker) (1994) *Nitrobenzol, BUA-Stoffbericht 133* (Also available in English translation). S. Hirzel Wissenschaftliche Verlagsgesellschaft, Stuttgart

Cattley RC, Everitt JI, Gross EA (1994) Carcinogenicity and toxicity of inhaled nitrobenzene in $B6C3F_1$ mice and F344 and CD rats. *Fundam Appl Toxicol 22(3)*: 328–340

Chandra M, Frith CH (1992) Spontaneous neoplasms in $B6C3F_1$ mice. *Toxicol Lett 60*: 91–98

Chiu CW, Lee LH, Wang CY, Bryan GT (1978) Mutagenicity of some comercially available nitro compounds for *Salmonella typhimurium*. *Mutat Res 58*: 11–22

CIIT (Chemical industry Institute of Toxicology) (1993) *A chronic inhalation toxicity study of nitrobenzene in F344 and CD rats, and $B6C3F_1$ mice*. In report issued by the Chemical Industry Institute of Toxicology, Research Triangle Park, NC

Dodd DE, Fowler EH, Snellings WM, Pitts IM, Tyl RW, Lyon JP, O'Neal FOO, Kimmerle G (1987) Reproduction and fertility evaluations in CD rats following nitrobenzene inhalation. *Fundam appl Toxicol 8*: 493–505

Favorsky MV (1939) New polyploidy-inducing chemicals. *C R Acad Sci URSS 25*: 71–74

Fu PP, Chou MW, Miller DW, White GL, Heflich RH, Beland FA (1985) The orientation of the nitro substituent predicts the direct-acting bacterial mutagenicity of nitrated polycyclic aromatic hydrocarbons. *Mutat Res 143*: 173–181

Garner RC, Nutman CA (1977) Testing of some azo dyes and their reduction products for mutagenicity using *Salmonella typhimurium* TA 1538. *Mutat Res 44*: 9–19

Henschler D, Lehnert G (Eds) (1989) Nitrobenzol. *Biologische Arbeitsstoff-Toleranz-Werte (BAT-Werte) und Expositionsäquivalente für krebserzeugende Arbeitsstoffe (EKA)*. 4th issue, and in English translation in *Biological Exposure Values for Occupational Toxicants and Carcinogens*, Volume 2, VCH Verlagsgesellschaft, Weinheim

IARC (International Agency for Research on Cancer) (1995) Printing process and printing inks, carbon black and some nitro compounds. *IARC Monographs on the evaluation of carcinogenic risks to humans, Vol. 65*, IARC Lyon, 381–407

Kligerman AD, Erexson GL, Wilmer JL, Phelps MC (1982) Analyses of cytogenetic damage in rat peripheral lymphocytes following exposure to ethylene oxide and nitrobenzene. CIIT Sept 7th: 3

Levan A (1947) Studies on the camphor reaction of the yeast. *Hereditas 33*: 457–514

McLaren TT, Foster PMD, Sharpe RM (1993) Identification of stage-specific changes in protein secretion by isolated seminiferous tubules from the rat following exposure to either *m*-dinitrobenzene or nitrobenzene. *Fundam Appl Toxicol 21*: 384–392

Medinsky MA, Irons RD (1985) Sex, strain and species differences in the response of rodents to nitrobenzene vapors. in Rickert DE (Ed.) *Toxicity of Nitroaromatic Compounds, 5th CIIT Conference on Toxicology*, Raleigh/NC, January 28–29 1982, Hemisphere Publishing Corporation, Washington

Mirsalis JC, Tyson CK, Butterworth BE (1982) Detection of genotoxic carcinogens *in vivo-in vitro* hepatocyte DNA repair assay. *Environ Mutagen 4*: 553–562

Nohmi T, Yoshikawa K, Nakadate M, Miyata R, Ishidate M jr (1984) Mutations in *Salmonella typhimurium* and inactivation of *Bacillis subtilis* transforming DNA induced by phenylhydroxylamine derivatives. *Mutat Res 136*: 159–168

Purchase IFH, Longstaff E, Ashby J, Styles JA, Anderson D, Lefevre PA, Westwood FR (1978) An evaluation of 6 short term tests for detecting organic chemical carcinogens. *Br J Cancer 37*: 873–903

Simonet M, Guinochet M (1939) L'action de quelques dérivés halogenés et nitrés des carbures cycliques sur les mitoses, dans les cellules végétales (The action of some halogenated and nitrated derivatives of cyclic hydrocarbons on mitosis in plant cells) (French). *C R Soc Biol 132*: 455–457

Solleveld HA, Haseman JK, McConnel EE (1984) Natural history of body weight gain, survival, and neoplasia in the F344 rat. *J Natl Cancer Inst 72*: 929–940

Suzuki J, Koyama T, Suzuki S (1983) Mutagenicities of mono-nitrobenzene derivatives in the presence of norharman. *Mut Res 120*: 105–110

Tarone RE, Chu KC, Ward JM (1981) Variability in the rates of some naturally occurring tumors in Fischer 344 rats and (C57BL/6N×C3H/HeN)F_1 (B6C3F_1) mice. *J Nat Cancer Inst 66*: 1175–1181

Tyl RW, France KA, Fisher LC, Dodd DE, Pitts IM, Lyon JP, O'Neal FOO, Kimmerle G (1987) Development toxicity evaluation of inhaled nitrobenzene in CD rats. *Fundam appl Toxicol 8*: 482–492

Vance WA, Levin DE (1984) Structural features of nitroaromatics that determine mutagenic activity in *Salmonella typhimurium*. *Environ Mutag 6*: 797–811

completed 31.03.1998

Nitroethane

MAK value (1958)	**100 ml/m^3 (ppm) ≙ 310 mg/m^3**
Peak limitation (2000)	**Category II, excursion factor 4**
Absorption through the skin	**–**
Sensitization	**–**
Carcinogenicity	**–**
Prenatal toxicity (2000)	**see Section IIc of the *List of MAK and BAT Values***
Germ cell mutagenicity	**–**
BAT value	**–**
Synonyms	mononitroethane
Chemical name (CAS)	nitroethane
CAS number	79-24-3
Structural formula	$H_3C–CH_2–NO_2$
Molecular formula	$C_2H_5NO_2$
Molecular weight	75.07
Melting point	–90°C
Boiling point	114°C
Density at 20°C	1.051 g/cm^3
Vapour pressure at 20°C	20.8 hPa
log P_{ow}*	0.18
1 ml/m^3 (ppm) ≙ 3.11 mg/m^3	**1 mg/m^3 ≙ 0.321 ml/m^3 (ppm)**

The MAK value of 100 ml/m^3 for nitroethane was established in 1958 in analogy to the TLV value at that time.

* *n*-octanol/water distribution coefficient

1 Toxic Effects and Mode of Action

Ingested in large doses by small children, nitroethane causes vomiting, sleepiness and, with a latency period of 10 to 14 hours, an increase in the methaemoglobin level with breathlessness and cyanosis. Nitroethane is a weak irritant of skin and mucous membranes. In animal studies, exposure to a nitroethane concentration of 480 ml/m^3 is tolerated without recognizable symptoms of irritation. One animal study yielded evidence that the substance has neurotoxic and hepatotoxic potential. From a long-term study with rats, a NOEL (no observed effect level) of 200 ml/m^3 may be derived for nitroethane. There is no evidence of genotoxic or carcinogenic effects.

There are no data available for the mechanism of action of nitroethane.

2 Toxicokinetics and Metabolism

When anaesthetized F344 rats were exposed to nitroethane in a concentration of 100 ml/m^3, about 50 % of the inhaled substance was absorbed in the respiratory tract. In the isolated upper respiratory tract 30 % to 70 % of the substance was absorbed and in the isolated lower respiratory tract about 50 % (Stott and McKenna 1984).

After intravenous injection into rabbits, nitroethane was rapidly metabolized and eliminated to some extent with the exhaled air. The elimination was complete within 30 hours (no other details) (Machle *et al.* 1942).

3 Effects in Man

The odour threshold for nitroethane was found to be 2.1 ml/m^3 (Amoore and Hautala 1983).

Ingestion of at most one ounce (about 28 g) of nitroethane by a 20-month-old boy caused vomiting and sleepiness within 10 hours and later breathlessness, cyanosis and a methaemoglobin level of 39 %. After treatment with methylene blue, the methaemoglobin level decreased to 5.7 %. The child recovered without any problems (Hornfeldt and Rabe 1994). A 13-month-old girl who had drunk at most 90 ml of nitroethane vomited 7 hours later and became lethargic, cyanotic and short of breath (44 inhalations/min). After another 7 hours, severe methaemoglobinaemia was diagnosed and treated with methylene blue. A short time later without any other treatment, the child was free of symptoms (Osterhoudt *et al.* 1995).

4 Animal Experiments and *in vitro* Studies

4.1 Acute toxicity

4.1.1 Inhalation

In rats exposed for 6 hours to 2000 ml/m^3 no symptoms of intoxication were seen. After exposure to 13000 ml/m^3 for 6 to 7 hours, all animals died (no other details given in the cited review) (ACGIH 1992).

Groups of 2 rabbits and 2 guinea pigs were exposed to nitroethane in concentrations between 1500 and 90000 mg/m^3 (480–29000 ml/m^3). The animals were exposed for periods between 0.5 and 12 hours (maximum 6 hours daily). Shortly after the start of exposure to concentrations above 1500 mg/m^3 (480 ml/m^3), the animals became restless and showed signs of respiratory tract irritation with dyspnoea and occasional rales. Irritation of the conjunctiva, erythema of the eyelids and lacrimation were recorded; the animals kept their eyes closed. At higher (unspecified) concentrations, salivation and jerking movements of the head and extremities were observed. Exposure to concentrations of 15000 mg/m^3 (4800 ml/m^3) or more was lethal. In animals exposed for 5 to 6 hours to 3000 mg/m^3 (960 ml/m^3) or for 3 hours to 30000 mg/m^3 (9600 ml/m^3) or for 1 hour to 90000 mg/m^3 (29000 ml/m^3), stupor, narcosis or general anaesthesia developed (Machle *et al.* 1940). From this study a NOEL of 480 ml/m^3 can be deduced for the acute irritant effects of nitroethane in the rabbit and guinea pig.

4.1.2 Ingestion

The oral LD_{50} for the rat was found to be 1625 mg/kg body weight (Martin and Baker 1967) and for the CD-1 mouse 2150 mg/kg body weight (Hite and Skeggs 1979). For the rabbit, a dose of 500 to 750 mg/kg body weight was lethal (Machle *et al.* 1940).

In rats given a nitroethane dose of 275 mg/kg body weight, the turnover of norepinephrine and serotonin in the *medulla oblongata* was accelerated; this was interpreted as a neurotoxic effect. The levels of other neurotransmitters of monoamine type or their metabolites in the hippocampus, midbrain and hypothalamus were not affected (Kanada *et al.* 1994).

4.1.3 Intraperitoneal injection

Four hours after intraperitoneal injection of a nitroethane dose of 200 mg/kg body weight to 3-month-old male Wistar rats, the "acid proteinase" activity in the brain was increased. In the hepatic microsome fraction, increased activities of epoxide hydrolase and UDP-glucuronosyl transferase persisted for up to 48 hours *post applicationem* and the activity of 7-ethoxycoumarin-O-deethylase was decreased. At the same time in the liver cells, proliferation of the smooth endoplasmic reticulum and degranulation and disorganization of the rough endoplasmic reticulum were observed (Zitting *et al.* 1982). An intraperitoneal nitroethane dose of 9 mmol/kg body weight (681 mg/kg body weight) was not hepatotoxic for male or female BALB/c mice (5–6 animals per group), as shown by determination of the activities of sorbitol dehydrogenase, alanine aminotransferase (ALT) and aspartate aminotransferase (AST) and by histopathological examination of the liver after 1, 2, 3 and 4 days (Dayal *et al.* 1989).

4.2 Subacute, subchronic and chronic toxicity

Inhalation of nitroethane in a concentration of 1500 mg/m^3 (480 ml/m^3) (6 hours/day, 5 days/week) for 5 to 7 days was tolerated by 2 rabbits and 2 guinea pigs without any signs of intoxication (Machle *et al.* 1940).

In a long-term study, groups of 40 male and 40 female Long-Evans rats were exposed for 2 years (7 hours/day, 5 days/week) to nitroethane in concentrations of 100 and 200 ml/m^3; no substance-related effects were detected. Body weights and the weights of brain, liver, kidneys, lungs and heart were recorded. The haematological parameters studied were leukocyte and erythrocyte counts, mean erythrocyte volume, and haemoglobin level. Neither the gross pathological nor the histological examination of all organs revealed substance-related alterations. No changes were found in the assayed serum parameters: total bilirubin, total protein, blood urea nitrogen, creatinine, sodium, potassium, AST, ALT (Griffin *et al.* 1988). A NOEL of 200 ml/m^3 for the repeated exposure of rats to nitroethane can be deduced from this study.

4.3 Local effects on skin and mucous membranes

Nitroethane is a weak irritant for rabbit skin (ACGIH 1992) and for the conjunctiva and respiratory tract (Machle *et al.* 1940). The severity of the irritation was about the same as that produced by the weak irritant diethyl ether (Machle *et al.* 1940).

There are no data available for allergenic effects of nitroethane.

4.4 Reproductive and developmental toxicity

Exposure of 25 pregnant mice to nitroethane in a concentration of 14.3 ± 2.0 ml/m^3 on days 6 to 17 of gestation (8.25 ± 2.25 hours daily) had neither toxic effects on the dams nor teratogenic, embryotoxic or foetotoxic effects (Beliles *et al.* 1978).

4.5 Genotoxicity

The few genotoxicity studies which have been carried out with nitroethane do not suggest that the substance could have genotoxic potential. This is in accordance with the results of numerous studies which have shown that primary nitroalkanes, unlike secondary nitroalkanes, are generally neither genotoxic nor carcinogenic (see e.g. Andrae *et al.* 1988, Conaway *et al.* 1991a, 1991b, Fiala *et al.* 1987, George *et al.* 1989).

4.5.1 *In vitro*

Nitroethane was not mutagenic in the *Salmonella* mutagenicity test with the strains TA92, TA98, TA100, TA102 and TA1537 either with or without a metabolic activation system. Concentrations between 11 and 15 µl/plate (Dayal *et al.* 1989, Hite and Skeggs 1979) or 20 and 50 µl/plate (Löfroth *et al.* 1986) could be studied. Higher concentrations were cytotoxic.

4.5.2 *In vivo*

Single oral nitroethane doses of 250, 500 or 1000 mg/kg body weight were administered to groups of 8 male and 8 female CD-1 mice; no micronucleus formation was detected in bone marrow erythrocytes prepared 6 hours *post applicationem* (Hite and Skeggs 1979).

4.6 Carcinogenicity

Groups of 40 male and 40 female Long-Evans rats were exposed for 2 years (7 hours daily, 5 days/week) to nitroethane in concentrations of 100 and 200 ml/m^3. No increased tumour incidences were observed (Griffin *et al.* 1988) (see Section 4.2).

5 Manifesto (MAK value/classification)

There are no studies appropriate for the assessment of the effects of nitroethane in man. In a long-term inhalation study in which Long-Evans rats were exposed to nitroethane concentrations of 100 and 200 ml/m^3, substance-related toxic effects were not observed. The MAK value of 100 ml/m^3 (310 mg/m^3) can, therefore, be retained.

At concentrations equivalent to twice the MAK value, no effects were seen in long-term studies. The weak irritant effects of nitroethane on skin and mucous membranes are like those of diethyl ether (> 400 ml/m^3). Therefore the substance is classified in Peak limitation category II with an excursion factor of 4.

The germ cell mutagenic and carcinogenic potential of nitroethane cannot be evaluated. Because of lack of data, it cannot be decided whether designation with "S" or "H" is necessary.

The only reproductive toxicity study available is inappropriate for the assessment of the prenatal toxicity of nitroethane because the exposure concentration used was much too low. Therefore nitroethane is listed in Section IIc of the *List of MAK and BAT Values*.

6 References

ACGIH (American Conference of Governmental Industrial Hygienists) (1992) Nitroethane. in: *Documentation of TLVs and BEIs*, ACGIH, Cincinnati, OH, USA

Amoore JE, Hautala E (1983) Odor as an aid to chemical safety: odor thresholds compared with threshold limit values and volatilities for 214 industrial chemicals in air and water dilution. *J Appl Toxicol 3*: 272–290

Andrae U, Homfeldt H, Vogt L, Lichtmannegger J, Summer KH (1988) 2-Nitropropane induces DNA repair synthesis in rat hepatocytes *in vitro* and *in vivo*. *Carcinogenesis 9*: 811–815

Beliles RP, Makris SL, Ferguson F, Putman C, Sapanski W, Kelly N, Partymiller K, Heicklen J (1978) Teratology study in mice subjected to inhalation of diethylhydroxylamine, nitroethane, and diethylamine hydrogen sulfite. *Environ Res 17*: 165–176

Conaway CC, Hussain NS, Way BM, Fiala ES (1991a) Evaluation of secondary nitroalkanes, their nitronates, primary nitroalkanes, nitrocarbinols, and other aliphatic nitro compounds in the Ames *Salmonella* assay. *Mutat Res 261*: 197–208

Conaway CC, Nie G, Hussain NS, Fiala ES (1991b) Comparison of oxidative damage to rat liver DNA and RNA by primary nitroalkanes, secondary nitroalkanes, cyclopentanone oxime, and related compounds. *Cancer Res 51*: 3143–3147

Dayal R, Gescher A, Harpur ES, Pratt I, Chipman JK (1989) Comparison of the hepatotoxicity in mice and the mutagenicity of three nitroalkanes. *Fundam Appl Toxicol 13*: 341–348

Fiala ES, Czerniak R, Castonguay A, Conaway CC, Rivenson A (1987) Assay of 1-nitropropane, 2-nitropropane, 1-azoxypropane and 2-azoxypropane for carcinogenicity by gavage in Sprague-Dawley rats. *Carcinogenesis 8*: 1947–1949

George E, Burlinson B, Gatehouse D (1989) Genotoxicity of 1- and 2-nitropropane in the rat. *Carcinogenesis 10*: 2329–2334

Griffin TB, Stein AA, Coulston F (1988) Chronic inhalation exposure of rats to vapors of nitroethane. *Ecotoxicol Environ Saf 16*: 11–24

Hite M, Skeggs H (1979) Mutagenic evaluation of nitroparaffins in the *Salmonella typhimurium*/mammalian-microsome test and the micronucleus test. *Environ Mutagen 1*: 383–389

Hornfeldt CS, Rabe WH (1994) Nitroethane poisoning from an artificial fingernail remover. *J Toxicol Clin Toxicol 32*: 321–324

Kanada M, Miyagawa M, Sato M, Hasegawa H, Honma T (1994) Neurochemical profile of effects of 28 neurotoxic chemicals on the central nervous system in rats. (1) Effects of oral administration on brain contents of biogenic amines and metabolites. *Ind Health 32*: 145–164

Löfroth G, Nilsson L, Andersen JR (1986) Structure-activity relationship of nitroalkane-induced mutagenicity in the Ames *Salmonella* assay. *Prog Clin Biol Res 209 B*: 149–155

Machle W, Scott EW, Treon J (1940) The physiological response of animals to some simple mononitroparaffins and to certain derivatives of these compounds. *J Ind Hyg Toxicol 22*: 315–332

Machle W, Scott EW, Treon J (1942) The metabolism of mononitroparaffins. I. Recovery of nitroethane from the animal organism. *J Ind Hyg Toxicol 24*: 5–9

Martin JL, Baker jr PJ (1967) Nitroparaffins. *Enc Chem Tech 13*: 877

Osterhoudt KC, Wiley CC, Dudley R, Sheen S, Henretig FM (1995) Rebound severe methemoglobinemia from ingestion of a nitroethane artificial-fingernail remover. *J Pediatr 126*: 819–821

Stott WT, McKenna MJ (1984) The comparative absorption and excretion of chemical vapors by the upper, lower, and intact respiratory tract of rats. *Fundam Appl Toxicol 4*: 594–602

Zitting A, Nickels J, Savolainen H (1982) Comparison of acute toxic effects of intraperitoneally injected nitromethane and nitroethane in rats. *Toxicol Lett 13*: 189–194

completed 24.02.2000

Nitromethane

MAK value	**–**
Peak limitation	**–**
Absorption through the skin (2000)	**H**
Sensitization	**–**
Carcinogenicity (2000)	**Category 3B**
Prenatal toxicity	**–**
Germ cell mutagenicity	**–**
BAT value	**–**
Synonyms	mononitromethane nitrocarbol
Chemical name (CAS)	nitromethane
CAS number	75-52-5
Structural formula	$H_3C–NO_2$
Molecular formula	CH_3NO_2
Molecular weight	61.04
Melting point	–28.55°C
Boiling point	101.2°C
Density at 15°C	1.32 g/cm^3
Vapour pressure at 20°C	37.05 hPa
log P_{ow}*	0.17
Solubility in water at 20°C	9.5 g/100 ml
1 ml/m^3 (ppm) ≙ 2.53 mg/m^3	**1 mg/m^3 ≙ 0.395 ml/m^3 (ppm)**

The MAK value of 100 ml/m^3 for nitromethane which was valid until 2000 was established in 1958 in analogy to the TLV value at that time. The present documentation is based largely on the review of the toxicological data for nitromethane by the American Conference of Governmental Industrial Hygienists (ACGIH 1996).

---

* *n*-octanol/water distribution coefficient

1 Toxic Effects and Mode of Action

Nitromethane has narcotic effects in man and is a weak irritant in the respiratory tract and on the skin. The main effects of nitromethane in the rat are nervous system effects characterized by paralysis of the hind limbs, and effects on the respiratory and olfactory epithelia. Thyroid gland and blood parameters were also affected. Nitromethane causes mammary tumours in female rats and hepatocellular adenomas and carcinomas and lung adenomas and carcinomas in mice. The mechanism responsible for development of these tumours is unclear. The available data do not suggest that nitromethane has genotoxic effects.

2 Mechanism of Action

Nitromethane binds covalently to the active centre of histidase and so causes inactivation of the enzyme (ACGIH 1996).

3 Toxicokinetics and Metabolism

3.1 Absorption, distribution, elimination

The dermal penetration rate for nitromethane was calculated to be 0.6 mg/cm^2 and hour; in persons exposed to a concentration equivalent to the TLV value of 25 ml/m^3, significant absorption through the skin is to be expected (Fiserova-Bergerova *et al.* 1990). 300 µl of a 5.5 % v/v nitromethane solution (47.1 mg nitromethane of which 7.1 mg was radioactively labelled) was applied occlusively to the shaved skin of each of 2 female rhesus monkeys for a period of 12 hours. During the 72 hours after the end of the exposure, radioactivity was determined in blood, urine and faeces. The highest rate of elimination was seen after 24 hours. After 72 hours, 15.4 µg nitromethane had been excreted, 74.3 % in the urine and 25.7 % in the faeces. The highest level in plasma (39 ng/ml blood) was found during the period from 20 minutes to 6 hours after the exposure. After 72 hours only very little nitromethane (3.5 µg) was detected in the skin and none in the subcutaneous adipose tissue. The authors stated that only very little absorption through the skin took place. They point out, however, that neither exhalation nor the formation of metabolites was investigated. In addition, the amount of test substance in the patch after 12 hours was very small; the authors ascribed this to the very high volatility of nitromethane (Angus Chemie 1990).

3.2 Metabolism

Nitromethane was incubated under aerobic conditions with rat liver microsomes. The metabolites detected were nitrite and formaldehyde. An unstable complex of cytochrome P450 with NO was formed; this complex has only been detected in incubations with nitromethane and not with other nitroalkanes. Much less nitrite was formed from nitromethane than from 2-nitropropane (Sakurai *et al.* 1980). *In vitro* studies with microsomes from the nasal epithelium of the rat revealed no formaldehyde formation and with liver microsomes only very little (Dahl and Hadley 1983).

4 Effects in Man

The lethal dose given for man is in the range from 0.5 to 5 g/kg body weight. Evidence of methaemoglobin formation has not been described. Nitromethane is a weak narcotic and causes slight irritation of the respiratory tract. On the skin it causes slight irritation (ACGIH 1996).

5 Animal Experiments and *in vitro* Studies

There are no data available for allergenic effects of nitromethane nor from reproductive and developmental toxicity studies.

5.1 Acute toxicity

5.1.1 Inhalation

The predominant symptoms seen after acute inhalation exposure to nitromethane are central nervous effects and signs of slight irritation in the respiratory tract. Histopathological examination revealed mainly changes in the kidney and the liver characterized by subcapsular damage, focal necrosis, fatty deposits, congestion and oedema. For the rabbit, lethal concentrations of 2500 ml/m^3 after exposure for 12 hours and 5000 ml/m^3 after 6 hours have been reported. Exposure to 5000 ml/m^3 for 3 hours was lethal for the guinea pig (ACGIH 1996).

In rats exposed for 6 hours to nitromethane in concentrations of 500, 2500 or 13000 ml/m^3, the methaemoglobin levels were increased only slightly, by a maximum of 0.25 % (Dequidt *et al.* 1973).

5.1.2 Ingestion

The oral LD_{50} for the mouse is 1440 mg/kg, for the rat 1210 mg/kg body weight. The lowest dose with toxic effects (no other details in the cited review) was given as 750 mg/kg body weight for the rabbit and 125 mg/kg body weight for the dog. The dog was more sensitive to the effects of nitromethane than the rat or mouse (ACGIH 1996).

5.1.3 Dermal absorption

300 μl of a 5.5 % v/v nitromethane solution (47.1 mg nitromethane of which 7.1 mg was radioactively labelled) was applied occlusively to the shaved skin of each of 2 female rhesus monkeys for a period of 12 hours. Signs of toxicity were not seen in either monkey; food and water consumption were unaffected (Angus Chemie 1990).

5.1.4 Intraperitoneal injection

Groups of 3 to 6 male and female BALB/c mice were given nitromethane in doses of 9.0 mmol/kg body weight (about 2200 mg/kg body weight) by intraperitoneal injection. The animals were killed after 24, 48, 72 or 96 hours and the activities of sorbitol dehydrogenase, alanine aminotransferase and aspartate aminotransferase in serum were determined. The liver was not subjected to histopathological examination. No differences between the treated and the control animals were found (Dayal *et al.* 1989).

In rats given nitromethane doses of 200 mg/kg body weight by intraperitoneal injection, increased proteinase activity and glutathione concentration and decreased cytochrome C reductase activity were found in the liver. Fatty deposits in the liver were not observed (ACGIH 1996).

5.2 Subacute, subchronic and chronic toxicity

5.2.1 Inhalation

Groups of 10 male and 10 female F344 rats and $B6C3F_1$ mice were exposed to nitromethane in concentrations of 94, 188, 375, 750 or 1500 ml/m^3, 6 hours/day, 5 days/week for 13 weeks. Survival was unaffected in both species. No changes in organ weights were seen in the rats. At the lowest concentration the haematocrit and haemoglobin values were reduced at some sampling times, the erythrocyte cell volume decreased. In animals exposed to 375 ml/m^3 or more, the effects on the blood parameters were more pronounced and degeneration of the olfactory epithelium and the spinal cord were observed. Paralysis of the hind limbs was seen at concentrations of 750 ml/m^3 or more. In the mouse, the exposures had no effects on body weights. At the lowest concentration the relative kidney weights were increased in the males and hyaline droplets were seen in the respiratory epithelium in two female animals. At concentrations of 188 ml/m^3 or more, the first signs of degeneration of the olfactory epithelium were detected in the male mice. At the highest concentration in both males and females, minimal extramedullary haematopoiesis was detected in the spleen. From this study a NOAEL cannot be deduced for either the rat or the mouse (NTP 1997). The results are shown in detail in Table 1.

Table 1. 13-week study with F344 rats and B6C3F$_1$ mice exposed to nitromethane (NTP 1997)

Concentration (ml/m^3)	Rat	Mouse
94	♂ + ♀: decreased haematocrit, haemoglobin (at some sampling times), and erythrocyte cell volume	♂: increased relative kidney weights 2 ♀: hyaline droplets in respiratory epithelium
188	♂: decreased erythrocyte count 1 ♀: mild bone marrow hyperplasia	♂: minimal nasal lesions ♀: increased absolute kidney weights, ♂ + ♀: first signs of degeneration of the olfactory epithelium, hyaline droplets in the respiratory epithelium
375	♂: increased methaemoglobin level, decreased T_3, T_4, free T_4 (reversible) ♂ + ♀: decreased haematocrit and haemoglobin, degeneration of the olfactory epithelium, slight degeneration of the spinal cord and sciatic nerve	♂: increased relative liver weight, decreased sperm motility ♀: increased relative kidney weights, increased cycle duration ♂ + ♀: minimal extramedullary haematopoiesis in the spleen
750	♂: Heinz bodies, mild bone marrow hyperplasia, decreased sperm motility ♀: decreased T_3, T_4, free T_4 (reversible), decreased erythrocyte count, increased methaemoglobin ♂ + ♀: hind limb paralysis, hyaline droplets in the respiratory epithelium and goblet cell hyperplasia, nasal lesions	♂: increased absolute liver weights
1500	♂: decreased body weights, increased testis weights, all the above-mentioned effects more pronounced	all the above-mentioned effects more pronounced

T_3 triiodothyronine
T_4 thyroxine

Groups of 40 male and 40 female Long-Evans rats were exposed 7 hours/day, 5 days weekly for 2 years to nitromethane in concentrations of 100 and 200 ml/m^3. Survival and body weights of the male animals were not different from the control values. The body weights of the female animals were significantly reduced after one year. Haematological and biochemical parameters, and organ weights were unaffected. Non-neoplastic changes included bronchitis, glomerulosclerosis, calcification of the kidneys, vacuolation of the adrenal cortex and fibrocystomas in the mammary gland; these changes also occurred at high incidence in the control animals and so cannot be used in the present evaluation. The authors considered that the effects were not a result of the exposure to nitromethane. Histopathological examination revealed no evidence of benign or malignant tumours (Griffin *et al.* 1996).

Groups of 50 male and 50 female Sprague-Dawley rats and 15 rabbits were exposed 7 hours/day, 5 days/week for up to 24 weeks to nitromethane in concentrations of 98 and 745 ml/m^3. In neither species were effects seen on erythrocyte count, methaemoglobin concentration, prothrombin time or serum glutamate pyruvate transaminase activity. In the rats, the relative lung weights were slightly reduced at all examination times (after 2 and 10 days and 1, 3, or 6 months exposure) but the differences from the control values were not statistically significant. At 745 ml/m^3 the thyroid weights, body weights, haematocrit and haemoglobin levels were reduced. Microscopic alterations were not observed. In the rabbits, gross pathological changes were not seen but, in the lungs of animals from both concentration groups after exposure for one month, focal haemorrhage and congestion of the alveoli and alveolar walls were detected. In these areas, interstitial oedema and alveolar wall necrosis were also present. Thyroid weights were increased, haemoglobin and thyroxine levels decreased (Lewis *et al.* 1979).

5.2.2 Ingestion

Rats were given nitromethane in the drinking water in doses of 80 or 150 mg/kg body weight for a period of 15 weeks; 4/10 rats in the 80 mg/kg group and 3/10 in the 150 mg/kg group died. Body weights were reduced, liver damage was observed (ACGIH 1996).

5.2.3 Intraperitoneal injection

Methaemoglobin formation was not detected in rats given 8 or 15 nitromethane doses of 110 mg/kg body weight by intraperitoneal injection (Dequidt *et al.* 1973).

5.3 Local effects on skin and mucous membranes

Nitromethane (47.1 mg, of which 7.1 mg was radioactively labelled) was applied occlusively to the shaved skin of each of 2 female rhesus monkeys for a period of 12 hours. Histological examination of the skin revealed neither damage nor irritation (Angus Chemie 1990).

5.4 Genotoxicity

5.4.1 *In vitro*

In the *Salmonella* mutagenicity test with the strains TA100, TA1535, TA1537, TA97 and TA98, nitromethane in concentrations of 6 to 1000 µg/plate was not mutagenic either with or without metabolic activation (Chiu *et al.* 1978, Kawai *et al.* 1987, Löfroth *et al.* 1986, Mortelmans *et al.* 1986). Likewise, in a modified *Salmonella* mutagenicity test with S9 mix from hamster and flavin mononucleotide, nitromethane in concentrations of 0.3 to 100 μmol/plate was not mutagenic (Dellarco and Prival 1989).

Nitromethane in concentrations of 5 to 6 μg/ml in DMSO or 3500 to 5000 μg/ml in culture medium did not increase the incidence of micronucleus formation in hamster embryo cells (Gibson *et al.* 1997). Nitromethane in concentrations up to 5000 μg/ml did not increase the incidence of sister chromatid exchange in CHO cells (a cell line from Chinese hamster ovary) (NTP 1997). In a cell transformation test with hamster embryo cells exposed to nitromethane in concentrations of 2000 to 5000 μg/ml, positive results were obtained (Kerckaert *et al.* 1996).

Nitromethane in concentrations of 1 to 100 μg/ml did not increase the intracellular p53 concentration in cells of the mouse fibroblast cell line NCTC 929 (Duerksen-Hughes *et al.* 1999).

5.4.2 *In vivo*

In mice exposed for 13 weeks to nitromethane in concentrations of 94 to 1500 ml/m^3, no increase in the incidence of micronucleus formation was found (NTP 1997).

The lack of genotoxicity of nitromethane is in accordance with the general finding that primary nitroalkanes, unlike secondary nitroalkanes, have no genotoxic activity. The inactivity of primary nitroalkanes is a result of the fact that the anionic forms do not exist at significant levels in the physiological pH range (Conaway *et al.* 1991). Thus they are not substrates for the sulfotransferases which convert the nitronates of secondary nitroalkanes into DNA-damaging species (Fiala *et al.* 1995).

5.5 Carcinogenicity

Groups of 50 male and 50 female F344 rats were exposed for 2 years to nitromethane in concentrations of 94, 188 and 375 ml/m^3 (Table 2). Survival of the animals was unaffected; the body weights of the females of the highest dose group were increased. Unlike in the 13-week study, in this study the animals did not develop paralysis of the hind limbs. The incidence of fibroadenomas and that for the sum of fibroadenomas, adenomas and carcinomas of the mammary glands of female rats were increased in a concentration-dependent manner, significantly at concentrations of 188 ml/m^3 or more. Two carcinomas produced metastases in the lungs.

All the male animals, both exposed and control animals, developed mild to moderate nephropathy. The number of adenomas in the kidneys was slightly increased in the 94 ml/m^3 and 375 ml/m^3 groups but not in the 188 ml/m^3 group and the differences from the control group were not significant.

Groups of 50 male and 50 female B6C3F$_1$ mice were exposed for 2 years to nitromethane in concentrations of 188, 375 or 750 ml/m^3. Survival and body weights were not different from those in the control group. Clinical examination revealed swelling of the periorbital tissue and exophthalmos in male and female animals in which harderian gland tumours had formed. Degeneration and metaplasia of the olfactory epithelium and hyaline degeneration of the respiratory epithelium were more pronounced in the exposed male and female animals than in the controls. In the olfactory epithelium, loss of neurons, atrophy of nerves and enlargement of Bowman's glands were seen. Suppurative inflammation of the tear ducts was seen in the male animals of the two high dose groups but not in the control animals.

Table 2. Incidences of tumours in F344 rats and B6C3F$_1$ mice after long-term inhalation of nitromethane (NTP 1997)

F344 rat					
concentration (ml/m^3)		**0**	**94**	**188**	**375**
survivors	♂	**13/50**	**16/50**	**14/50**	**8/50**
	♀	**28/50**	**19/50**	**30/50**	**23/50**
mammary gland					
hyperplasia	♀	0/50	0/50	1/50	2/50
fibroadenomas	♀	19/50	21/50	33/50 [1]	36/50 [1]
adenomas	♀	2/50	0/50	0/50	2/50
carcinomas	♀	2/50	7/50	1/50	11/50 [2]
fibroadenomas + adenomas + carcinomas	♀	21/50	25/50	34/50	41/50 [1]
kidney					
nephropathy	♂	50/50	50/50	50/50	50/50
hyperplasia	♂	6/50	8/50	6/50	12/50
adenomas	♂	2/50	5/50	2/50	5/50
B6C3F$_1$ mouse					
concentration (ml/m^3)		**0**	**188**	**375**	**750**
survivors	♂	**31/50**	**36/50**	**30/50**	**29/50**
	♀	**25/50**	**28/50**	**26/50**	**36/50**
harderian gland adenomas + carcinomas	♂	10/50	11/50	25/50 [1]	37/50 [1]
	♀	6/50	9/50	20/50 [1]	21/50 [1]
hepatocellular adenomas + carcinomas	♂	29/50	24/50	22/50	26/50
	♀	19/50	34/49 [1]	22/49	40/50 [1]
alveolar/bronchiolar carcinomas	♂	2/50	3/50	3/50	11/50 [1]
	♀	0/50	3/50	5/49	3/50
alveolar/bronchiolar adenomas + carcinomas	♂	13/50	13/50	12/50	20/50
	♀	3/50	6/50	6/49	12/50 [3]

[1] significantly different from the control value ($p < 0.01$; logistic regression test)
[2] significantly different from the control value ($p = 0.011$; logistic regression test)
[3] significantly different from the control value ($p = 0.015$; logistic regression test)

The incidence of adenomas and that for the sum of adenomas and carcinomas of the harderian gland were increased in a concentration-dependent manner in male and female rats, and the increase was significant at concentrations of 375 ml/m^3 or more. In the female animals, the incidence of hepatocellular adenomas and that for the sum of hepatocellular adenomas and carcinomas were increased significantly in animals exposed to 188 or 750 ml/m^3. The incidence of multiple hepatocellular adenomas was also increased in the two exposure groups. In the female animals of the 750 ml/m^3 group, the sum of the incidences of lung adenomas and alveolar/bronchiolar carcinomas was significantly increased; in the males, the incidence of alveolar/bronchiolar carcinomas on its own was significantly different from the control value. Cellular infiltration of histiocytes was seen in female animals of the 375 ml/m^3 group significantly more often than in the controls. The tumour incidences are shown in Table 2.

In the organs in which tumours developed, there were no signs of toxic effects. The authors of the study also point out that no signs of neurotoxicity or other toxic effects were observed (NTP 1997).

Another long-term inhalation study with Long-Evans rats (Griffin *et al.* 1996; see Section 5.2.1) exposed to nitromethane in concentrations up to 200 ml/m^3 yielded no evidence of carcinogenic effects of nitromethane.

6 Manifesto (MAK value/classification)

There is to date no mechanistic explanation for the development of mammary tumours in the female F344 rat and of liver carcinomas, lung carcinomas and tear duct tumours in the B6C3F$_1$ mouse exposed to nitromethane. The mammary tumours appear to be specific to the F344 strain because they did not develop in Long-Evans rats exposed under the same conditions. Because nitromethane could not be shown to have genotoxic activity either *in vitro* or *in vivo*, a genotoxic mechanism does not seem to be responsible for the tumour development. The incidences of spontaneous liver and lung tumours are high in this strain of mouse; thus it is unclear whether the exposure to nitromethane was involved at all in the development of these tumours. At least the incidence of liver tumours was not associated with the exposure concentration. It is also unclear whether the increased incidence of harderian gland tumours in the mouse is of significance for man. Therefore nitromethane has been classified provisionally in Carcinogen category 3B of Section III of the *List of MAK and BAT Values*. A MAK value could be established if it were possible to derive a NOEL for inhalation exposure to the substance. However, effects were seen even at the lowest tested concentration (94 ml/m^3). The MAK value of 100 ml/m^3 which was valid until 2000 has therefore been withdrawn.

The potential of the substance for germ cell mutagenicity has not been investigated, but in view of the lack of genotoxic effects is not to be expected.

Because of the calculated dermal penetration rate of 0.6 mg/cm^2 and hour and the results obtained in studies with rhesus monkeys, it must be assumed that the substance is absorbed through the skin in relevant amounts. Because it is also a suspected carcinogen, nitromethane is designated with an “H”. It cannot be decided whether designation with “Sa” or “Sh” is necessary because there are no data.

7 References

ACGIH (American Conference of Governmental Industrial Hygienists) (1996) Nitromethane. in: *Documentation of TLVs and BEIs*, ACGIH, Cincinnati, OH, USA

Angus Chemie (1990) *Skin absorption and metabolism/toxicokinetic study of ^{14}C—nitromethane in female rhesus monkeys.* Final Report, Project 890408, Coulston International (unpublished study)

Chiu CW, Lee LH, Wang CY, Bryan GT (1978) Mutagenicity of some commercially available nitro compounds for *Salmonella typhimurium. Mutat Res 58*: 11–22

Conaway CC, Nie G, Hussain NS, Fiala ES (1991) Comparison of oxidative damage to rat liver DNA and RNA by primary nitroalkanes, secondary nitroalkanes, cyclopentanone oxime, and related compounds. *Cancer Res 51*: 3143–3147

Dahl AR, Hadley WM (1983) Formaldehyde production promoted by rat nasal cytochrome P-450-dependent monooxygenases with nasal decongestants, essences, solvents, air pollutants, nicotine, and cocaine as substrates. *Toxicol Appl Pharmacol 67*: 200–205

Dayal R, Gescher A, Harpur ES, Pratt I, Chipman K (1989) Comparison of the hepatotoxicity in mice and the mutagenicity of three nitroalkanes. *Fundam Appl Toxicol 13*: 341–348

Dellarco VL, Prival MJ (1989) Mutagenicity of nitro compounds in *Salmonella typhimurium* in the presence of flavin mononucleotide in a preincubation assay. *Environ Mol Mutagen 13*: 116–127

Dequidt PJ, Vasseur P, Potencier J (1973) Etude toxicologique expérimentale de quelques nitro-paraffines (4) Etude du nitrométhane (Toxicology of some nitroparaffins (4) Nitromethane. *Bull Soc Pharm (Lille) 1*: 29–35

Duerksen-Hughes PJ, Yang J, Ozcan O (1999) p53 induction as a genotoxic test for twenty-five chemicals undergoing *in vivo* carcinogenicity testing. *Environ Health Perspect 107*: 805–812

Fiala ES, Sodum R, Hussain NS, Rivenson A, Dolan L (1995) Secondary nitroalkanes: Induction of DNA repair in rat hepatocytes, activation by aryl sulfotransferase and hepatocarcinogenicity of 2-nitrobutane and 3-nitropentane in male F344 rats. *Toxicology 99*: 89–97

Fiserova-Bergerova V, Pierce JT, Droz PO (1990) Dermal absorption potential of industrial chemicals: criteria for skin notation. *Am J Ind Med 17*: 617–635

Gibson DP, Brauninger R, Shaffi HS, Kerckaert GA, LeBoeuf RA, Isfort RJ, Aardema MJ (1997) Induction of micronuclei in Syrian hamster embryo cells: comparison to results in the SHE cell transformation assay for national toxicology program test chemicals. *Mutat Res 392*: 61–70

Griffin TB, Coulston F, Stein AA (1996) Chronic inhalation exposure of rats to nitromethane. *Ecotoxicol Environ Saf 34*: 109–117

Kawai A, Goto S, Matsumoto Y, Matsushita H (1987) Mutagenicity of aliphatic and aromatic nitro compounds (Japanese). *Jpn J Ind Health 29*: 34–54

Kerckaert GA, Brauninger R, LeBoeuf RA, Isfort RJ (1996) Use of the Syrian hamster embryo cell transformation assay for carcinogenicity prediction of chemicals currently being tested by the national toxicology program in rodent bioassays. *Environ Health Perspect 104*: 1075–1084

Löfroth G, Nilsson L, Andersen J (1986) Structure-activity relationship of nitroalkane-induced mutagenicity in the Ames *Salmonella* assay. *Prog Clin Biol Res 209B*: 149–155

Lewis TR, Ulrich CE, Busey WM (1979) Subchronic inhalation toxicity of nitromethane and 2-nitropropane. *J Environ Pathol Toxicol 2*: 233–249

Mortelmans K, Haworth S, Lawlor T, Speck W, Tainer B, Zeiger E (1986) *Salmonella* mutagenicity tests: II. Results from the testing of 270 chemicals. *Environ Mutagen 8*: 1–119

NTP (National Toxicology Program) (1997) *Toxicology and carcinogenesis studies of nitromethane in F344/N rats and B6C3F$_1$ mice (inhalation studies).* NIH Publication No. 97-3377, US Department of Health and Human Services, National Institutes of Health, Bethesda, MD, USA

Sakurai H, Hermann G, Ruf HH, Ullrich V (1980) The interaction of aliphatic nitro compounds with the liver microsomal monooxygenase system. *Biochem Pharmacol 29*: 341–345

completed 24.02.2000

Rotenone

MAK value	**not yet established, see Section IIb of the *List of MAK and BAT Values***
Peak limitation	**–**
Absorption through the skin (2000)	**H**
Sensitization	**–**
Carcinogenicity	**–**
Prenatal toxicity	**–**
Germ cell mutagenicity	**–**
BAT value	**–**
Synonyms	–
Chemical name (CAS)	[2*R*-(2*α*,6a*α*,12a*α*)]-1,2,12,12a-tetra hydro-8,9-dimethoxy-2-(1-methyl ethenyl)-[1]benzopyrano[3,4-*b*]furo [2,3-*h*][1]benzopyran-6(6a*H*)-one
CAS number	83-79-4
Structural formula	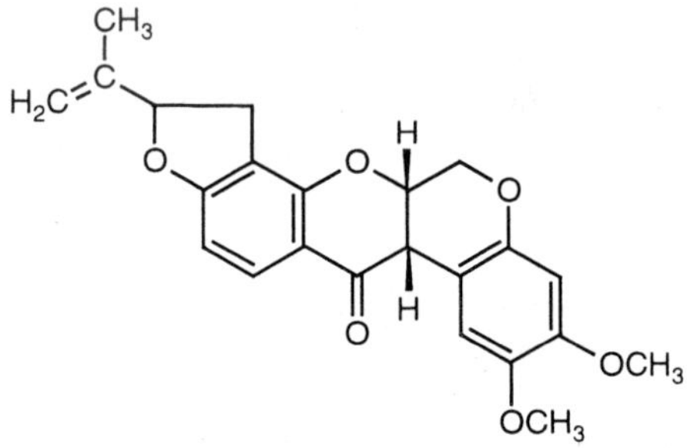
Molecular formula	$C_{23}H_{22}O_6$
Molecular weight	394.4
Melting point	165–166°C
Solubility	insoluble in water (15 mg/l at 100°C), soluble in organic solvents

Rotenone is a versatile pesticide which is used against insects in agriculture, fruit growing and horticulture, against ectoparasites in farm animals, and against unwanted species of fish in fresh-water fish farming. The substance is obtained from the roots of various plants (Leguminosae) of the subtropics and tropics. In practice, powdered ground roots, which contain not only rotenone but also other rotenoids, and various formulations of the pure substance are used. Pure rotenone is colourless and practically insoluble in water but soluble in most organic solvents. Exposed to air and light, it is converted to yellow, orange and red oxidation products.

The MAK value of 5 mg/m^3 (inhalable fraction) for rotenone which was valid until 2000 was established in 1957 in line with the TLV value at that time. The present documentation is based in part on the review of the toxicological data for rotenone by the American Conference of Governmental Industrial Hygienists (ACGIH 1992).

1 Toxic Effects and Mode of Action

Rotenone in the form of inhalable dust is of high acute toxicity, presumably because it inhibits the electron transport chain in the mitochondria. The oral toxicity depends on how readily the preparation is absorbed, that is, on the particle size and the vehicle. The initial symptom of intoxication is respiratory stimulation; it is followed by respiratory depression, incoordination, convulsions, muscle tremor and respiratory failure.

On the skin rotenone is a mild irritant but in the rabbit eye it causes severe irritation.

The results of genotoxicity studies are inconsistent. In bacterial test systems, rotenone does not induce gene mutations; in mouse L5178Y cells a positive result of unclear relevance has been obtained. Sister chromatid exchange (SCE) is not induced by rotenone or at least not unequivocally; tests for DNA damage by detection of unscheduled DNA synthesis (UDS) or by alkaline elution yielded no evidence of genotoxic effects.

Rotenone is a spindle poison. Numerical chromosomal aberrations are found in cultures of Chinese hamster cells but no structural aberrations. In cultures of human lymphocytes, micronuclei are found but also no structural chromosomal aberrations.

Increased incidences of mammary tumours in female rats given rotenone orally and by intraperitoneal injection could not be reproduced in several other studies. In male rats an increased incidence of tumours of the parathyroid gland was observed but the statistical significance of the finding was doubtful. In male mice, a reduction in the incidence of hepatocellular tumours was seen; thus it has been suggested that rotenone has anticarcinogenic activity. As a whole, the studies do not suggest that rotenone has carcinogenic potential.

2 Mechanism of Action

Rotenone is one of the most powerful known inhibitors of the mitochondrial electron transport chain; it inhibits NADH-ubiquinone reductase and so causes ATP deficiency (Ueno *et al.* 1996).

In addition, it acts as a spindle poison because it inhibits microtubule formation (Barham and Brinkley 1976). Rotenone and colchicine act competitively in their inhibition of microtubule formation (Marshall and Himes 1978).

Mechanistic studies have been carried out to clarify the question of an anticarcinogenic activity of rotenone. The results showed that at the concentration which was used in the long-term study with male B6C3F$_1$ mice (600 and 1200 ppm in the diet), rotenone inhibits liver cell proliferation in the animals by over 90 %. Rotenone had, however, no effect on peroxisomal β-oxidation (cyanide-insensitive palmitoyl-CoA oxidation) or on liver catalase activity (Cunningham *et al.* 1995).

In another study, male B6C3F$_1$ mice were first treated with diethylnitrosamine twice weekly for 8 weeks to induce liver foci and then given either a normal diet, a diet containing rotenone (600 ppm), a diet containing the peroxisome proliferater WY-14,643 (1000 ppm), or a diet containing both substances. On its own, rotenone did not affect the number or volume of preneoplastic liver foci; WY-14,643, however, increased the values significantly above the control. In animals treated with rotenone as well as WY-14,643, this increase did not take place. WY-14,643 increased DNA synthesis in normal liver tissue, but when the substance was administered together with rotenone, this increase did not take place. In the liver foci of animals fed WY-14,643, the level of DNA synthesis was increased but simultaneous administration of rotenone also prevented this increase (Isenberg *et al.* 1997). Prevention of an increase in proliferation rate and inhibition of DNA synthesis are plausible mechanisms of an anticarcinogenic effect.

Rotenone and the closely related rotenoids, deguelin and tephrosin, inhibit the induction of ornithine decarboxylase by phorbol ester in an epidermal mouse cell line ("308-cells") and in human MCF-7 cells at non-toxic concentrations (Fang and Casida 1998, Gerhäuser *et al.* 1995).

3 Toxicokinetics and Metabolism

3.1 Absorption, distribution, elimination

There are no data for the absorption of rotenone after inhalation or dermal exposure. Four hours after oral administration of a non-toxic dose of ^{14}C-rotenone to mice (12 µg rotenone dissolved in DMSO per 18 g mouse, that is, 0.67 mg/kg body weight) the ^{14}C-activity was distributed as follows: brain 0.04 %, kidney 0.6 %, liver 4.4 %, small intestine 21.6 %. Rats (dose not specified) excreted 19.5 % of the ^{14}C dose in the urine within 22 hours, mice 20 % within 24 hours, mainly as metabolites (Fukami *et al.* 1969).

Male albino mice and rats were given oral or intraperitoneal doses of ^{14}C-rotenone (0.29 mg/kg body weight, ^{14}C-label in a methyl ether group); within 50 hours, mice exhaled 27 % and rats 12.5 % of the ^{14}C-dose as $^{14}CO_2$ independent of the administration route; 7 % to 17 % of the radioactivity was excreted in the urine (Unai *et al.* 1973).

3.2 Metabolism

Rotenone is metabolized in mammals, fish and insects by cytochrome P450-dependent monooxygenases. Metabolism involves oxidative dealkylation at the methyl ether groups (Unai *et al.* 1973) and hydroxylation of aromatic C atoms. Metabolites identified in the mouse, rat and carp are 8-hydroxyrotenone, 6,7-dihydro-6,7-dihydroxyrotenone and various rotenolones (Fukami *et al.* 1967, 1969).

In sunlight and in the presence of oxygen, rotenone is converted to photo-oxidation products including epoxides (Cheng *et al.* 1972); these have not been tested for mutagenicity. Whether or not they are also formed during metabolic activation of rotenone in the liver is not known.

4 Effects in Man

In rotenone production workers exposed in the 1930s to fine rotenone dust, dermatitis developed in the genital region, ulcerative rhinitis with anosmia, and irritation of the skin wherever perspiration resulted in dust accumulation (ACGIH 1992).

Powdered rotenone was applied to the armpits of four persons twice daily for 30 days. One person developed a mild transient rash, another noted a very mild smarting. Application of 10 % rotenone in petrolatum to the forearm did not cause irritation. When derris (a plant preparation containing rotenone) or an aqueous derris extract was taken into the mouth briefly by volunteers, the persons experienced a slight sensation of numbness and a metallic taste which persisted for 3 to 4 hours (Ray 1991).

Accidental ingestion of a liquid insecticide formulation containing rotenone was fatal. A girl aged three and a half years drank about 10 ml of the formulation. The initial symptoms, vomiting and sleepiness, were followed by unconsciousness. Emergency medical treatment was begun about 2 hours after the poison had been ingested. At this time neither spontaneous breathing nor heart beat could be detected. The child died from respiratory failure about 8 hours after ingestion of the poison. Post-mortem analyses revealed a rotenone concentration of 1260 mg/kg in the stomach contents. The concentrations in blood, kidney and liver tissue were between 2 and 4 mg/kg; rotenone was not detectable in the thymus, brain or muscle. The ingested lethal dose was estimated as 40 mg/kg body weight (De Wilde *et al.* 1986).

5 Animal Experiments and *in vitro* Studies

5.1 Acute toxicity

Because rotenone is insoluble in water, its absorption and the resulting toxicity is decisively affected by the form in which it is administered (particle size, dispersion or solution, solvent). For the rat, oral LD_{50} values in the range from 25 to 132 mg/kg body weight have been found, for the guinea pig 13 to 130 mg/kg body weight. Dermal LD_{50} values in the range from 100 to 200 mg/kg body weight have been reported for the rabbit (Ray 1991). LD_{50} values of 0.2 and 0.3 mg/kg body weight were found after intravenous injection into rats (Ray 1991). The symptoms of acute poisoning include respiratory stimulation followed by respiratory depression, incoordination, clonic or tonic convulsions and muscle tremor. Death results from respiratory failure (ACGIH 1992).

5.2 Subacute, subchronic and chronic toxicity

In the only available studies, the substance was administered orally.

Groups of 5 F344 rats and 5 $B6C3F_1$ mice were given diets containing rotenone in concentrations of 0, 50, 100, 200, 400 or 600 ppm (about 0–60 mg/kg body weight and day), in a second study 0, 300, 600, 1200, 2400, 4800 ppm (about 0–480 mg/kg body weight and day) for periods of 14 days. In the second study in animals given doses of 120 mg/kg body weight (1200 ppm in the diet) or more, the body weights of the rats decreased in a dose-dependent manner. 2/5 male rats in the 240 mg/kg group and 4/5 in the 480 mg/kg group died. Food consumption was reduced in animals given doses of 120 mg/kg body weight or more, the hair coats were rough, the faeces hard, and the animals sat in a hunched position. The clinical and pathological examinations revealed no effects. No signs of toxicity were seen in the mice (NTP 1988).

In a 13-week study, groups of 10 rats were given diets containing rotenone in concentrations of 0, 75, 150, 300, 600 and 1200 ppm (about 7.5, 15, 30, 60 and 120 mg/kg body weight and day), and groups of 10 mice were given 0, 600, 1900, 5000, 16000 and 50000 ppm (about 90, 285, 750, 2400, 7500 mg/kg body weight and day). In the highest dose group all the male rats and 6 female rats died. Delayed body weight gains were seen in male rats given 30 mg/kg body weight or more and female rats given 15 mg/kg body weight or more. Relative liver weights were increased at doses of 30 mg/kg body weight and more. Atrophy of the bone marrow and inflammation and hyperplasia of the forestomach were seen in male rats at 30 mg/kg or more and in females at 15 mg/kg body weight or more. From these results a NOEL of 7.5 mg/kg body weight can be derived for female rats and of 15 mg/kg body weight for male rats.

In the mice, all males and females in the highest dose group died and only one male and two females survived the 2400 mg/kg dose. Body weights were reduced after doses of 750 mg/kg body weight and more. Relative liver weights were decreased at doses of

90 mg/kg body weight and more. The clinical and pathological examinations revealed no effects. From this study a NOEL for the mouse cannot be deduced (NTP 1988).

In a two-year study, groups of 50 F344 rats and 50 B6C3F$_1$ mice were given rotenone in the diet. The concentrations in the rat diet were 38 and 75 ppm (1.8 and 3.6 mg/kg body weight and day), in the mouse diet 600 and 1200 ppm (♂: 111 and 242 mg/kg, ♀: 124 and 265 mg/kg body weight and day). The average daily doses for the mice were thus about 70 times those for the rats. The neoplastic changes found in this study are discussed in Section 5.6. In the rats, the treatment had no effect on average body weight, food consumption or survival. In the male rats of the highest dose group, the incidence of focal hyperplasia in the pituitary gland was increased but not in a dose-dependent manner. In the mice, food consumption was the same in the three groups, but in the two groups of animals given rotenone the average body weights were significantly lower than in the control group. Survival of the male animals of the high dose group was significantly higher ($p < 0.05$) than that of the controls. The gross pathological and histological examinations revealed no evidence of systemic toxic effects (Abdo *et al.* 1988, NTP 1988; see also Section 5.6). From this study, a NOEL of 1.8 mg/kg body weight and day for non-neoplastic effects can be deduced for the rat. For the mouse, no NOEL can be derived because reduced body weights were also seen at the lowest tested dose of 111 mg/kg body weight and day.

In an early study of the chronic toxicity of derris in the rat, a NOEL for this preparation in the diet of 156 ppm (about 15.6 mg/kg body weight) was determined (ACGIH 1992). Administration of rotenone to rats in the drinking water at a concentration of 2.5 ppm (about 0.25 mg/kg body weight and day) with 2.5 ppm dimethylsulfoxide resulted in a marked delay in body weight gains (ACGIH 1992).

5.3 Local effects on skin and mucous membranes

Rotenone in the form of derris powder caused slight irritation on the skin of experimental animals (no other details given in the cited reviews); on rabbit skin the substance caused severe irritation and pus formation (ACGIH 1992, Ray 1991).

There are no data available for allergenic effects of rotenone.

5.4 Reproductive and developmental toxicity

5.4.1 Fertility

Incubation of human sperm with 25 μM rotenone *in vitro* resulted in ATP depletion and inhibition of sperm motility (DeLamirande and Gagnon 1992).

5.4.2 Developmental toxicity

Technical grade rotenone (87 % rotenone, 13 % other extract components) was administered orally in corn oil in daily doses of 2.5, 5 and 10 mg/kg body weight to groups of 20 Wistar rats from day 6 to day 15 of gestation. The high dose killed 60 % of the dams, caused a significant reduction in body weight gains, reduced the number of living foetuses per dam and increased the number of resorptions. The incidence of skeletal malformations was not increased at any dose. In the middle dose group body weight gains were reduced significantly; in this group the incidence of skeletal variations and anomalies (extra ribs, delayed ossification of the sternebrae, missing sternebrae) was increased significantly. At the low dose no significant effects were seen (Khera *et al.* 1982).

When Sprague-Dawley rats were given rotenone in the diet in concentrations of 10, 100, 200, 400, 600 or 800 ppm (about 1, 10, 20, 40, 60, 80 mg/kg body weight and day) from day 6 to day 15 of gestation, maternal toxicity was seen at the two highest doses but no spontaneous abortions or resorptions (ACGIH 1992).

5.5 Genotoxicity

In a non-quantitative reversion test (spot test) with *Escherichia coli* WP2, no mutagenic effects of rotenone could be detected (Ashwood-Smith *et al.* 1972). In concentrations up to 10 mg/plate, rotenone was not mutagenic in the *Salmonella* mutagenicity test in the strains TA98, TA100, TA1535 or TA1537 either in the presence or absence of a metabolic activation system (liver S9 from rat and hamster liver) (Moriya *et al.* 1983, Zeiger *et al.* 1987).

In the thymidine kinase test with mouse lymphoma L5178Y cells with and without a metabolic activation system, concentration-dependent mutagenicity of rotenone was demonstrated with toxic concentrations in the range from 0.25 to 4 µg/ml. Small and large mutant colonies were not distinguished and therefore it is not possible to differentiate between chromosomal aberrations and gene mutations. Tests with a metabolic activation system were not carried out (Myhr *et al.* 1990). In the interpretation of this result, the high toxicity must be taken into account and also the fact that the thymidine kinase test with mouse lymphoma cells has only a low specificity for carcinogenic effects because it detects mutagenic activity in 60 % of the non-carcinogens which have been tested (Zeiger *et al.* 1990).

Tests for SCE in CHO cells (a cell line from Chinese hamster ovary) with and without metabolic activation yielded negative results (concentration not specified) (Tomkins *et al.* 1980). Likewise, in two other studies without a metabolic activation system the incidence of SCE in CHO cells was not increased (concentrations 0.0001–0.0016 µg/ml), in one of two studies with a metabolic activation system (concentrations 0.05–1.6 µg/ml) the incidence of SCE was increased slightly but not in a concentration-dependent manner (Anderson *et al.* 1990). In the NTP report this result was called "equivocal" (NTP 1988).

Chromosomal aberrations were not induced in CHO cells by rotenone in concentrations up to 100 µg/ml in the absence of a metabolic activation system or up to 200 µg/ml in the presence of an activation system (Anderson *et al.* 1990). In the Chinese hamster cell line CHL, rotenone (0.005–10 µg/ml) did not induce structural chromosomal aberrations but induced numerical aberrations (hypodiploidy, hyperdiploidy, polyploidy, endoreduplication) (Matsumoto and Ohta 1991, 1993).

Cultures of human lymphocytes were treated with rotenone (0.1–1 µg/ml; 0.25–2.5 µM) with and without metabolic activation and then examined for SCE, chromosomal aberrations and micronuclei. Neither SCE nor chromosomal aberrations were induced by the treatment but the incidence of micronuclei in the tests without metabolic activation was increased in a concentration-dependent manner to up to ten times the spontaneous value. With metabolic activation, the micronucleus induction was weaker. The absence of structural chromosomal aberrations in the presence of clear micronucleus induction suggests that rotenone disturbs mitotic chromosome distribution (Guadano *et al.* 1998).

UDS was not induced by rotenone (1–1000 µM or 0.39–390 µg/ml, 8 hours) in SV-40 transformed human fibroblasts (Ahmed *et al.* 1977). Likewise in cultured primary rat hepatocytes, no UDS induction was observed (no other details) (Probst and Hill 1980).

DNA damage detectable in the alkaline elution test was not induced by treatment of primary rat hepatocytes with rotenone (15–60 µM, 3 hours) (Storer *et al.* 1996).

Rotenone (12 µM, 4.73 µg/ml) inhibited all phases of the cell cycle of CHO cells. At this concentration, cell respiration of the CHO cells was partially inhibited. In a cell-free system, rotenone (12 µM), like colchicine, inhibited the aggregation of tubulin to form microtubules. This is conceivably the cause of the mitosis inhibition. The inhibition of other phases of the cell cycle could result from a reduced ATP supply (Barham and Brinkley 1976).

5.6 Carcinogenicity

Carcinogenicity studies have been carried out with rotenone administered perorally and by intraperitoneal injection. One research group demonstrated increased incidences of mammary tumours in female rats both after oral administration and intraperitoneal injection; other kinds of tumour did not develop. In these studies the substance was administered for only 42 to 60 days and the animals were then observed for 17 or 18 months (Gosalvez 1983, Gosalvez and Merchan 1973). In later studies under identical conditions, the increased incidence of mammary tumours could not be confirmed (Freudenthal *et al.* 1981, Greenman *et al.* 1993). Because once more the substance was applied for only 42 to 58 days, the studies are not useful for assessing the carcinogenic effects of rotenone.

In the two-year feeding study described in Section 5.2 above (Abdo *et al.* 1988, NTP 1988) the incidence of parathyroid adenomas was increased in male rats but the increase above the concurrent control value was not statistically significant (control 1/41, low dose 0/44, high dose 4/44). However, the difference from the historical control value was significant. Tumours of this kind are rare in the rat; the NTP historical control value is 4/1314 for untreated male rats. The increased incidence at the high rotenone dose was considered to be "equivocal evidence" of a carcinogenic effect. The results obtained with female rats and with mice yielded no evidence of carcinogenicity. Conspicuous was the negative trend in the incidence of hepatocellular adenomas and carcinomas in male mice (control 12/47, low dose 12/49, high dose 1/50). The incidence in the high dose group is unusually low and conceivably a result of the treatment with rotenone (NTP 1988).

In summary, it may be concluded that the increased incidence of mammary tumours could not be reproduced in comparative studies, even with higher doses and longer duration of administration, and that the increased incidence of parathyroid tumours in male rats was of doubtful statistical significance. Seen as a whole, the studies do not suggest that rotenone has carcinogenic potential.

6 Manifesto (MAK value/classification)

Rotenone, one of the most powerful known inhibitors of the mitochondrial electron transport chain, is of high acute toxicity.

Rotenone does not interact with DNA, is not mutagenic and causes no structural chromosomal aberrations. However, it has been shown to be a spindle poison and to induce numerical chromosomal aberrations. This mechanism is probably responsible for the observed micronucleus induction. A non-linear dose-response relationship may be assumed for this process.

The reported induction of mammary tumours could not be reproduced in later studies. The NTP study demonstrated an increased incidence of tumours of the parathyroid in male rats, but the increase was of doubtful statistical significance. In the mice, which were treated with much higher doses of rotenone, no increased tumour incidences were seen. In all, the studies do not suggest that rotenone has carcinogenic potential.

Results of studies of exposed persons and inhalation studies with experimental animals which would be suitable for the derivation of a MAK value are not available. Because of the irritant effects of rotenone, the feeding study cannot be used to establish a MAK value. The MAK value which was valid until 2000 has therefore been withdrawn. Rotenone is listed in Section IIb of the *List of MAK and BAT Values*.

Because of the low dermal LD_{50} value in the rabbit, the substance is designated with an "H". Since there are no data for allergenic effects of the substance, rotenone is not designated with an "S".

The germ cell mutagenicity of the substance has not been investigated.

7 References

Abdo KM, Eustis SL, Haseman J, Huff JE, Peters A, Persing R (1988) Toxicity and carcinogenicity of rotenone given in the feed to F344/N rats and $B6C3F_1$ mice for up to two years. *Drug Chem Toxicol 11*: 225–235

ACGIH (American Conference of Governmental Industrial Hygienists) (1992) Rotenone. in: *Documentation of TLVs and BEIs*, ACGIH, Cincinnati, OH, USA

Ahmed FE, Hart RW, Lewis NJ (1977) Pesticide induced DNA damage and its repair in cultured human cells. *Mutat Res 42*: 161–174

Anderson BE, Zeiger E, Shelby MD, Resnick MA, Gulati DK, Ivett JL, Loveday KS (1990) Chromosome aberration and sister chromatid exchange test with 42 chemicals. *Environ Mol Mutagen 16, Suppl 18*: 55–137

Ashwood-Smith MJ, Trevino J, Ring R (1972) Mutagenicity of dichlorvos. *Nature 240*: 418–420

Barham SS, Brinkley BR (1976) Action of rotenone and related respiratory inhibitors on mammalian cell division. I. Cell kinetics and biochemical aspects. *Cytobios 15*: 85–96

Cheng HM, Yamamoto I, Casida JE (1972) Rotenone photodecomposition. *J Agric Food Chem 20*: 850–856

Cunningham ML, Soliman MS, Badr MZ, Matthews HB (1995) Rotenone, an anticarcinogen, inhibits cellular proliferation but not peroxisome proliferation in mouse liver. *Cancer Lett 95*: 93–97

De Wilde AR, Heyndrickx A, Carton D (1986) A case of fatal rotenone poisoning. *J Forensic Sci 31*: 1492–1498

DeLamirande, Gagnon E (1992) Reactive oxygen species and human spermatozoa. II. Depletion of adenosine triphosphate plays an important role in the inhibition of sperm motility. *J Androl 13*: 379–386

Fang N, Casida JE (1998) Anticancer action of cube insecticide: correlation for rotenoid constituents between inhibition of NADH: ubiquinone oxidoreductase and induced ornithine decarboxylase activities. *Proc Natl Acad Sci USA 95*: 3380–3384

Freudenthal R, Leber A, Thake D (1981) *Carcinogenic potential of rotenone: subchronic oral and peritoneal administration to rats and chronic dietary administration to Syrian golden hamsters.* EPA 600/S1-81-037. U.S. Environmental Protection Agency, Health Effects Research Laboratory, Research Triangle Park, NC

Fukami J, Yamamoto I, Casida JE (1967) Metabolism of rotenone *in vitro* by tissue homogenates from mammals and insects. *Science 155*: 713–716

Fukami J, Shishido T, Fukunaga K, Casida JE (1969) Oxidative metabolism of rotenone in mammals, fish and insects and its relation to selective toxicity. *J Agric Food Chem 17*: 1217–1226

Gerhäuser C, Mar W, Lee SK, Suh N, Luo Y, Kosmeder J, Luyengi L, Fong HH, Kinghorn AD, Moriarty RM, Metha RG, Constantinou A, Moon RC, Pezzuto JM (1995) Rotenoids mediate potent cancer chemopreventive activity through transcriptional regulation of ornithine decarboxylase. *Nat Med 1*: 260–266

Gosalvez M (1983) Minireview: Carcinogenesis with the insecticide rotenone. *Life Sci 32*: 809–816

Gosalvez M, Merchan J (1973) Induction of rat mammary adenomas with the respiratory inhibitor rotenone. *Cancer Res 33*: 3047–3050

Greenman DL, Allaben WT, Burger GT, Kodell RL (1993) Bioassay for carcinogenicity of rotenone in female Wistar rats. *Fundam Appl Toxicol 20*: 383–390

Guadano A, González-Coloma A, De la Pena E (1998) Genotoxicity of the insecticide rotenone in cultured human lymphocytes. *Mutat Res 414*: 1–7

Isenberg JS, Kolaja KL, Ayoubi SA, Watkins JB 3rd, Klaunig JE (1997) Inhibition of WY-14,643 induced hepatic lesion growth in mice by rotenone. *Carcinogenesis 18*: 1511–1519

Khera KS, Whalen C, Angers G (1982) Teratogenicity study on pyrethrum and rotenone (natural origin) and ronnel in pregnant rats. *J Toxicol Environ Health 10*: 111–119

Marshall LE, Himes RH (1978) Rotenone inhibition of tubulin self-assembly. *Biochim Biophys Acta 543*: 590–594

Matsumoto K, Ohta T (1991) Rotenone induces aneuploidy, polyploidy and endoreduplication in cultured Chinese hamster cells. *Mutat Res 263*: 173–177

Matsumoto K, Ohta T (1993) Mitosis of rotenone-induced endoreduplication in Chinese hamster cells. *Jpn J Genet 6*: 185–194

Moriya M, Ohta K, Watanabe T, Miyazawa T, Kato K, Shirasu Y (1983) Further mutagenicity studies on pesticides in bacterial reversion assay systems. *Mutat Res 116*: 185–216

Myhr B, McGregor D, Bowers L, Riach C, Brown AG, Edwards I, McBride D, Martin R, Caspary WJ (1990) L5178Y mouse lymphoma cell mutation assay results with 41 compounds. *Environ Mol Mutagen 16, Suppl 18*: 138–167

NTP (National Toxicology Program) (1988) *Toxicology and carcinogenesis studies of rotenone (CAS No. 83-79-4) in F344/N rats and B6C3F$_1$ mice (feed studies)*, Technical Report 320, US Department of Health and Human Services, National Institutes of Health, Bethesda, MD, USA

Probst G, Hill L (1980) Chemically-induced DNA-repair synthesis in primary rat hepatocytes: a correlation with bacterial mutagenicity. *Ann NY Acad Sci 349*: 405–406

Ray DE (1991) Pesticides derived from plants and other organisms. in: Hayes WJ, Laws ER (Eds) *Handbook of Pesticide Toxicology Vol. 2*, Academic Press, New York, 599–603

Storer RD, McKelvey TW, Kraynak AR, Elia MC, Barnum JE, Harmon LS, Nichols WW, DeLuca JG (1996) Revalidation of the *in vitro* alkaline elution/rat hepatocyte assay for DNA damage; improved criteria for assessment of cytotoxicity and genotoxicity and results for 81 compounds. *Mutat Res 368*: 59–101

Tomkins D, Kwok E, Douglas G (1980) Testing of pesticides for induction of sister-chromatid exchange in Chinese hamster ovary cells (Abstract). *Can J Genet Cytol 22*: 681

Ueno H, Miyoshi H, Inoue M, Niidome Y, Iwamura H (1996) Structural factors of rotenone required for inhibition of various NADH-ubiquinone oxidoreductases. *Biochim Biophys Acta Bioenergetics 1276*: 195–202

Unai T, Cheng HM, Yamamoto I, Casida JE (1973) Chemical and biological *O,O*-demethylation of rotenone derivatives. *Agric Biol Chem 37*: 1937–1944

Zeiger E (1990) Mutagenicity of 42 chemicals in *Salmonella*. *Environ Mol Mutagen 16, Suppl 18*: 32–54

Zeiger E, Anderson B, Haworth S, Lawlor T, Mortelmans K, Speck W (1987) *Salmonella* mutagenicity tests. III. Results from the testing of 255 chemicals. *Environ Mutagen 9, Suppl 9*: 1–109

completed 07.10.1999

Strychnine

MAK value (1999)	**not yet established, see Section IIb of the *List of MAK and BAT Values***
Peak limitation	**–**
Absorption through the skin	**–**
Sensitization	**–**
Carcinogenicity	**–**
Prenatal toxicity	**–**
Germ cell mutagenicity	**–**
BAT value	**–**
Synonyms	–
Chemical name (CAS)	strychnidin-10-one
CAS number	57-24-9
Structural formula	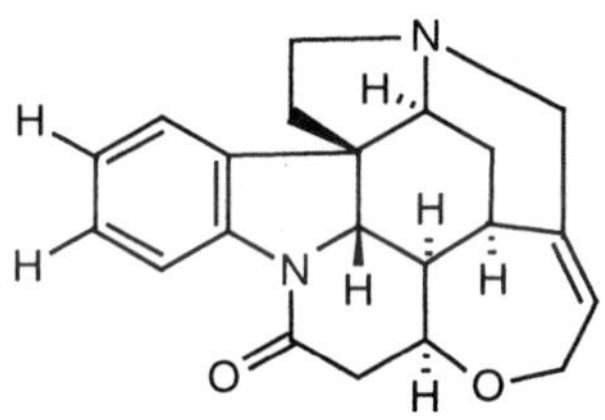
Molecular formula	$C_{21}H_{22}N_2O_2$
Molecular weight	334.40
Melting point	268–290°C
Solubility in water	143–156 mg/l at 20°C

The present documentation is based mainly on a review of the toxicological data for strychnine prepared by the Health Council for the Netherlands (HCN 1999). Strychnine is a non-flammable, colourless, odourless, bitter tasting crystalline solid. The alkaloid is obtained from the dried seeds of the Indian tree *Strychnos nux-vomica* and other *Strychnos* species (family Loganiaceae). Strychnine salts (nitrate, sulfate and phosphate) have been used as rodenticides and in medicine for the treatment of non-ketotic hyperglycinaemia, impotence and sleep apnoea (ACGIH 1991, Gitzelmann *et al.* 1977, 1978).

1 Toxic Effects and Mode of Action

Ingested strychnine is rapidly absorbed, concentrated in the liver and oxidatively detoxified by cytochrome P450 enzymes. Of a strychnine dose, 20 % is excreted unchanged in urine and faeces. In man, the half-time for the elimination of strychnine from the blood is 10 to 16 hours. Strychnine blocks the postsynaptic inhibitory effect of the neurotransmitter glycine in the spinal cord. Symptoms of strychnine poisoning include abnormal eye movements, photophobia, stiff joints, myalgia, muscle cramps and, as the cause of death, respiratory paralysis. Doses of 15 to 30 mg strychnine are generally lethal for man. More recently, deaths have been documented after doses of less than 7 mg strychnine. Persons differ widely in their individual sensitivity to the substance and children are more sensitive than adults. The oral LD_{50} of strychnine for the rat is between 2.35 and 16 mg/kg body weight. From a 28-day study with rats, a no observed effect level (NOEL) could not be derived because deaths were recorded even after the lowest strychnine dose of 2.5 mg/kg body weight. Strychnine has effects on genetic recombination.

2 Mechanism of Action

Strychnine causes nervous system stimulation by selective and competitive blocking of the postsynaptic inhibitory effect of the neurotransmitter glycine in the spinal cord. This reduces the efflux of sodium ions from the spinal nerves which in turn reduces the hyperpolarization of the postsynaptic membranes and thus the threshold for the evoking of action potentials. The result is a cerebrospinal hyperexcitability, tetanic cramps in the diaphragm and the striated muscles, hyperexcitability of the sympathetic nervous system with tachycardia and hypertension and even respiratory arrest. The persons suffering tonic spasms are fully conscious because the normal function of the cerebral cortex and subcortical centres is unaffected (Forth *et al.* 1992). Sensory stimuli trigger and amplify the seizures via an over-sensitive reflex arc. In the early phase of a strychnine intoxication, symptoms such as trembling, twitching, and stiffness of the facial muscles and the legs appear after 15 to 30 minutes followed by painful spasms and increased sensory perception. The seizures become more frequent and death is caused by exhaustion or anoxia during a spasm. Only few poisoned persons survived more than 5 seizures with concurrent respiratory arrest. Secondary effects include lactic acid acidosis, renal failure, hyperthermia and the destruction of striated muscle fibres (rhabdomyolysis).

3 Toxicokinetics and Metabolism

After ingestion, strychnine is rapidly absorbed from the gastrointestinal tract. Parenterally injected strychnine is also rapidly absorbed. Strychnine binds to plasma protein hardly at all. In poisoned persons the concentration of strychnine in the blood was mostly above 2 mg/l and in the liver above 4 mg/kg liver. After lethal intoxications of persons and dogs, the highest concentrations of strychnine were found in the blood, liver, bile and kidneys. Much lower levels were found in muscles, brain and even in the spinal cord, the main site of action of the substance.

About 80 % of a dose of strychnine is oxidized in the liver to non-toxic metabolites. In rats given strychnine in a concentration of 0.15 mg/ml in the drinking water for 5 days, a marked induction of the cytochrome P450 isoenzymes CYP2B1 and CYP2B2 with a maximum after 3 days was seen. In addition, the glutathione-*S*-transferase and UDP-glucuronosyl transferase activities were shown to be increased.

Pretreatment with phenobarbital induced oxidative metabolism more markedly than did that with 3-methylcholanthrene. Female rats reacted more sensitively to strychnine than did male rats whose hepatic strychnine oxidation was 2.5-times faster. This difference could no longer be demonstrated in castrated animals. In mice, no sex-related differences in strychnine metabolism have been described. After subcutaneous injection of a strychnine dose of 0.5 mg/kg body weight into rats, the following urinary metabolites were identified: strychnine-21,22-epoxide as the main metabolite, and strychnine-*N*-oxide, 21α,22α-dihydroxy-22-hydrostrychnine, 21α,22β-dihydroxy-22-hydrostrychnine, 2-hydroxystrychnine and 16-hydroxystrychnine (for the structures see HCN 1999).

With microsomal fractions from the livers of the mouse, rat, rabbit, guinea pig and dog, the same strychnine metabolites were found *in vitro* as were known from the *in vivo* studies with the rat and, in addition, 11,12-dehydrostrychnine, 22-hydroxystrychnine and 18-oxostrychnine. There were marked differences between the species in the pattern of metabolites obtained *in vitro*. Strychnine-21,22-epoxide was not detected in the rabbit fractions. 16-Hydroxystrychnine was the main metabolite *in vitro* for the rat and mouse, 2-hydroxystrychnine for the guinea pig and rabbit and strychnine-*N*-oxide for the dog. Strychnine-21,22-epoxide and 18-oxostrychnine were minor metabolites. The metabolic activity of the liver microsomes from the guinea pig was much higher than that of other species. Apart from in the dog, there was a good inverse correlation between the observed metabolic activity and the acute toxicity of strychnine.

The metabolite strychnine-*N*-oxide could be reduced by the flora of the human intestine under anaerobic conditions to yield strychnine and 16-hydroxystrychnine.

Depending on the dose taken up, between 1 % and 20 % of the absorbed substance is excreted unchanged in human urine. About 70 % of the dose was recovered within the first 6 hours. One hour after ingestion of a lethal dose, however, no strychnine could be detected in the urine. Traces of the substance can be detected in the urine for up to five days after dosing. Strychnine can be detected in exhumed bodies years after their death.

Determined in serum between 0.5 and 43 hours after ingestion of the substance, the concentration of strychnine sank from 3.8 mg/l to less than 0.1 mg/l. The elimination is best described by a single compartment model with saturated Michaelis-Menten kinetics.

Assuming 100 % absorption, the following parameters were calculated: absorption rate $k_a = 2.68$/hour, absorption half-time $T_{1/2} = 15.5$ minutes, $V_{max} = 3.71$ mg/kg and hour, $k_m = 1.46$ mg/l and distribution volume 13 l/kg. These data for man have been confirmed essentially by two other studies. In a person who had ingested strychnine, the maximum level in serum was 2.1 mg/l after 3 hours. The elimination half-time was 10 to 15.9 hours.

Rats given a ^{3}H-strychnine dose of 0.5 mg/kg body weight excreted 30 % of the dose with the urine and 65 % with the faeces within 7 days; 6 % and 3 %, respectively, were unchanged substance. About 80 % of the radioactivity was recovered within 24 hours.

4 Effects in Man

No data are available for irritation, sensitization, reproductive toxicity, genotoxicity, or carcinogenicity of strychnine in man.

4.1 Single exposures

In persons exposed to strychnine, the following symptoms have been observed: abnormal eye movements, photophobia, stiff joints, myalgia, muscle cramps and respiratory paralysis as the cause of death. The median lethal oral dose for man has been given as 1.5 to 2.0 mg/kg body weight. Other sources have stated that doses of 15 to 30 mg strychnine (about 0.22–0.42 mg/kg body weight) are generally lethal for man. However, deaths after ingestion of only 5 to 10 mg strychnine (about 0.07–0.14 mg/kg body weight) have been reported and one patient survived a dose of 3750 mg (about 53.57 mg/kg body weight). Children are more sensitive than adults.

The results of the pathological examination varied. Generally the findings included petechial bleeding and hyperaemia of the organs as a result of major seizures. Necrosis in the brain was caused by oxygen depletion. Extensive formation of lipofuscin pigments in the liver parenchyma and also some hyaline droplet formation have been reported. It has been suggested that high doses caused death without any seizures and low doses later death after typical symptoms of intoxication. Complete recovery has been observed 7 to 10 days after intoxication. Other patients who had survived strychnine poisoning suffered from severe loss of visual acuity and left-sided paresis caused by the severe acidosis and prolonged hypoxia.

4.2 Repeated exposures

In women and men given doses of up to 7 mg strychnine for therapeutic purposes (no other details) stiffness of the muscles especially in the neck, the little fingers and the jaw was reported. Strychnine used to be used in the treatment of ketotic hyperglycinaemia, a rare metabolic disorder, as a glycine receptor antagonist. Children were given doses of about 0.9 mg/kg body weight and day, divided into several single doses. Symptoms of intoxication have not been described. However, these children have higher strychnine-tolerance because of their pathologically increased levels of glycine which cause postsynaptic inhibition of neurones in the spinal cord (Gitzelmann *et al.* 1977, 1978).

5 Animal Experiments and *in vitro* Studies

No data are available for local effects of strychnine on skin and mucous membranes, for allergenic effects, reproductive and developmental toxicity or carcinogenicity.

5.1 Acute toxicity

The LD_{50} values for strychnine in various species after administration by various routes are shown in Table 1.

Table 1. LD_{50} values for strychnine

Administration route	Species	LD_{50} (mg/kg body weight)
oral	rat	2.35–16
	mouse	2
	dog	0.5–1.1
intraperitoneal	rat	0.9–2.8
	mouse	1.4–1.8
subcutaneous	rat	1.2–4.0
	mouse	0.47
	dog	0.46
intravenous	rat	0.57–0.96
	guinea pig	0.39

Strychnine doses which cause seizures are doses in the lethal range. In adult CBA mice given strychnine doses of 0.75, 1.25, 1.50 mg/kg body weight as single subcutaneous injections, spasms were seen in 8/20, 19/20 and 21/21 animals with average latency periods of 6.1 to 10.3 minutes. 4/20, 17/20 and 19/21 animals died. Sex-related differences were not observed. Tonic spasms could be induced in male albino mice by intraperitoneal injection of strychnine doses between 0.5 and 4 mg/kg body weight. The autopsy findings tend to be unspecific and suggestive of massive spasms and anoxia. Congestion and minor bleeding were detected in the brain and viscera. Spasm-inducing doses of strychnine (strychnine sulfate, 2.5 mg/kg body weight, subcutaneous) caused a marked reduction in hormone release from the hypothalamus and posterior pituitary gland (neurohypophysis).

5.2 Subacute, subchronic and chronic toxicity

In a 28-day study, groups of 12 male rats were given daily oral strychnine doses of 5 or 10 mg/kg body weight and 12 female rats doses of 2.5 mg/kg body weight (2 % aqueous solution of strychnine hydrochloride). 10 to 20 minutes after each dose, slight trembling developed in the animals and gradually regressed within the subsequent hour. Of the male rats, in the low dose group one animal died after the fifth dose and in the high dose group 5 animals died, the first after only the third dose. One female rat died as well. In the surviving animals, no changes were detected in body weight or organ weight gains, in biochemical blood or urine parameters, or in the eyes. The histological examination also yielded negative results. Even after the animals had been treated for 28 days, their behaviour and their performance in the rota-rod test were normal. The ECG was normal (Seidl and Zbinden 1982). Although no substance-related changes were detected in this study, one rat in the group given the lowest strychnine dose of 2.5 mg/kg body weight died. Therefore a NOEL cannot be derived from the results of this study.

5.3 Genotoxicity

Strychnine was not mutagenic in the white-ivory reversion test in *Drosophila melanogaster*. Strychnine induced a high incidence of Trp^f duplications in *Salmonella typhimurium* but under the same test conditions did not cause reversion of the hisG46, hisD3052 or hisC3076 alleles (no other details). This result suggests that strychnine affects genetic recombination but is not mutagenic. Increases in the incidence of recombinations have also been reported in *Drosophila melanogaster* and in yeasts.

6 Manifesto (MAK value/classification)

The MAK value for strychnine which applied until 1999 was 0.15 mg/m^3 (about 0.02 mg/kg body weight and day) and was based on data which indicated that therapeutic doses of strychnine sulfate up to 7 mg were tolerated by man without adverse effects on health. However, a recent study indicated that deaths had occurred among persons given doses in this range. The 28-day study with rats did not yield a NOEL. As no other data for the effects of long-term exposure of persons or experimental animals are available, the MAK value has been withdrawn and strychnine listed in Section IIb of the *List of MAK and BAT Values*.

7 References

ACGIH (American Conference of Governmental Industrial Hygienists) (1991) Strychnine. in: *Documentation of TLVs and BEIs*, ACGIH, Cincinnati, OH, USA

Forth W, Henschler D, Rummel W (Eds) (1992) *Allgemeine und spezielle Pharmakologie und Toxikologie*, Bibliographisches Institut, Mannheim, Wien, Zürich, 464–465

Gitzelmann R, Steinmann B, Otten A, Dumermuth G, Herdan M, Reubi JC, Cuenod M (1977) Nonketotic hyperhyperglycinemia treated with strychnine, a glycine receptor antagonist. *Helv Paediatr Acta 32*: 517–525

Gitzelmann R, Steinmann B, Cuenod M (1978) Strychnine for the treatment of nonketotic hyperhyperglycinemia. *N Engl J Med 298*: 1424

HCN (Health Council of the Netherlands) (1999) *Strychnine (Draft document)*. Committee on Updating of Occupational Exposure Limits, Health Council of the Netherlands, Den Haag

Seidl I, Zbinden G (1982) Subchronic oral toxicity study of strychnine in rats. *Arch Toxicol 41*: 267–272

completed 02.03.1999

Toxaphene

MAK value	–
Peak limitation	–
Absorption through the skin (1966)	**H**
Sensitization	–
Carcinogenicity (1998)	**Category 2**
Prenatal toxicity	–
Germ cell mutagenicity	–
BAT value	–
Synonyms	camphechlor camphochlor chlorinated camphene chlorocamphene octachlorocamphene polychlorinated camphenes polychlorocamphene
Chemical name (CAS)	toxaphene
CAS number	8001-35-2
Structural formula	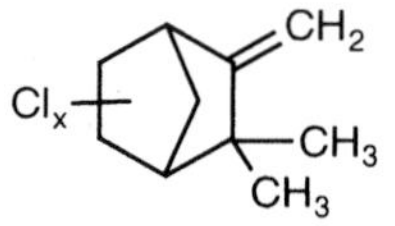 x = 4–12, on average 8 $C_{10}H_{10}Cl_8$ (approximate formula for the mixture) (chlorine content 60 %)
Molecular weight	average 414
Melting point	65–90°C
Boiling point (decomposition)	120–155°C
Vapour pressure at 20°C	0.27–0.53 hPa
log P_{ow}*	6.44 (reported for technical grade mixture) Saleh (1991)

* *n*-octanol/water distribution coefficient

Toxaphene (chlorinated camphene) with a chlorine content of 67 % to 69 % is a wax-like mixture of congeners which is practically insoluble in water and has a weak resinous odour. Large-scale production of toxaphene by chlorination of technical grade camphene began in 1948; the substance was used under a variety of trade names as a contact insecticide in agriculture and horticulture and as a rodenticide. In Germany, the use of the substance has been forbidden since 1980; also in the EC, the use of toxaphene as a pesticide is no longer permitted; its use is forbidden in the USA. Toxaphene was applied as a powder, as granules, oil or emulsion. It contains at least 177, and according to other reports, more than 670 compounds. None of the components makes up more than 5 % of the mixture. Of the components, 22 individual compounds have been chemically characterized and identified as heptachlorobornane, octachlorobornane, nonachlorobornane and decachlorobornane. Because of its high vapour pressure, toxaphene is distributed globally. Occupational exposure to toxaphene can still occur in Germany during waste disposal, residue removal, etc.

Toxaphene has been reviewed by Saleh (1991) and ATSDR (1996).

1 Toxic Effects and Mode of Action

Toxaphene is a mixture of numerous polychlorinated lipophilic compounds with pesticidal activity. It is readily absorbed from the digestive tract and in animals exposed to the substance is found mainly in the adipose tissue. Toxaphene is more rapidly and extensively broken down by metabolism than are other polychlorinated pesticides; reductive dechlorination and dehydrochlorination yields polar products and chloride. The chlorobornane metabolites are mostly excreted with the faeces, to a lesser extent with the urine.

Acute toxaphene poisoning in man caused by direct swallowing or careless handling of the substance was manifested in jerking movements in the arms and legs and severe convulsions which occasionally resulted in death from suffocation. Cases of chronic toxaphene poisoning, on the other hand, have not been documented. The target organs of toxaphene in animals are the central nervous system and the liver (liver hypertrophy with increased microsomal enzyme activity) and also the kidneys and the thyroid gland.

In feeding studies, three or five generations of mice and rats treated with the substance showed no signs of prenatal toxicity. However, in other studies, foetal mortality was slightly but not dose-dependently increased in mice and slight, also not dose-dependent effects on foetal weights and ossification centres were seen in rats; occasional effects were also seen in behavioural tests.

Toxaphene is genotoxic. It causes mutations in *Salmonella typhimurium* and sister chromatid exchange (SCE) in cultured mammalian cells. In an NTP study, toxaphene administered to mice and rats in the diet caused hepatocellular carcinomas in the mice and thyroid tumours in the rats.

2 Mechanism of Action

Because toxaphene is a mixture of many substances, it is conceivable that its action involves several mechanisms.

In an NTP carcinogenicity study, thyroid tumours were observed in rats and liver tumours in mice (see Section 5.6). Hormone analysis and the histological findings in rats (see Section 5.2.1) together with data from the literature led the authors to conclude that toxaphene is a cytochrome P450 inducer of phenobarbital type and causes induction of enzymes such as glucuronosyl transferases. This results in an increased breakdown of thyroxine (T_4) (Waritz *et al.* 1996). As rats do not have the specific T_4-binding globulins (Capen and Martin 1989, Curran and DeGroot 1991) which are essential for the storage of thyroid hormones, they are unable to compensate fluctuations in the T_4 level. For this reason rats react very sensitively to losses of T_4. Reduced T_4 levels lead, via a feedback mechanism, to increased formation of thyroid stimulating hormone (TSH) in the pituitary gland, which in turn stimulates growth of the thyroid gland. Via this mechanism, prolonged exposure of rats to toxaphene can cause thyroid tumours which are of practically no relevance for man. Toxaphene is, however, mutagenic *in vitro*. The possibility that a genotoxic mechanism is involved in induction of the thyroid tumours in rats and the liver tumours in mice can therefore not be excluded. Studies of the mechanism of induction of the liver tumours in mice are not available. Toxaphene, like a series of other polychlorinated and tumour-promoting compounds, induces protein kinase C *in vitro* (Moser and Smart 1989). Thus it is conceivable that toxaphene acts as a tumour promoter by this mechanism.

There is also evidence that toxaphene causes inhibition of metabolic cooperation in V79 cells and inhibition of the intercellular communication mediated via gap junctions. In the opinion of the authors, these effects could also play a role in a tumour-promoting activity of toxaphene and be responsible for the observed neurotoxicity (Saleh 1991).

Other mechanisms which have been suggested to be involved in the acute neurotoxicity of toxaphene are binding to a brain-specific receptor and effects on a γ-aminobutyric acid (GABA)-regulated chloride ion channel (Lawrence and Casida 1984) and inhibition of a Ca^{2+}-ATPase and calcium transport in rat brain (Moorthy *et al.* 1987, Rao *et al.* 1986).

The induction of hepatic microsomal enzymes including cytochrome P450 has been described in a series of publications (Chu *et al.* 1986, Peakall 1976, Pollock *et al.* 1983, Schwabe and Wendling 1967, Trottman and Desaiah 1980). Other effects at the molecular level which could cause toxicity include the inhibition of various ATPases in the liver, heart, kidney and brain of the rat and mouse (Mourelle *et al.* 1985, Trottman und Desaiah 1979, 1983, Trottman *et al.* 1985) and interaction with oestrogen receptors (Bonefeld Jørgensen *et al.* 1997, Soto *et al.* 1994).

3 Toxicokinetics and Metabolism

3.1 Absorption, distribution, elimination

Three hours after oral administration of ^{14}C-labelled toxaphene in a dose of 16 mg/kg body weight to Sprague-Dawley rats, high levels of radioactivity were found in brown adipose tissue, in the adrenal cortex, bone marrow, liver and kidney. 24 hours after the treatment, radioactivity was still detectable in liver and kidney but the highest level was found in white adipose tissue. The ^{14}C-activity was determined by whole body autoradiography (Mohammed *et al.* 1985).

Male Sprague-Dawley rats were given oral doses of 8.5 mg/kg body weight of ^{14}C-labelled toxaphene. 14 days later, the residual levels of toxaphene or its metabolites in the animals were determined. To this end, the radioactivity in the various organs and tissues and in blood was determined and expressed as toxaphene equivalents. The residual levels were 0.52 µg/g adipose tissue, 0.12 µg/g liver, 0.17 µg/g kidney, 0.14 µg/g blood and 0.02–0.09 µg/g tissue in bone, brain, heart, lung, muscle, spleen and testis (Ohsawa *et al.* 1975).

At various times after intravenous injection of ^{14}C-labelled toxaphene (^{14}C on the exocyclic methylene group of the camphene) in doses of 16 mg/kg body weight into NMRI mice, the distribution of the radioactivity in the animals was determined by scintillation counting or whole body autoradiography. The highest levels of ^{14}C-activity were found after 20 minutes in the adrenal cortex, liver and lung, after 4 hours in the intestinal contents and in abdominal adipose tissue. Then the activity decreased rapidly and continuously in all tissues. 16 days after the injection only traces of radioactivity were found, the highest level in abdominal adipose tissue. After 32 days, radioactivity was no longer detectable (Mohammed *et al.* 1983).

In a one-year feeding study, 2 male and 2 female macaques were given toxaphene in doses of 1 mg/kg body weight and day. The levels of toxaphene in blood and adipose tissue were determined during the study in terms of the amounts of 22 defined congeners of chlorinated camphene. After one year, the levels of toxaphene in blood were about 40 ng/g, in adipose tissue about 4000 ng/g. Two congeners were predominant in all the assayed samples and together accounted for 60 % of the recovered toxaphene (Andrews *et al.* 1996).

16 cows were given feed containing toxaphene in concentrations of 0, 2.5, 5, 10, 15 or 20 mg/kg feed for 32 or 77 days. The concentration of toxaphene in the milk at the end of the feeding period was in the range between 0.04 and 0.2 µg/g milk and increased with the dose. When the animals were given normal feed again, the toxaphene concentration in the milk from cows of all dose groups fell within 14 days to values under the detection limit (< 0.02 µg/g milk) (Zweig *et al.* 1963).

Albino rats given a gavage dose of ^{36}Cl-labelled toxaphene of 20 mg/kg body weight excreted 52 % of the administered ^{36}Cl-radioactivity within 9 days, 15 % with the urine and 37 % with the faeces. About 70 % of the excreted ^{36}Cl-radioactivity was in the form of chloride (Crowder and Dindal 1974).

Sprague-Dawley rats given a single oral dose of ^{14}C-labelled toxaphene of 8.5 or 19.0 mg/kg body weight eliminated 57 % to 59 % of the ^{14}C-radioactivity within 14 days, 21 % or 32 % with the urine and 35 % or 27 % with the faeces and about 1 % in the exhaled air. The faeces contained in addition about 30 % of the administered radioactivity in methanol-insoluble form, so that in all more than 80 % of the ^{14}C was eliminated. The rats given an oral dose of ^{36}Cl-labelled toxaphene (14.2 mg/kg body weight) excreted 76 % of the administered ^{36}Cl-radioactivity within 14 days, 49 % with the urine and 27 % with the faeces. Here too, some of the radioactivity (10 %) remained in the methanol-insoluble faeces fraction. Half of the total administered ^{36}Cl-activity was recovered in the form of the chloride ion. The elimination half-time for the ^{36}Cl-ion was 2 to 3 days (Ohsawa *et al.* 1975).

In another study, the elimination of radioactivity within the 7 days after an oral dose (10 mg/kg body weight) of ^{14}C-labelled toxaphene was followed in Sprague-Dawley rats; 58 % of the total dose (as ^{14}C) was excreted, with the urine 22 % and with the faeces 36 %. It was not determined whether or not a fraction of the excreted radioactivity remained in the methanol-insoluble fraction of the faeces, as shown by Ohsawa *et al.* (1975). The adipose tissue contained 6.4 μg toxaphene equivalents per g tissue, other tissues (brain, heart, lung, liver, kidney, spleen) less than 0.2 μg (Pollock and Kilgore 1980).

3.2 Metabolism

Quantitative metabolism studies are difficult because toxaphene is such a complex mixture of substances. The results of studies with ^{14}C-labelled and ^{36}Cl-labelled toxaphene are summarized below. Within 14 days, about 3 % of an oral dose of ^{14}C-labelled toxaphene (12.7 mg/kg body weight, average dose) was excreted unchanged in the faeces by Sprague-Dawley rats. The only metabolite identified was the chloride ion. About half of the total administered ^{36}Cl-radioactivity appeared in the urine in this form. Partially dechlorinated metabolites accounted for 27 % of the ^{14}C-activity; between 5 % and 10 % was recovered as chlorine-free metabolites, conceivably mercapturic acids, terpene carboxylic acids and terpene alcohols with the corresponding glucuronides and sulfates (Ohsawa *et al.* 1975).

Oral toxaphene doses of about 13 mg/kg body weight were administered to animals of various species (chicken, mouse, rat, hamster, guinea pig, rabbit, monkey). Three days later, samples of adipose and liver tissue and the faeces of the animals were analysed by means of gas chromatography. The chromatograms obtained for the adipose tissue and faeces of all species except the monkey contained 29 peaks characteristic of toxaphene. In contrast, the chromatograms of the liver samples contained no peaks in common with the toxaphene chromatograms; the peaks which were found were thought to be mainly toxaphene metabolites. Toxaphene was metabolized most extensively by monkeys and rats and least extensively by chickens. A chemically characterized toxaphene component (2,2,5-*endo*,6-*exo*,8,9,10-heptachlorobornane) was also metabolized most by monkeys and least by chickens during the 3 days after oral administration of about 3 mg/kg body weight (Saleh *et al.* 1979).

Toxaphene was taken up rapidly by perfused bovine liver (75 % uptake in 15 minutes). Toxaphene metabolites were found in the perfusate, presumably as a result of dechlorination and dehydrochlorination reactions (Maiorino *et al.* 1984).

4 Effects in Man

There are several reports of children who died of toxaphene poisoning after swallowing the substance. Several hours after ingestion of toxaphene, involuntary jerking movements of the extremities and convulsions began. Death was from suffocation (McGee *et al.* 1952). During the course of the evening after an evening meal at which they ate collard greens which had shortly before been sprayed intensively with toxaphene, three members of a family suddenly collapsed with convulsions. They recovered overnight in hospital. The four other family members had no or only very slight symptoms. The vegetable contained toxaphene in a concentration of about 3300 mg/kg. In this case and in another in which a family was poisoned by a meal containing toxaphene, the symptoms did not include abdominal pains, spontaneous vomiting or diarrhoea (McGee *et al.* 1952). In the attempted suicide of a 26-year-old man, ingestion of an unknown dose of toxaphene resulted in repeated convulsions. With medical treatment, he recovered overnight (Wells and Milhorn 1983).

Cases of suspected toxaphene poisoning resulting from occupational exposure have been described. Two persons whose job involved spraying toxaphene in cotton plantations in Egypt suffered after 1.5 to 2 months exposure from dyspnoea, weariness and weakness. The symptoms and x-ray findings were interpreted as non-infectious allergic bronchopneumonia. The suggestion that toxaphene was the causative agent was, however, based entirely on the time of appearance of the symptoms (Warraki 1963).

In a controlled laboratory study, 25 persons were exposed for 30 minutes daily on 10 consecutive days to an aerosol containing toxaphene in concentrations up to 500 mg/m^3. Three weeks later the same persons were exposed again for 30 minutes daily on 3 days. The dermatological and internal examinations of the test persons yielded no abnormal findings (ATSDR 1996, Keplinger 1963). The low information content of the publication and the unusual exposure conditions make this study difficult to interpret.

Eight women who worked without protection on a hot, windless, humid day in a field which had been sprayed from the air with toxaphene developed slight or moderate symptoms of intoxication 4 to 5 hours after starting work (no other details). Lymphocytes obtained from the eight exposed women and from seven healthy women were cultured for the study of chromosomal aberrations. In the exposed group 13.1 % of metaphases were aberrant (183/1400), in the control group 1.6 % (24/900). The aberrations in the exposed group were mainly acentric fragments and chromatid exchanges (Samosh 1974).

An increased incidence of lung cancer found in an epidemiological study of 316 persons exposed long term (6 to 23 years) at work to numerous pesticides including toxaphene, other insecticides such as DDT and hexachlorocyclohexane, herbicides, organophosphates, organonitro compounds and fungicides cannot be interpreted as causally related to the toxaphene exposure because the persons were exposed to many substances (Barthel 1976, IARC 1979).

5 Animal Experiments and *in vitro* Studies

5.1 Acute toxicity

Symptoms of acute toxaphene poisoning include tonic-clonic cramps, salivation, restlessness, vomiting and hyperexcitability (IARC 1979, Saleh 1991). Table 1 shows the published LD_{50} values.

Table 1. Acute toxicity of toxaphene

Species, sex	Administration route	LD_{50} (mg/kg body weight)	References
rat, ♂	oral	90	Gaines (1969)
rat, ♀	oral	80	Gaines (1969)
rat, sex not specified	oral	50–70	Schwabe and Wendling (1967)
rat, ♂	oral	220	Boyd and Taylor (1971)
rat, ♂	intraperitoneal	228	Pollock *et al.* (1983)
rat, ♂	dermal	1075	Gaines (1969)
rat, ♀	dermal	780	Gaines (1969)
mouse, ♂	intraperitoneal	42	Khalifa *et al.* (1974)

Two pure substances isolated from technical grade toxaphene, an octachloro compound called toxicant A and a heptachlorobornane called toxicant B, with intraperitoneal LD_{50} values of 3.1 and 6.6 mg/kg body weight, respectively, are much more toxic for mice than is toxaphene (Khalifa *et al.* 1974).

5.2 Subacute, subchronic and chronic toxicity

There are no data available for the effects of long-term inhalation of toxaphene, nor for its allergenicity.

5.2.1 Ingestion

In a 28-day study, effects of toxaphene on the thyroid gland were studied specifically. Forty male Crl:Cd BR rats were given toxaphene daily by gavage. An initial dose of 100 mg/kg body weight was reduced after three days to 75 mg/kg body weight because some of the animals (2/40) had died. An untreated control group was included. TSH levels were significantly higher in the treated animals than in the control animals after 1, 2 and 4 weeks. The levels of thyroid hormones (T_4, T_3 and rT_3) were slightly and not significantly lower in the treated animals than in the controls. Histological examination of the thyroid gland at the end of the study revealed follicle cell hypertrophy and hyperplasia in all treated animals. The results are indicative of toxaphene-induced hyperactivity of the thyroid (Waritz *et al.* 1996).

In a 13-week feeding study with male and female Sprague-Dawley rats given toxaphene in the diet in concentrations of 0, 4, 20, 100 or 500 mg/kg diet (about 0, 0.4, 2, 10 and 50 mg/kg body weight), food consumption and body weight gains were not different from the control values in any of the dose groups. In the 500 mg/kg group, relative liver weights were increased in both sexes and relative kidney weights in the male animals. No effects on haematological or biochemical serum parameters were found. The histopathological examination revealed dose-related effects in the liver, kidney and thyroid gland; the effects in the liver and thyroid were mostly described as slight and adaptive. The NOAEL was 4 mg/kg diet (0.35 mg/kg body weight and day) (Chu *et al.* 1986).

In a reproductive toxicity study, 30 female and 15 male Sprague-Dawley rats per dose group were given toxaphene in the diet in concentrations of 0, 4, 20, 100 or 500 mg/kg diet (about 0, 0.4, 2, 10 and 50 mg/kg body weight) for 27 weeks. In parent animals given 20 mg/kg diet or more, histological examination revealed changes in the liver, thyroid gland and kidneys. Reduced weight gains, increased serum cholesterol levels, increased absolute and relative liver and kidney weights and increased enzyme activities in the liver microsomes were seen mainly in the highest dose group. A NOAEL of 4 mg/kg diet (0.29–0.38 mg/kg body weight and day) was reported (Chu *et al.* 1988).

In a feeding study with male and female Sherman rats given toxaphene in the diet in concentrations of 50 or 200 mg/kg diet (about 5 and 20 mg/kg body weight), after 2, 4, 6 and 9 months food consumption, body weights and relative liver weights were not different from the control values. Histological examination revealed centrilobular hypertrophy and fatty cytoplasmic inclusion bodies in the liver after 9 months (Ortega *et al.* 1957).

In a long-term NTP study, male and female Osborne-Mendel rats were given toxaphene in the diet in average concentrations of 556 or 1112 mg/kg diet (males) and 540 or 1080 mg/kg diet (females) for a period of 80 weeks. The initial dose was halved in the course of the study, for the male rats twice. The recovery period was 28 or 30 weeks. In the female animals of both dose groups, body weights were reduced. The symptoms observed frequently in the treated rats included abdominal distension, diarrhoea, dyspnoea and unkempt fur, and in the animals of the high dose group generalized tremor in week 53 was followed by paralysis of the legs, ataxia, bleeding from the nose and vagina and blood in the urine (NCI 1979).

In the NTP study, male and female $B6C3F_1$ mice were given a diet containing toxaphene in average concentrations of about 100 or 200 mg/kg diet for a period of 80 weeks.

After the end of exposure, the animals were observed for another 10 or 11 weeks. Body weights were reduced only in the male animals of the high dose group. As in the treated rats, the symptoms seen in mice included abdominal distension, diarrhoea, dyspnoea and unkempt fur (NCI 1979).

In a 13-week study, groups of 6 male and 6 female beagle dogs were given toxaphene in gelatine capsules in doses of 0, 0.2, 2 and 5 mg/kg body weight and day. The highest dose on the first two days of the study was 10 mg/kg body weight; some animals reacted with seizures, salivation and vomiting. When the dose was reduced to 5 mg/kg body weight, such symptoms did not develop. All animals survived; food consumption and weight gains were not different from the control values. The serum alkaline phosphatase levels were increased in both sexes in a dose-dependent manner; the increase was significant only in the 5 mg/kg group. Relative liver weights were increased in males given 5 mg/kg body weight and in females at both 2 and 5 mg/kg body weight. Slight to moderate, dose-dependent effects were seen in the histopathological examination in the liver and to a lesser extent in the kidney and thyroid. A NOAEL of 0.2 mg/kg body weight was found (Chu *et al.* 1986).

5.2.3 Dermal absorption

In an early review in which neither experimental details nor literature citations were given, it is pointed out that toxaphene may be taken up through the skin and cause systemic effects. Dogs were rather insensitive but rabbits very sensitive. Formulations of toxaphene in oil or as an emulsion seem to be most dangerous from the point of view of absorption through the skin (Stormont and Conley 1952).

5.3 Local effects on skin and mucous membranes

In a review of toxaphene, the substance is described as slightly to moderately irritating when applied in concentrated form to the skin of laboratory animals (Stormont and Conley 1952).

5.4 Reproductive and developmental toxicity

In a feeding study, Sprague-Dawley rats were given toxaphene in the diet in concentrations of 25 or 100 mg/kg diet (about 2.5 and 10 mg/kg body weight and day) for 72 days before mating and then continuously for three generations. The parent animals (F_0 generation) were subjected to histological examination after 42 weeks, the F_1 and F_2 generations after 39 weeks. The only pathological alteration found was cytoplasmic vacuolation of the liver at the high dose. Fertility, litter size, survival and weight at weaning were normal (Kennedy *et al.* 1973).

In another study, 25 pregnant Sprague-Dawley rats were given highly toxic doses of toxaphene (32 mg/kg body weight and day) by gavage on days 6 to 15 of gestation.

Mortality among the dams was 50 %. At the end of the study on day 20, the incidence of foetuses with supernumerary ribs was significantly higher in the group treated with toxaphene; other developmental disorders were not seen (Chernoff *et al.* 1990).

Groups of 30 female and 15 male Sprague-Dawley rats were given toxaphene in the diet in concentrations of 0, 4, 20, 100 or 500 mg/kg diet for 27 weeks. Toxicity was observed in the parent animals of the highest dose group; changes were seen in the histological examination also in animals given concentrations down to 20 mg/kg. A NOAEL of 4 mg/kg diet was determined (see Section 5.2.1). Two litters from each animal were studied. Fertility, litter size, birth weights and survival were unaffected (Chu *et al.* 1988).

Behavioural changes were studied in Sprague-Dawley rats exposed perinatally and postnatally to toxaphene. For this purpose, either the dams were given toxaphene doses of 6 mg/kg body weight by gavage, daily from day 7 of gestation, or the same doses were given to male and female rats from age 5 to 6 weeks for 21 days. The toxaphene treatment did not affect body weight gains or mortality of the animals. The perinatally treated rats were subjected to behavioural tests from the age of 7 days. In one of three tests, a test of the righting reflex (the ability to land on all fours when dropped upside down), the treated animals performed significantly worse than the controls. At the age of 14 to 16 weeks, the learning ability of perinatally and postnatally exposed rats was studied. No significant differences from the controls were found (Crowder *et al.* 1980).

In a feeding study, pregnant rats were given toxaphene doses of 0.05 mg/kg body weight and day. In the perinatally exposed pups, the development of swimming ability was delayed. Two components of toxaphene, "toxicant A" and "toxicant B", in doses of 0.002 mg/kg body weight produced the same effect (Olson *et al.* 1980).

Toxaphene was administered orally to pregnant CD rats and CD-1 mice in daily doses of 0, 15, 25 or 35 mg/kg body weight during the period of organogenesis, days 7 to 16 of gestation. In mice foetal mortality was increased but the increase was neither dose-dependent nor significant. In the rats, foetal mortality was in the normal range but foetal weights were reduced, significantly in the middle dose group. At the same dose, the numbers of sternal and caudal ossification centres were significantly reduced. In the mice, pups with encephalocele (brain hernia) were found in 5 of 61 litters in the highest dose group but not in any other groups (Chernoff and Carver 1976).

Administration of toxaphene to mice in concentrations of 25 mg/kg diet during 5 generations produced "no embryotoxicity or teratogenicity" (no other details) (IARC 1979, Keplinger *et al.* 1968).

5.5 Genotoxicity

5.5.1 *In vitro*

Toxaphene is mutagenic in *Salmonella typhimurium* strain TA100 and weakly mutagenic in TA98 without metabolic activation. In TA100 it was demonstrated that addition of liver S9 mix halved the mutagenic activity. The mutagenic potency in TA100 of various commercially available samples without added S9 mix ranged from 310 to 1270 revertants per mg. The individual mutagenic components were not identified; heptachlorobornane-I (2,2,5-*endo*,6-*exo*,8,9,10-heptachlorobornane), a toxic component, was not mutagenic in TA98, TA100, TA1535, TA1537 or TA1538 with or without S9 mix (Hooper *et al.* 1979). A pre-incubation test confirmed the mutagenic activity of toxaphene in TA100 and TA98 (Mortelmans *et al.* 1986). In a test for the induction of λ phage in a lysogenic strain of *Escherichia coli*, toxaphene had phage-inducing activity with and without S9 mix (Houk and DeMarini 1987).

SCE was induced by toxaphene in concentrations of 10^{-5} M and 10^{-4} M (about 4 and 40 µg/ml) in cells of the human lymphoid cell line LAZ-007 (Sobti *et al.* 1983). Likewise, in cells of the Chinese hamster cell line DON, toxaphene in concentrations of 5, 10 and 15 µg/ml significantly increased the incidence of SCE. The incidence of SCE increased with the concentration (Steinel *et al.* 1990).

A test for morphological transformation of C3H/10T½ cells is incompletely described and yielded an unclear result (Nesnow *et al.* 1985).

5.5.2 *In vivo*

In a screening study of 49 chemicals known to be carcinogenic in rodents, female Sprague-Dawley rats were given oral toxaphene doses of 12 or 36 mg/kg body weight at 21 hours and again at 4 hours before they were killed. With the alkaline elution method, no liver DNA damage could be detected in these animals. Various other carcinogens, on the other hand, produced DNA damage (Kitchin and Brown 1994).

Dominant lethal mutations were studied in ICR/Ha Swiss mice. Male mice were given toxaphene either as single intraperitoneal injections of 36 or 180 mg/kg body weight or by daily oral dosing on five days with 40 or 80 mg/kg body weight. The doses were in the lethal range. Nine of the twelve male animals given the high oral dose and two of the nine given the high intraperitoneal dose died. The surviving male animals were then mated during a period of 8 weeks with untreated female mice. The preimplantation losses and numbers of dead implants provided no evidence of a genotoxic effect of toxaphene (Epstein *et al.* 1972).

5.6 Carcinogenicity

In an NTP study, the carcinogenic activity of toxaphene was studied in male and female Osborne-Mendel rats (male: 0, 556, 1112 mg/kg diet; female: 0, 540, 1080 mg/kg diet) and male and female B6C3F$_1$ mice (0, 100, 200 mg/kg diet) (NCI 1979, IARC 1979). In the male rats, the incidence of thyroid adenomas and carcinomas was increased above that in the pooled control (see Table 2); the increase was dose-dependent and significant.

In the high dose group female rats, the incidence of thyroid adenomas was significantly increased (Table 2) (NCI 1979, IARC 1979). The incidence of hepatocellular carcinomas in the male mice of the low and high dose groups and the female mice in the high dose group was increased significantly above that in both control groups (Table 2). In addition, the incidence of neoplastic nodules in the livers of the exposed mice was increased (NCI 1979, IARC 1979). When the histological evaluation of the liver preparations was repeated at a later date, the results revealed a slightly lower incidence of hepatocellular carcinomas but on the whole confirmed the original findings (ATSDR 1996).

Table 2. Carcinogenicity studies with toxaphene

Author:	NCI 1979
Species:	rat (Osborne-Mendel), 5 weeks old; groups of 50 ♂ and 50 ♀ treated, 10 ♂, 10 ♀ control animals (concurrent control), and about 55 ♂, 55 ♀ control animals (pooled control; including animals from other long-term studies)
Administration:	in the diet
Dose:	**for ♂ rats:** **about 550 mg/kg diet:** 1280 mg/kg for 2 weeks, 640 mg/kg for 53 weeks, from week 54 320 mg/kg (about 41 mg/kg body weight and day, assuming 60 g diet per kg body weight) **about 1110 mg/kg diet:** 2560 mg/kg for 2 weeks, 1280 mg/kg for 53 weeks, from week 54 640 mg/kg (about 83 mg/kg body weight and day, assuming 60 g diet per kg body weight) **for ♀ rats:** **540 mg/kg diet:** 640 mg/kg for 55 weeks, from week 56 320 mg/kg (about 41 mg/kg body weight and day, assuming 60 g diet per kg body weight) **1080 mg/kg diet:** 1280 mg/kg for 55 weeks, from week 56 640 mg/kg (about 81 mg/kg body weight and day, assuming 60 g diet per kg body weight)
Duration:	80 weeks, then until the end of the study 28 weeks without treatment for ♂ and 30 weeks for ♀
Toxicity:	**540 mg/kg diet:** decreased body weights in ♀ **1080 mg/kg diet:** decreased body weights in ♀ (see Section 5.2)
Tumours:	

Concentration (mg/kg diet)		0[1]	0[2]	about 550	about 1100
thyroid gland tumours					
adenomas	♂	1/7 (14 %)	2/44 (5 %)	7/41 (17 %)	7/35 (20 %)
adenomas	♀	0/6 (0 %)	1/46 (2 %)	1/43 (2 %)	7/42 ** (17 %)
carcinomas	♂	0/7 (0 %)	0/44 (0 %)	0/41 (0 %)	2/35 (6 %)
adenomas and carcinomas	♂	1/7 (14 %)	2/44 (5 %)	7/41 (17 %)	9/35 * (26 %)

[1] concurrent control (matched controls)
[2] pooled control
* incidence significantly higher than in the pooled control, $p < 0.01$
** incidence significantly higher than in the pooled control, $p < 0.05$

Table 2. continued

Author:	NCI 1979
Species:	mouse (B6C3F$_1$), 5 weeks old; groups of 50 ♂ and 50 ♀ treated, 10 ♂, 10 ♀ control animals (concurrent control) and about 50 ♂, 50 ♀ control animals (pooled control; including animals from other long-term studies)
Administration:	in the diet
Dose:	**about 100 mg/kg diet:** 160 mg/kg for 19 weeks, from week 20 80 mg/kg (about 15 mg/kg body weight and day, assuming 150 g diet per kg body weight) **about 200 mg/kg diet:** 320 mg/kg for 19 weeks, from week 20 160 mg/kg (about 30 mg/kg body weight and day, assuming 150 g diet per kg body weight)
Duration:	80 weeks, then until the end of the study 10 or 11 weeks without treatment
Toxicity:	**100 mg/kg diet:** increased mortality in ♂ **200 mg/kg diet:** decreased body weights in ♂, increased mortality in ♂ and ♀ (see Section 5.2)
Tumours:	

Concentration (mg/kg diet)		0[1]	0[2]	100	200
liver tumours					
carcinomas	♂	0/10 (0 %)	4/48 (8 %)	34/49 * (69 %)	45/46 * (98 %)
	♀	0/9 (0 %)	0/48 (0 %)	5/49 (10 %)	34/49 * (69 %)
neoplastic nodules	♂	2/10 (20 %)	not specified	6/49 (12 %)	not specified
	♀	0/9 (0 %)	not specified	13/49 (27 %)	6/49 (12 %)

[1] concurrent control (matched controls)
[2] pooled control
* incidence significantly higher than in the concurrent and pooled controls, $p < 0.001$

In female A/J mice given toxaphene in concentrations of 100 or 200 mg/kg diet (about 15 and 30 mg/kg body weight and day) for only 12 weeks, no induction of lung tumours was observed; only 6 to 7 mice per group were treated. Benzo[*a*]pyrene administered in two oral doses (3 mg on each of days 7 and 21) was shown to induce lung tumours in the same test but with larger groups of animals, 36 to 48 mice; the incidence of the tumours was reduced slightly but significantly by the low dose of toxaphene. At the high toxaphene dose, however, no reduction in tumour incidence was seen (Triolo *et al.* 1982).

Another carcinogenicity study has been described in a review. Male mice were given toxaphene with the diet in a concentration of 50 mg/kg diet for a period of 18 months (no other details) and were then observed for another 6 months. The incidence of liver tumours (adenomas and carcinomas) was increased significantly (Saleh 1991).

5.8 Other effects

Effects on the immune system

Immune reactions were studied in young female Swiss-Webster mice given toxaphene in the diet in concentrations of 0, 10, 100 or 200 mg/kg diet (about 0, 1, 10 and 20 mg/kg body weight) for 8 weeks. Whereas the cellular immune response was not affected at any dose, at both high doses IgG antibody production was reduced (Allen *et al.* 1983). In another experiment, animals were given toxaphene in the diet in concentrations of 0, 10, 100 or 200 mg/kg diet (about 0, 1, 10 and 20 mg/kg body weight). Three weeks after feeding began, the mice were mated and maintained on the diets until 3 weeks after parturition, at which time the pups were weaned. After weaning the pups were given a normal diet and examined at age 8 weeks. Antibody production was reduced in the 100 mg/kg group but increased in the 200 mg/kg group. The cell-mediated immune response was less in the 100 mg/kg group but unchanged in the 10 and 200 mg/kg groups. The phagocytic activity of the macrophages was significantly reduced in the progeny of all treated groups, and reduced most in the 100 mg/kg group (Allen *et al.* 1983).

6 Manifesto (MAK value/classification)

In vitro, toxaphene induces mutations in the *Salmonella* mutagenicity test and SCE in mammalian cells. In the dominant lethal test in the mouse, negative results were obtained with toxaphene.

Administered in the diet, toxaphene produced liver tumours in male and female mice and thyroid tumours in male and female rats. It is conceivable that the thyroid tumours are not relevant for man (see Section 2). However, because of the *in vitro* mutagenicity of toxaphene, the possibility that a genotoxic mechanism is involved in the formation of both kinds of tumour cannot be excluded. Therefore toxaphene is classified in Carcinogen category 2 of Section III of the *List of MAK and BAT Values* and the MAK value has been withdrawn.

Because of the high toxicity of the substance after dermal application, the designation with an "H" has been retained. There are no data for a sensitization potential of toxaphene.

7 References

Allen AL, Koller LD, Pollock GA (1983) Effect of toxaphene exposure on immune responses in mice. *J Toxicol Environ Health 11*: 61–69

Andrews P, Headrick K, Pilon J-C, Bryce F, Iverson F (1996) Capillary GC-ECD and ECNI GCMS characterization of toxaphene residues in primate tissues during a feeding study. *Chemosphere 32*: 1043–1053

ATSDR (Agency for Toxic Substances and Disease Registry) (1996) *Toxicological profile for toxaphene*. US Department of Health & Human Services

Barthel E (1976) High incidence of lung cancer in persons with chronic professional exposure to pesticides in agriculture. *Z Erkr Atmungsorgane 146*: 266–274

Bonefeld Jørgensen EC, Autrup H, Hansen JC (1997) Effect of toxaphene on estrogen receptor functions in human breast cancer cells. *Carcinogenesis 18*: 1651–1654

Boyd EM, Taylor FI (1971) Toxaphene toxicity in protein-deficient rats. *Toxicol Appl Pharmacol 18*: 158–167

Capen CC, Martin SL (1989) The effects of xenobiotics on the structure and function of thyroid follicular and C-cells. *Toxicol Pathol 17*: 266–293

Chernoff N, Carver BD (1976) Fetal toxicity of toxaphene in rats and mice. *Bull Environ Contam Toxicol 15*: 660–664

Chernoff N, Setzer RW, Miller DB, Rosen MB, Rogers JM (1990) Effects of chemically induced maternal toxicity on prenatal development in the rat. *Teratology 42*: 651–658

Chu I, Villeneuve DC, Sun C-W, Secours V, Procter B, Arnold E, Clegg D, Reynolds L, Valli VE (1986) Toxicity of toxaphene in the rat and beagle dog. *Fundam Appl Toxicol 7*: 406–418

Chu I, Secours V, Villeneuve DC, Valli VE, Nakamura A, Colin D, Clegg DJ, Arnold EP (1988) Reproduction study of toxaphene in the rat. *J Environ Sci Health B 23*: 101–126

Crowder LA, Dindal EF (1974) Fate of ^{36}Cl-toxaphene in rats. *Bull Environ Contam Toxicol 12*: 320–327

Crowder LA, Lanzaro GC, Whitson RS (1980) Behavioral effects of methyl parathion and toxaphene exposure in rats. *J Environ Sci Health B 15*: 365–378

Curran PG, DeGroot LJ (1991) The effect of hepatic enzyme-inducing drugs on thyroid hormones and the thyroid gland. *Endocr Rev 12*: 135–150

Epstein SS, Arnold E, Andrea J, Bass W, Bishop Y (1972) Detection of chemical mutagens by the dominant lethal assay in the mouse. *Toxicol Appl Pharmacol 23*: 288–325

Gaines TB (1969) Acute toxicity of pesticides. *Toxicol Appl Pharmacol 14*: 515–534

Hooper NK, Ames BN, Saleh MA, Casida JE (1979) Toxaphene, a complex mixture of polychloroterpenes and a major insecticide, is mutagenic. *Science 205*: 591–593

Houk VS, DeMarini DM (1987) Induction of prophage lambda by chlorinated pesticides. *Mutat Res 182*: 193–201

IARC (International Agency for Research on Cancer) (1979) Toxaphene (polychlorinated camphenes). *IARC monographs on the evaluation of carcinogenic risks to humans, Vol. 20*, IARC, Lyon, 327–348

Kennedy Jr GL, Frawley JP, Calandra JC (1973) Multigeneration reproductive effects of three pesticides in rats. *Toxicol Appl Pharmacol 25*: 589–596

Keplinger ML (1963) Use of humans to evaluate safety of chemicals. *Arch Environ Health 6*: 342–349

Keplinger ML, Deichmann WB, Sala F (1968) Effects of combinations of pesticides on reproduction in mice. *Ind Med Surg 37*: 525

Khalifa S, Mon TR, Engel JL, Casida JE (1974) Isolation of 2,2,5-endo,6-exo,8,9,10-heptachlorobornane and an octachloro toxicant from technical toxaphene. *J Agric Food Chem 22*: 653–657

Kitchin KT, Brown JL (1994) Dose-response relationship for rat liver DNA damage caused by 49 rodent carcinogens. *Toxicology 88*: 31–49

Lawrence LJ, Casida JE (1984) Interactions of lindane, toxaphene and cyclodienes with brain-specific *t*-butylbicyclophosphorothionate receptor. *Life Sci 35*: 171–178

Maiorino RM, Whiting FM, Brown WH, Reid BL, Stull JW (1984) Metabolism of toxaphene by the isolated perfused bovine liver. *Arch Environ Contam Toxicol 13*: 565–571

McGee LC, Reed HL, Fleming JP (1952) Accidental poisoning by toxaphene. *J Am Med Assoc 149*: 1124–1126

Mohammed A, Andersson Ö, Biessmann A, Slanina P (1983) Fate and specific tissue retention of toxaphene in mice. *Arch Toxicol 54*: 311–321

Mohammed A, Hallberg E, Rydström J, Slanina P (1985) Toxaphene: accumulation in the adrenal cortex and effect on ACTH-stimulated corticosteroid synthesis in the rat. *Toxicol Lett 24*: 137–143

Moorthy KS, Trottman CH, Spann CH, Desaiah D (1987) *In vivo* effects of toxaphene on calmodulin-regulated calcium-pump activity in rat brain. *J Toxicol Environ Health 20*: 249–259

Mortelmans K, Haworth S, Lawlor T, Speck W, Tainer B, Zeiger E (1986) *Salmonella* mutagenicity tests: II. Results from the testing of 270 chemicals. *Environ Mutagen 8, Suppl 7*: 1–119

Moser GJ, Smart RC (1989) Hepatic tumor-promoting chlorinated hydrocarbons stimulate protein kinase C activity. *Carcinogenesis 10*: 851–856

Mourelle M, Garcia M, Aguilar C (1985) Adenosine triphosphatase activities in plasma liver membranes of rats treated with DDT and toxaphene. *J Appl Toxicol 5*: 39–41

NCI (National Cancer Institute) (1979) *Bioassay of toxaphene for possible carcinogenicity (CAS No. 8001-35-2). Techn Report Series No. 37*, DHEW (NIH) Publication No. 79-837, National Institutes of Health, Bethesda, MD

Nesnow S, Curtis G, Garland H (1985) Tests with the C3H/10T1/2 clone 8 morphological transformation bioassay. in: Ashby J, de Serres FJ, Draper M, Ishidate jr M, Margolin BH, Matter BE, Shelby MD (Eds) *Progress in Mutation Research, Vol. 5*, Elsevier, Amsterdam, 659–664

Ohsawa T, Knox JR, Khalifa S, Casida JE (1975) Metabolic dechlorination of toxaphene in rats. *J Agric Food Chem 23*: 98–106

Olson KL, Matsumura F, Boush GM (1980) Behavioral effects on juvenile rats from perinatal exposure to low levels of toxaphene, and its toxic components, toxicant A, and toxicant B. *Arch Environ Contam Toxicol 9*: 247–257

Ortega P, Hayes Jr WJ, Durham WF (1957) Pathologic changes in the liver of rats after feeding low levels of various insecticides. *AMA Arch Pathol 64*: 614–622

Peakall DB (1976) Effects of toxaphene on hepatic enzyme induction and circulating steroid levels in the rat. *Environ Health Perspect 13*: 117–120

Pollock GA, Kilgore W (1980) Excretion and storage of [^{14}C] toxaphene and two isolated [^{14}C] toxaphene fractions. *J Toxicol Environ Health 6*: 127–40

Pollock GA, Krasnec JP, Niemann BR (1983) Rat hepatic microsomal enzyme induction by pretreatment with toxaphene and toxaphene fractions. *J Toxicol Environ Health 11*: 355–363

Rao KS, Trottman CH, Morrow W, Desaiah D (1986) Toxaphene inhibition of calmodulin dependent calcium ATPase activity in rat brain synaptosomes. *Fundam Appl Toxicol 6*: 648–653

Saleh MA (1991) Toxaphene: chemistry, biochemistry, toxicity and environmental fate. *Rev Environ Contam Toxicol 118*: 1–85

Saleh MA, Skinner RF, Casida JE (1979) Comparative metabolism of 2,2,5-endo,6-exo,8,9,10-heptachlorobornane and toxaphene in six mammalian species and chickens. *J Agric Food Chem 27*: 731–737

Samosh LV (1974) Chromosome aberrations and character of satellite associations after accidental exposure of the human body to polychlorocamphene (Russian). *Tsitol Genet 8*: 24–27

Schwabe U, Wendling I (1967) Beschleunigung des Arzneimittel-Abbaus durch kleine Dosen von DDT und anderen Chlorkohlenwasserstoff-Insekticiden (Increased breakdown of medicines caused by small doses of DDT and other chlorinated hydrocarbon insecticides) (German). *Arzneim-Forsch 17*: 614–618

Sobti RC, Krishan A, Davies J (1983) Cytokinetic and cytogenetic effect of agricultural chemicals on human lymphoid cells *in vitro*. II. Organochlorine pesticides. *Arch Toxicol 52*: 221–231

Soto AM, Chung KL, Sonnenschein C (1994) The pesticides endosulfan, toxaphene, and dieldrin have estrogenic effects on human estrogen-sensitive cells. *Environ Health Perspect 102*: 380–383

Steinel HH, Arlauskas A, Baker RSU (1990) SCE induction and cell-cycle delay by toxaphene. *Mutat Res 230*: 29–33

Stormont and Conley BE (1952) Pharmacologic properties of toxaphene, a chlorinated hydrocarbon insecticide. *J Am Med Assoc 149*: 1135–1137

Triolo AJ, Lang WR, Coon JM, Lindstrom D, Herr DL (1982) Effect of the insecticides toxaphene and carbaryl on induction of lung tumors by benzo[*a*]pyrene in the mouse. *J Toxicol Environ Health 9*: 637–649

Trottman CH, Desaiah D (1979) Adenosine triphosphatase activities in brain, kidney and liver of mice treated with toxaphene. *J Environ Sci Health B 14*: 393–404

Trottman CH, Desaiah D (1980) Induction of rat hepatic microsomal enzymes by toxaphene pretreatment. *J Environ Sci Health B 15*: 121–134

Trottman CH, Desaiah D (1983) Effect of toxaphene on the binding of ^{3}H-labeled ouabain and dopamine to rat brain synaptosomes. *Toxicol Lett 18*: 323–330

Trottman CH, Rao KS, Morrow W, Uzodinma JE, Desaiah D (1985) *In vitro* effects of toxaphene on mitochondrial calcium ATPase and calcium uptake in selected rat tissues. *Life Sci 36*: 427–433

Waritz RS, Steinberg M, Kinoshita FK, Kelly CM, Richter WR (1996) Thyroid function and thyroid tumors in toxaphene-treated rats. *Regul Toxicol Pharmacol 24*: 184–192

Warraki S (1963) Respiratory hazards of chlorinated camphene. *Arch Environ Health 7*: 137–140

Wells WL, Milhorn jr HT (1983) Suicide attempt by toxaphene ingestion: A case report. *J M State Med Assoc 24*: 329–330

Zweig G, Pye EL, Sitlani R, Peoples SA (1963) Residues in milk from dairy cows fed low levels of toxaphene in their daily ration. *J Agric Food Chem 11*: 70–72

completed 30.03.1998

Xylidine (isomers)

2,3-Xylidine, 2,5-Xylidine, 3,4-Xylidine, 3,5-Xylidine

MAK value	–
Peak limitation	–
Absorption through the skin (1966)	**H**
Sensitization	–
Carcinogenicity (1998)	**Category 3A**
Prenatal toxicity	–
Germ cell mutagenicity	–
BAT value	–

2,4-Xylidine, 2,6-Xylidine

MAK value	–
Peak limitation	–
Absorption through the skin (1966)	**H**
Sensitization	–
Carcinogenicity (1998)	**Category 2**
Prenatal toxicity	–
Germ cell mutagenicity	–
BAT value	–

Chemical names and CAS numbers	see next page
Molecular weight	121.18
Structural formula	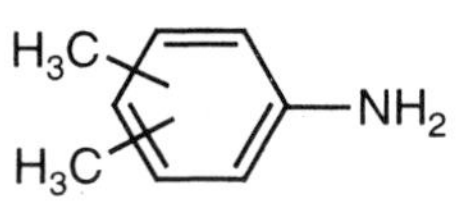
Molecular formula	$C_8H_{11}N$
Physical properties	see next page
1 ml/m³ (ppm) ≙ 5.03mg/m³	**1 mg/m³ ≙ 0.200 ml/m³ (ppm)**

Chemical names:

Substance	Chemical name (CAS)	Synonyms	CAS number
2,3-xylidine	2,3-dimethyl-benzenamine	1-amino-2,3-dimethylbenzene 3-amino-*o*-xylene 2,3-dimethylaniline 2,3-dimethylphenylamine	87-59-2
2,4-xylidine	2,4-dimethyl-benzenamine	4-amino-1,3-dimethylbenzene 4-amino-3-methyltoluene 4-amino-1,3-xylene 2,4-dimethylaniline 2-methyl-*p*-toluidine	95-68-1
2,5-xylidine	2,5-dimethyl-benzenamine	1-amino-2,5-dimethylbenzene 2-amino-*p*-methyltoluene 2-amino-*p*-toluidine 2-amino-1,4-xylene 2,5-dimethylaniline 5-methyl-*o*-toluidine	95-78-3
2,6-xylidine	2,6-dimethyl-benzenamine	1-amino-2,6-dimethylbenzene 2-amino-*m*-methyltoluene 2-amino-1,3-xylene 2,6-dimethylaniline 3-methyl-*o*-toluidine	87-62-7
3,4-xylidine	3,4-dimethyl-benzenamine	1-amino-3,4-dimethylbenzene 4-amino-1,2-xylene 4-amino-2-methyltoluene 3,4-dimethylaniline 3,4-dimethylphenylamine 3-methyl-*p*-toluidine	95-64-7
3,5-xylidine	3,5-dimethyl-benzenamine	1-amino-3,5-dimethylbenzene 5-amino-*m*-methyltoluene 5-amino-1,3-xylene 3,5-dimethylaniline 5-methyl-*m*-toluidine	108-69-0

Physical properties:

Substance	Melting point (°C)	Boiling point (°C)	Density at 20°C (g/ml)	Vapour pressure (hPa)	log P_{ow} *
2,3-xylidine	–15 to +3.5	221–224	0.993	20 (106°C)	1.89–2.34
2,4-xylidine	–16 to –14	212–217	0.972–0.995	0.11 to < 0.13 (20°C)	1.68–2.13 (experimental)
2,5-xylidine	14.2–16	213–218	0.9735	13 (96°C)	1.89–2.34
2,6-xylidine	11	205–218	0.979–0.984	0.18–0.20 (20°C)	1.71–2.21
3,4-xylidine	49–51	225–226	1.076	1.076 (20°C)	1.84–2.21
3,5-xylidine	8–10	218–222	0.969–0.972	about 1.3 (20°C)	1.90–2.21

* *n*-octanol/water distribution coefficient

The xylidine isomers have been reviewed in detail in a BUA report (BUA 1995).

1 Toxic Effects and Mode of Action

2,4-Xylidine, 2,5-xylidine, 2,6-xylidine and 3,4-xylidine are readily absorbed from the gastrointestinal tract and excreted mainly in the form of metabolites with the urine. After repeated ingestion or inhalation, the main effects of the substances are nephrotoxicity and hepatotoxicity (2,4-xylidine, 2,5-xylidine, 2,6-xylidine). The isomers all cause methaemoglobin formation and, after metabolic activation, are mutagenic in bacterial test systems. Feeding studies suggest that 2,4-xylidine and 2,5-xylidine may have carcinogenic potential. Carcinogenicity studies with 2,6-xylidine yielded positive results with both mouse and rat.

2 Mechanism of Action

There are only few mechanistic studies of the acute and chronic toxicity of the isomers. The lipid accumulation seen in dogs treated with 2,6-xylidine is ascribed to competitive inhibition of coenzyme A during acetylation of the reactive amino group. In the rat, the 3-methyl-4-aminobenzoic acid formed as a metabolite of 2,4-xylidine is said to be a folic acid antagonist (Weinberg and Sahini 1981).

The methaemoglobin formation is a result of cooxidation of oxyhaemoglobin and the *N*-hydroxy metabolites. *N*-Hydroxy derivatives and their esters are thought to be responsible for the genotoxicity and chronic toxicity of the xylidine isomers.

3 Toxicokinetics and Metabolism

That ingested xylidines are readily absorbed from the gastrointestinal tract can be concluded from the metabolites detected in the urine of the rat and dog. Toxic doses of xylidines may be taken up through the lungs and skin (see Section 5.2).

Metabolism of the xylidines is characterized by acetylation or oxidation of the amino group, oxidation of a methyl group to a carboxylic acid and conjugation with glycine, oxidation of an unsubstituted carbon in the aromatic ring to yield a phenol followed by conjugate formation. Small amounts of *N*-methylation products are also found. Isomer-specific and species-specific differences in metabolism are observed. Most metabolic studies were carried out many years ago. With the methods available today, more detailed analyses would be possible. For 2,4-xylidine (Birner and Neumann 1988) and 2,6-xylidine (Sabbioni 1992, 1994) in experimental animals and in man, hydrolysable haemoglobin adducts have been found; this is evidence for metabolic activation of these isomers to the *N*-hydroxy derivatives.

2,4-Xylidine

In the rat and dog, the main metabolite of 2,4-xylidine is *N*-acetyl-4-amino-3-methylbenzoic acid (Lindstrom 1961, Short *et al.* 1989b). Only this metabolite was detected after low doses (5 mg/kg body weight). Evidence for *N*-oxidation of 2,4-xylidine was provided by the detection of 2,4-dimethylnitrosobenzene in liver homogenates from male rats given two 2,4-xylidine doses of 125 mg/kg body weight by intraperitoneal injection (Kimmel *et al.* 1986). The *in vivo* methaemoglobin formation is also evidence of *N*-oxidation. After incubation of 2,4-xylidine with S9 mix from the livers of male rats which had been induced with Kanechlor® KC-400, 2,4-dimethylphenylhydroxylamine was detected and considered to be the active metabolite in the *Salmonella* mutagenicity test (Nohmi *et al.* 1984).

2,5-Xylidine

The main metabolite of 2,5-xylidine in the rat was 4-hydroxy-2,5-xylidine (Lindstrom *et al.* 1963).

2,6-Xylidine

After administration of 2,6-xylidine to the rat and dog, mainly 4-hydroxy-2,6-xylidine was detected (Short *et al.* 1989b).

3,4-Xylidine

The 3,4-xylidine metabolites identified in the rat were 4-acetamido-2-methylbenzoic acid, 2-amino-4,5-dimethylphenyl sulfate and 4-amino-2-methylbenzoic acid as glucuronide (Boyland and Sims 1959).

Xylidines occur as components of a variety of products from which they can be released metabolically: 2,4-xylidine from amitraz (acaricide), 2,6-xylidine from lidocaine (local anaesthetic), etidocaine (local anaesthetic), tocainide (antiarrhythmic) and D2624 (anticonvulsant) (Hollunger 1960, Knowles and Benezet 1981, Martin *et al.* 1997, Parker *et al.* 1996, Venkataramanan *et al.* 1982).

4 Effects in Man

Methaemoglobinaemia was seen in patients treated with lidocaine (Burne and Doughty 1964, Weiss *et al.* 1987). As lidocaine is broken down to 2,6-xylidine, the *N*-hydroxy metabolite of 2,6-xylidine can be considered responsible for the methaemoglobinaemia.

2,5-Xylidine

In the analogous experiment with 2,5-xylidine, the 25 male CD rats in the low dose group were given 6000 mg/kg diet for 5 months and 3000 mg/kg diet for 13 months, those in the high dose group 12000 mg/kg for 5 months and 6000 mg/kg diet for 13 months. Fibromas developed in the high dose group but the incidence was not increased above that in the concurrent controls.

In the corresponding groups of 25 male and 25 female CD-1 mice, the incidence of blood vessel tumours was significantly increased in both the low dose (6000 mg/kg diet for 18 months) and high dose (12000 mg/kg diet for 18 months) males (5/18 and 7/19 respectively, controls 2/16). In addition, the incidence of hepatomas was increased (5/16 and 2/20, controls 0/13) (Weisburger *et al.* 1978).

2,6-Xylidine

In a study of transplacental effects carried out as part of an NTP study, the parent generation CD rats were given 2,6-xylidine in concentrations of 300, 1000 or 3000 mg/kg diet (daily doses of about 15, 50 or 150 mg/kg body weight) before mating, during pregnancy and lactation (from age 5 weeks until mating 16 weeks later and then until postnatal day 21). Of the progeny, 56 males and 56 females were given the same doses after weaning for another 102 weeks. In the high dose group in both sexes the incidence of adenomas (males 10/56, females 6/56) and that of carcinomas of the nasal cavity (males 28/56, females 24/56) were significantly increased. In addition, in this group metastases were detected in the brain in 5 of 56 male rats and 7 of 56 female rats. No tumours were detected in the control animals. In all dose groups the incidence of subcutaneous fibromas and fibrosarcomas was increased and in the females that of hepatomas. Unusually, rhabdomyosarcomas were detected in the nasal cavities of 2 male and 2 female animals (Montgomery and Kornreich 1982).

6 Manifesto (MAK value/classification)

When the xylidine isomers have been studied separately they have shown similar toxic effects, although differences in their metabolism suggest that there could be quantitative differences. All tested isomers induce methaemoglobin production, reduce haemoglobin levels, are hepatotoxic and nephrotoxic, and cause the changes in the reticuloendothelial system typical of this class of substances. After metabolic activation, all the isomers are mutagenic *in vitro* under appropriate conditions in bacterial test systems. That the results of the animal studies may in principle be applied to man is indicated by the fact that in these species methaemoglobin and haemoglobin adducts are formed in the same way.

Structure-effect relationships for various substituted anilines do not reveal consistent correlations between the various parameters such as haemoglobin binding, mutagenicity and carcinogenicity. It is therefore not a contradiction that 2,6-xylidine produces less methaemoglobin than aniline but is more carcinogenic. The above considerations are also in line with the conclusion that the tumour development is a function of both genotoxic and chronic toxic effects. The structure-effect relationships demonstrate that the modification of the electronic properties of these substances by the substituents has only a slight influence on the potency and that therefore no qualitative differences between the effects of the isomers are to be expected.

Although most of the studies of carcinogenicity in experimental animals do not meet present-day requirements in their methods or documentation, they do suggest that the tested xylidine isomers could have carcinogenic potential. 2,4-Xylidine and 2,6-xylidine have been shown to be carcinogenic. The arguments which harden the suspicion that all the xylidine isomers are carcinogens and make it necessary to classify them as risk factors for cancer are listed below. After metabolic activation, all the isomers are mutagenic *in vitro* under appropriate conditions in bacterial test systems. They are all subject to metabolic *N*-oxidation and form the C-8 substituted guanine derivatives typical for this class of substances; the demonstration of binding to DNA in the liver and in the nasal cavity, the target organ, suggests that the reaction is also significant *in vivo*. Therefore it may be concluded that the xylidines have genotoxic properties and present a risk factor for cancer.

However, it is also clear that the high doses used in the carcinogenicity studies produced toxic effects in the target tissues which could be decisive for the location and timing of tumour development. The detection of inducing effects such as proliferation of the endoplasmic reticulum and increased enzyme activities suggests the involvement or saturation of receptor-mediated adaptive processes. Thus it is concluded that the xylidines are weak carcinogens with genotoxic properties; their contribution to human cancer risk is determined essentially by their toxic effects. Xylidine isomers must therefore be considered to be candidates for Carcinogen category 5. However, quantitative data and especially data for the dose-dependence of the observed effects are not available; therefore it is not possible to make species comparisons or to estimate the relative contribution of background exposures. It is therefore not yet possible to establish MAK values.

2,4-Xylidine, 2,6-xylidine and mixtures of isomers containing these substances are classified in Carcinogen category 2 in Section III of the *List of MAK and BAT Values*. 2,3-Xylidine, 2,5-xylidine, 3,4-xylidine and 3,5-xylidine are classified provisionally in Carcinogen category 3A.

Analysis of haemoglobin adducts of 2,4-xylidine and 2,6-xylidine in man has shown that there is significant iatrogenic and environmental background exposure to these substances.

Because of the symptoms of intoxication which have been described, it must be assumed that the isomers are absorbed not only through the respiratory tract but also through the skin. Therefore the xylidines are designated with an "H". Because of lack of data for allergenic effects, the sensitization potential of the xylidines cannot be assessed. They are not designated with an "S".

7 References

Bayer AG (1990) *Study for skin and eye irritation/corrosion in rabbits according to OECD guideline No. 404 and 405. Report No. T4034584*, Bayer AG Fachbereich Toxikologie, Wuppertal, unpublished

Beland FA, Melchior jr WB, Mourato LLG, Santos MA, Marques MM (1997) Arylamine-DNA adduct conformation in relation to mutagenesis. *Mutat Res 376*:13–19

Beyerbach A, Farmer PB, Sabbioni G (1996) Synthesis and analysis of DNA adducts of arylamines. *Biomarkers 1*: 9–20

BG Chemie (Berufsgenossenschaft der Chemischen Industrie) (1992) *Toxikologische Bewertung Nr. 64, 2,4-Xylidin* (Toxicological evaluation No. 64 2,4-Xylidine) (German). BG Chemie, Heidelberg

Birner G, Neumann HG (1988) Biomonitoring of aromatic amines II: hemoglobin binding of some monocyclic aromatic amines. *Arch Toxicol 62*: 110–115

Boyland E, Sims S (1959) The biochemistry of aromatic amines. *Biochem J 73*: 377–390

Bryant MS, Vineis P, Skipper PL, Tannenbaum SR (1988a) Hemoglobin adducts of aromatic amines: association with smoking status and type of tobacco. *Proc Nat Acad Sci USA 85*: 9788–9791

Bryant MS, Vineis P, Skipper PL, Tannenbaum SR (1988b) Hemoglobin adducts of aromatic amines in people exposed to cigarette smoke. *IARC Sci Publ 89*: 133–136

Bryant MS, Simmons HF, Harrell RE, Hinson JA (1994) 2,6-Dimethylaniline-hemoglobin adducts from lidocaine in humans. *Carcinogenesis 15*: 2287–2290

BUA (Beratergremium für umweltrelevante Altstoffe) (1995) *Xylidine. BUA-Stoffbericht 161* (also available in English translation). S. Hirzel Wissenschaftliche Verlagsgesellschaft, Stuttgart

Burne D, Doughty A (1964) Methaemoglobinaemia following lignocaine. *Lancet ii*: 971

Egorov YL, Shvedchenko VS, Mondoev GL, Malysheva MV, Zotova LV (1976) Criteria of the danger of uptake by the body of industrial poisons through the skin (Russian). *Gig Tr Prof Zabol 10*: 5–10

Hardy ML (1987) The hepatoxicity and urinary metabolites of 2,4- and 2,6-dimethylaniline in the rat and dog. *Diss Abstr Int 47*: 4440B–4441B

Hollunger G (1960) On the metabolism of lidocaine. II. The biotransformation of lidocaine. *Acta Pharmacol Toxicol 17*: 365–373

Kimmel EC, Casida JE, Ruzo LO (1986) Formamidine insecticides and chloroacetanilide herbicides: disubstituted anilines and nitrosobenzenes as mammalian metabolites and bacterial mutagens. *J Agric Food Chem 34*: 157–161

Knowles CO, Benezet HJ (1981) Excretion balance, metabolic fate and tissue residues following treatment of rats with amitraz and *N'*-(2,4-dimethylphenyl)-*N*-methylformamidine. *J Environ Sci Health B 16*: 547–555

Lindstrom HV (1961) The metabolism of FD&C red No. 1. The fate of 2,4-meta-xylidine in rats. *J Pharmacol Exp Ther 132*: 306–310

Lindstrom HV, Hansen WH, Nelson AA, Fitzhugh OG (1963) The metabolism of FD&C red No. 1. II. The fate of 2,5-para-xylidine and 2,6-meta-xylidine in rats and observation on the toxicity of xylidine isomers. *J Pharmacol Exp Ther 142*: 257–264

Lindstrom HV, Bowie WC, Wallace WC, Nelson AA, Fitzhug OG (1969) The toxicity and metabolism of mesidine and pseudocumidine in rats. *J Pharmacol Exp Ther 167*: 223–234. cited from Kiese M, *Methemoglobinemia: A comprehensive Treatise*, CRC Press, 1974

Magnusson G, Bodin NO, Hansson E (1971) Hepatic changes in dog and rats induced by xylidine isomers. *Acta Pathol Microbiol Scand Section A 79*: 639–648

Magnusson G, Majeed SK, Down WH, Sacharin RM, Jorgeson W (1979) Hepatic effects of xylidine isomers in rats. *Toxicology 12*: 63–74

Marques MM, Mourato LL, Santos MA, Beland FA (1996) Synthesis, characterization, and conformational analysis of DNA adducts from methylated anilines present in tobacco smoke. *Chem Res Toxicol 9*: 99–108.

Martin SW, Bishop FE, Kerr BM, Moor M, Moore M, Sheffels P, Rashed M, Slatter JG, Berthon-Cedille L, Lepage F, Descombe J-J, Picard M, Baille TA, Levy RH (1997) Pharmacokinetics and metabolism of the novel anticonvulsant agent *N*-(2,6-dimethyl-phenyl)-5-methyl-3-isoxazolecarboxamide (D2624) in rats and humans. *Drug Metab Dispos 25*: 40–46

McLean S, Starmer GA, Thomas J (1969) Methämoglobin formation by aromatic amines. *J Pharm Pharmacol 21*: 441–450. cited from Kiese M, *Methemoglobinemia: A comprehensive Treatise*, CRC Press, 1974

Mirsalis JC, Tyson CK, Steinmetz KL, Loh EK, Hamilton CM, Bakke JP, Spalding JW (1989) Measurement of unscheduled DNA synthesis and S-phase synthesis in rodent hepatocytes following *in vivo* treatment: testing of 24 compounds. *Environ Mol Mutagen 14*: 155–164

Montgomery CA, Kornreich M (1982) *NTP technical report on the carcinogenesis bioassay of 2,6-xylidine (2,6-dimethylmethylaniline) in Charles River CD rats (diet study) (NTP-82-94).* U.S. Department of Health and Human Services, 1–120

Nohmi T, Yoshikawa K, Nakadate M, Miyata R, Ishidate M (1984) Mutations in *Salmonella typhimurium* and inactivation of *Bacillus subtilis* transforming DNA induced by phenylhydroxylamine derivatives. *Mutat Res 136*: 159–168

Oettingen WF von, Sievers RF, Horecker BL (1947) Comparison of the acute effects of orally administered xylidine and aniline on the physiological functions of dogs. *Natl Inst Health Bull 188*: 7. cited from Kiese M, *Methemoglobinemia: A comprehensive Treatise*, CRC Press, 1974

Parker RJ, Collins JM, Strong JM (1996) Identification of 2,6-xylidine as a major metabolite in human liver slices. *Drug Metab Dispos 24*: 1167–1173

Parton JW, Probst GS, Garriott ML (1988) The *in vivo* effect of 2,6-xylidine on induction of micronuclei in mouse bone marrow cells. *Mutat Res 206*: 281–283

Parton JW, Beyers JE, Garriott ML, Tamura RN (1990) The evaluation of a multiple dosing protocol for the mouse bone-marrow micronucleus assay using benzidine and 2,6-xylidine. *Mutat Res 234*: 165–168

Przybojewska B (1997) An evaluation of the DNA damaging effect of selected aniline derivatives using the alkaline single cells gel electrophoresis ("comet") assay. *Mutat Res 394*: 53–57

Sabbioni G (1992) Hemoglobin binding of monocyclic aromatic amines: molecular dosimetry and quantitative structure activity relationships for the *N*-oxidation. *Chem Biol Interact 81*: 91–117

Sabbioni G (1994) Hemoglobin binding of arylamines and nitroarenes: molecular dosimetry and quantitative structure-activity relationships. *Environ Health Perspect 102, Suppl 6*: 61–67

Seiler JP (1977) Inhibition of testicular DNA synthesis by chemical mutagens and carcinogens. Preliminary results in the validation of novel short term test. *Mutat Res 46*: 305–310

Shardonofsky S, Krishnan K (1997) Characterization of methemoglobinemia induced by 3,5-xylidine in rats. *J Toxicol Environ Health 50*: 595–604

Short CR, King C, Sistrunk PW, Keer KM (1983) Subacute toxicity of several ring-substituted dealkylanilines in the rat. *Fundam Appl Toxicol 3*: 285–292

Short CR, Joseph M, Hardy ML (1989a) Covalent binding of [^{14}C]-2,6-dimethylaniline to DNA of rat liver and ethmoid turbinate. *J Toxicol Environ Health 27*: 85–94

Short CR, Hardy ML, Baker SA (1989b) The *in vivo* oxidative metabolism of 2,4- and 2,6-dimethylaniline in the dog and rat. *Toxicology 57*: 45–58

Sievers RF, Horecker BL (1947) Comparison of the acute effects of the cutaneous application of xylidine and aniline on dogs and cats. *Natl Inst Health Bull 188*: 10–13

Svirbely JL (1947) Determination of the minimal fatal concentration of xylidine in the air for mice and comparison with that of aniline. *Nat Inst Health Bull 188*: 13–16

Takahashi A, Omori Y, Takeuchi M (1974) Early biochemical morphological changes in the liver of mice after a single oral administration of 2,4-xylidine. *Jpn J Pharmacol 24*: 41

Treon JF, Deichmann WB, Sigmon HE, Wright H, Witherup SO, Heyroth FF, Kitzmiller KV, Keenan C (1949) The toxic properties of xylidine and monomethylaniline. I. The comparative toxicity of xylidine and monomethyaniline when administered orally or intravenously to animals or applied upon their skin. *J Ind Hyg Toxicol 31*: 1–20

Venkataramanan R, Abott FS, Axelson JE (1982) Metabolism of tocainide in the rat. *J Pharm Sci 71*: 491–494

Vernot EH, MacEwen JD, Haun CC, Kinkead ER (1977) Acute toxicity and skin corrosion data for some organic and inorganic compounds and aqueous solutions. *Toxicol Appl Pharmacol 42*: 417–423

Weinberg J, Sahini VE (1981) Xylidine isomers. VII. Structure-hepatotoxic activity relationships of some xylidines. *J Theor Biol 88*: 153–159

Weisburger EK, Russfield AB, Homburger F, Weisburger JH, Boger E, Van Dongen CG, Chu KC (1978) Testing of twenty-one environmental aromatic amines or derivates for long-term toxicity or carcinogenicity. *J Environ Pathol Toxicol 2*: 325–356

Weiss LD, Generalovich T, Heller MB, Paris PM, Stewart RD, Kaplan RM, Thompson DR (1987) Methemoglobin levels following intravenous lidocaine administration. *Ann Emerg Med 16*: 323–325

Zimmering S, Mason JM, Valencia R (1989) Chemical mutagenesis testing in *Drosophila*: VII. results of 22 coded compounds tested in larval feeding experiments. *Environ Mol Mutagen 14*: 245–251

completed 29.01.1998

Authors of Documents in Volume 19

Dr. R. Bartsch, Senatskommission der DFG zur Prüfung gesundheitsschädlicher Arbeitsstoffe, Technische Universität München, D-85350 Freising-Weihenstephan

Frau Dr. B. Brinkmann, Senatskommission der DFG zur Prüfung gesundheitsschädlicher Arbeitsstoffe, Technische Universität München, D-85350 Freising-Weihenstephan

Professor Dr. med. T. Brüning, Berufsgenossenschaftliches Forschungsinstitut für Arbeitsmedizin (BGFA), Bürkle-de-la-Camp-Platz 1, 44789 Bochum

Priv. Doz. Dr. G. A. Csanády, GSF-Forschungszentrum für Umwelt und Gesundheit GmbH, Neuherberg bei München, Institut für Toxikologie, 85758 Oberschleißheim

Professor Dr.rer.nat. Dr.med.habil. J.G. Filser, GSF-Institut für Toxikologie, Neuherberg bei München, D-85758 Oberschleißheim

Dr. S. Forderkunz, previously Senatskommission der DFG zur Prüfung gesundheitsschädlicher Arbeitsstoffe, Technische Universität München, D-85350 Freising-Weihenstephan

Professor Dr. Dr. H.-P. Gelbke, BASF Aktiengesellschaft, Abteilung Toxikologie, 67063 Ludwigshafen

Frau Dr. H.C. Henglein, previously GSF-Forschungszentrum für Umwelt und Gesundheit GmbH, Neuherberg bei München, Institut für Toxikologie, 85758 Oberschleißheim

Professor Dr. D. Henschler, Institut für Toxikologie und Pharmakologie der Universität Würzburg, Versbacher Str. 9, 97078 Würzburg

Professor Dr. med. R. Hess, Fluhweg 11, CH-4143 Dornach

Professor Dr.rer.nat. H. Kappus, Humbolt-Universität Berlin, Universitätsklinikum Rudolf Charité Virchow-Klinikum, Dermatologie, Augustenburger Platz 1, 13353 Berlin

Professor Dr. G. Kimmerle, Beethovenstraße 8a, 8038 Gröbenzell

Dr. H. Lessmann, Informationsverbund Dermatologischer Kliniken an der Universitätshautklinik, von-Siebold-Straße 3, 37075 Göttingen

Professor Dr. med. E. Löser, previously BAYER AG, Institut für Toxikologie, Friedrich-Ebert-Straße 217, 42117 Wuppertal

Professor Dr.rer.nat. H.-G. Neumann, Institut f. Toxikologie und Pharmakologie der Universität Würzburg, Versbacher Straße 9, 97078 Würzburg

Frau Dr. I. Neumann, Senatskommission der DFG zur Prüfung gesundheitsschädlicher Arbeitsstoffe, Technische Universität München, D-85350 Freising-Weihenstephan

Professor Dr. A.W. Rettenmeier, Universitätsklinikum der Gesamthochschule Essen, Institut für Hygiene und Arbeitsmedizin, Hufelandstraße 55, 45147 Essen

Frau Dr. U. Reuter, Senatskommission der DFG zur Prüfung gesundheitsschädlicher Arbeitsstoffe, Technische Universität München, D-85350 Freising-Weihenstephan

Professor Dr. Dr. H.-W. Thielmann, Deutsches Krebsforschungszentrum, Im Neuenheimer Feld 280, 69120 Heidelberg

Dr. D. Walter, Institut für Anorganische Chemie der TU-Berlin, Straße des 17. Juni 135, 10623 Berlin

Privatdozent Dr. D. Wild, Bundesanstalt für Fleischforschung, Institut für Mikrobiologie und Toxikologie, E.C.-Baumann-Straße 20, 95326 Kulmbach

Frau Dr. K. Ziegler-Skylakakis, c/- Senatskommission der DFG zur Prüfung gesundheitsschädlicher Arbeitsstoffe, Technische Universität München, D-85350 Freising-Weihenstephan

Index for Volumes 1–19